Hannes Isaak Reuter

Spatial crop and soil landscape processes under special consideration of relief information in a loess landscape

HORIZONTE Bd. 16

DER ANDERE VERLAG 2004

Bibliografische Information Der Deutschen Bibliothek
Die Deutsche Bibliothek verzeichnet diese Publikation in der Deutschen Nationalbibliografie; detaillierte bibliografische Daten sind im Internet über http://dnb.ddb.de abrufbar.

Der Andere Verlag, Kollegienwall 22A, 49074 Osnabrück
Tel. (0541) 20 20 125, Fax (0541) 20 20 127
E-Mail: talkto@der-andere-verlag.de
Internet: http://www.der-andere-verlag.de
ISBN 3-89959-238-7

HORIZONTE

HERRENHÄUSER FORSCHUNGSBEITRÄGE ZUR BODENKUNDE

BAND 16

Herausgeber:
Institut für Bodenkunde, Universität Hannover

In der Reihe
HORIZONTE - Herrenhäuser Forschungsbeiträge zur Bodenkunde
sind bisher erschienen:

Band 1: Puhlmann, Martina: Fällung und Lösung von Hydroxo-Sulfaten als Regulativ der Aziditäts- und Sulfat-Dynamik in sauren Sandböden.
Pb, 185 S. 57 Abb., 25 Tab. ISBN 3-934366-28-7 **€ 17,79 (1999)**

Band 2: Deurer, Markus: The dynamics of water and solute flow in the unsaturated zone of a coniferous forest: Measurement and numerical simulation.
Pb., 182 S., 66 Abb.,; 35 Tab. ISBN 3-934366-56-2 **€ 17,79 (2000)**

Band 3: Bauriegel, A.; Ilsemann, J.; Michel, C.; Van der Ploeg, R. R. (Hg.): Boden – Objekt menschlicher Nutzung und Forschung. Festschrift zu Ehren von Herrn Privatdozent Dr. Manfred Altermann.
Pb., 183 S. 45 Abb., 34 Tab. ISBN 3-934366-84-8 **€ 17,79 (2000)**

Band 4: Machulla, Galina Dr.: Mikrobielle Aktivität von Böden aus anthropogenen und natürlichen Substraten: Meß- und Schätzverfahren für die Standortbewertung.
Pb., 285 S., 54 Abb., 61 Tab. ISBN 3-935316-07-0 **€ 22,91 (2000)**

Band 5: Teichert, Andreas: Raum-Zeit-Strukturen von Messgrössen des Stoffhaushalts eines grundwassernahen sauren Sandbodens unter Kiefernwald.
Pb., 191 S., ISBN 3-935316-16-X **€ 21,88 (2001)**

Band 6: Hermsmeyer, Dirk: Soil Physical and Hydrological Evaluation of Aluminum Recycling By-Product as an Infiltration Barrier for Potash Mine Tailings.
Pb., 157 S., zahlreiche, teils farbige Abb., ISBN 3-935316-63-1 **€ 21,88 (2001)**

Band 7: Pamperin, Lars: Nitratverlagerung in Abhängigkeit von der Bodennutzung, den Standorteigenschaften und der Grundwasserneubildung eines stauwasserbeeinflussten Grundmoränenstandortes.
Pb., 245 S., 76 Abb., 31 Tab., ISBN 3-936231-31-1 **€ 25,90 (2002)**

Band 8: Pagel-Wieder, Sibylle: Estimation of the specific surface area of soil particles by adsorption of polyvinylalcohol in aqueous suspension.
Pb., 119 S., 44 Abb., 15 Tab., ISBN 3-936231-83-4 **€ 17,90 (2002)**

Band 9: Köhler, Kai: Ermittlung der Nitratauswaschung von Ackerstandorten bei unterschiedlicher N-Düngung und Bewirtschaftung aus Nmin-Messungen und Modellrechnungen.
PB., 220 S., 55 Abb., 47 Tab., ISBN 3-936231-91-5 **€ 23.90 (2002)**

Band 10: Spaethe, Andrea: Einfluss des Kationenmilieus auf Oberflächenladung und die hydraulischen und mechanischen Eigenschaften von Tonen.
Pb., 138 S., zahlr. Abb. u. Tab., ISBN 3-89959-006-6 **€ 19.90 (2002)**

Band 11: Böckenhoff, Klaus: Einfluss der Oberflächenladung auf das Flockungsverhalten von Tonfraktionen aus Schwarzerden des östlichen Harzer Vorlandes.
Pb., 104 S., zahlr. Abb. und Tab., ISBN 3-89959-044-9 € 16,50 (2003)

Band 12: Heumann, Sabine: Parameterizing net N mineralization in NW German sandy arable soils with different former land-uses.
Pb., 125 S., 17 Abb., 16 Tab., ISBN 3-89959-052-X € 17,90 (2003)

Band 13: Gieska, Matthias: Auswirkungen langjähriger Intensiv-Landwirtschaft auf das Gefüge von Lössböden - unter besonderer Berücksichtigung der Aggregatstabilität.
Pb., 150 S., 29 s/w- u. 9 farbige Abb., 35 Tab., ISBN 3-89959-057-0 € 25,90 (2003)

Band 14: Ringe, Horst, Böttcher, Jürgen und Wilhelmus H. M. Duijnisveld: Einfluss von Nadel- bzw. Mischforsten auf die Qualität der Grundwasserneubildung von Sandböden.
Pb., 136 S., 49 Abb., 76 Tab., ISBN 3-89959-093-7 € 19,90 (2003)

Band 15: Zimmerling, Berno: Beregnungsversuche zum Infiltrationsverhalten von Ackerböden nach Umstellung der konventionellen auf konservierende Bodenbearbeitung.
Pb., 210 S., 30 Abb., 26 Tab., ISBN 3-89959-164-X € 25,90 (2004)

Band 16: Reuter, Hannes Isaak: Spatial crop and soil landscape processes under special consideration of relief information in a loess landscape.
Pb., 286 S., 77 s/w- u. 21 farbige Abb., 57 Tab., ISBN 3-89959-238-7 € 45,90 (2004)

Spatial crop and soil landscape processes under special consideration of relief information in a loess landscape

Vom Fachbereich
Geowissenschaften und Geographie der Universität Hannover

zur Erlangung des Grades
Doktor der Naturwissenschaft
Dr. rer. nat.

genehmigte
Dissertation

von
Diplom-Geoökologe Hannes Isaak Reuter
Geboren am 21.05.1973 in Dessau

2004

Lieber Detlef!
Vielen Dank
für die
gemeinsame Zeit,
und manche
(viele) Hilfe
Hannes

Referent: Prof. Dr. J. Böttcher (Institut für Bodenkunde, Universität Hannover)
Koreferent: Prof. Dr. O. Wendroth (ZALF, now University of Kentucky)
weiteres Gutachten: Prof. D. R. Nielsen (UC Davis, California)

Tag der Promotion: 18.02.2004

Acknowledgements

First, I would like to thank Prof. Böttcher for the topic of this thesis, for being my advisor, our discussions and his comments.

Secondly, I would like to thank PD Dr. Ole Wendroth for the great possibility to work in his research group. He encouraged me to learn new things, starting with relief analysis, geostatistics, or state-space analysis. He always had a good scientific or personal advice in our discussions, always an open door and time to answer my questions.

Dr. K.C. Kersebaum for discussion about the HERMES model, for programming special features like incooperating the non-linear optimization model.

The German Research Foundation (DFG) for providing financial assistance (WE 1805/5-1, 5-2; EH 170/2-1, 2-2) for myself and the investigations.

The companies Südzucker, Agrocom and Amazone for the research project MOSAIK, performed at the research farm of the Südzucker company in Lüttewitz-Dreißig.

The family Steigerwald as the farmers for always providing support during the countless investigations at the field sites. I never will forget little Benny's comment: „Die lernens doch nie".

I enjoyed working together with my colleagues from the Institute for Agricultural Engineering Potsdam-Bornim in the MOSAIK project: A.Giebel, M.Heisig, P.Jürschik and J.Schwarz.

I have to thank all the people, which helped throughout field investigations, especially the people of the Department of Soil Landscape Research for your discussions, your hospitality, your friendship, the nice working environment. I will always remember having helping hands with counting straw (J.Biese, A.Griegoleit), taking soil samples (N.Wypler, I.Onasch), digitizing contour lines (L.Voelker), or servicing the weather station (M.Bähr).

The people of the ZALF at the central laboratory facility and at the plant laboratory of the research farm for analyzing so many soil and plant samples (D.Schulz and colleagues, R.Richter, P.Zablel).

The ZALF for providing a nice working environment (H.Rasch, T.Kühnert, A.Lange, R.Steinke).

Mr. Dave Fallow at the Department of Land Resource Science at the University of Guelph for having his time proofreading this thesis.

Some of my friends for reading, commenting, improving this thesis in various stages (A.Ledebuhr, E.Garlinski, M. Wilmking, R. Vanderspiegel, J. Fenske). The badminton group for the hours of exercises and many 'Apfelsaftschorles' afterwards.

Nadine for helping me improving the figures and the layout of my thesis.

My mother.

Table of Contents

Acknowledgements IX
Table of Contents XI
List of Figures XIV
List of Tables XX
Summary XXIII
Zusammenfassung XXIV

CHAPTER 1 *Introduction* 1

CHAPTER 2 *Literature Review* 4

CHAPTER 3 *Objective* 11

CHAPTER 4 *Site Description* 13

CHAPTER 5 *Methods* 17

Weather data 17
Soil data 18
Sampling Design 18
Global Positioning System 19
Sampling procedure 19
Soil Analysis 21
Sensor based sampling 23
Crop Yield Sampling and Data Processing 25
Hand Harvesting 25
Combine Harvesting 25
Topographic Analysis 26
Data Sets 26
Methods for topographic analysis 27
Methods for landform analysis 30
Solar Irradiance Modelling 33
SRAD 33
SOLARFLUX 36
Crop growth, nitrogen and soil water modelling 37
Non-linear optimization techniques 38
Statistical Methods 38

CHAPTER 6 *Results and Discussion for relief parameters* 41

General Description of Relief Parameters 41
Comparison of the Original and Extended Landform algorithm 45
Quality of relief parameters obtained from different data sources 47

Optimum resolution for different data sets 53
Semivariogram analysis of relief parameters 55

CHAPTER 7 *Results and Discussion for grain yield components* 58

Statistical and geostatistical grain yield analysis 58
Grain yield processing 60
Grain yield semivariogram analysis 65
Correlations between grain yield components 75
Temporal grain yield stability 79
Semivariogram analysis of grain yield components 82
Semivariogram analysis of ear length and length of internodes 83
Crosscorrelation analysis between grain yield and grain yield components for different relief parameters, and versus soil texture, penetration resistance and NDVI values 85
Grain yield components and plant properties for selected relief parameters 99
Grain yield development versus landform 99
Grain yield development versus profile curvature 105
Analysis of length of internodes and length of ears for landforms 107
Analysis of length of internodes and length of ears for profile curvature 109

CHAPTER 8 *Results and Discussion for soil properties* 111

Statistical and geostatistical analysis of soil properties 111
Spatial analysis for Nmin versus distance to tram lines 113
Semivariogram analysis for soil mineralized nitrogen 116
Spatial analysis for texture 117
Semivariogram analysis for soil Phosphorus and Potassium 118
Semivariogram analysis for soil water content 122
Crosscorrelation analysis between soil properties, different relief parameters and sensor values 125
Significant crosscorrelations of soil water content at different times 125
Significant crosscorrelations of soil surface water content and different parameters 126
Significant crosscorrelations of soil water content for different years and different parameters 129
Significant crosscorrelations of Nmin after harvest for different years and with different parameters 130
Significant crosscorrelations between soil textural properties and with different parameters 131
Significant crosscorrelations of soil penetration resistance with depth 132
Significant crosscorrelations between soil penetration resistance and relief / sensor values 132
Significant crosscorrelations for soil nutrients at different times and with different parameters 135
Soil properties for selected relief parameters 136
Soil water content distribution versus profile curvature and landform elements 136
Soil mineralized Nitrogen versus profile curvature and landform elements 141
Soil phosphorus and potassium content versus profile curvature and landform elements 145
Soil texture, soil organic matter and pH versus profile curvature and landform elements 151

CHAPTER 9 *Results and Discussion for interactions between grain yield, soil components and relief properties* 156

Evaluation of the topographic wetness index versus measured soil water content 156

Evaluation of topographic wetness index as additional model parameter 160

Evaluation of simulated solar irradiance against measured solar irradiance 162

Evaluation of crop growth modelling including simulated solar irradiance influenced by topography 166

CHAPTER 10 *Conclusions* 173

References 177

Appendix Abbreviations 193

Appendix Site Description 195

Appendix Statistics 198

Appendix AML 202

Appendix Geostatistics 232

Appendix Landform 241

Curriculum Vitae 251

List of Figures

Figure 1 View to the Southeast at the field site „Sportkomplex“. Image: Reuter. 13

Figure 2 Depth of ground water table (measured biweekly) for the time period 1928-1930 as published by Härtel (1931). See Figure 3d for well location No. 1034. 14

Figure 3 Height in m above sea level (a), satellite image of the investigation area (b), soil types of the german soil appraisal (c), and the geological map (d). Field borders with sampling grids are shown with black lines. Letters in (a) give the field names (AT Am Teich; KB Kleinbahn; BL Bei Lotte; GH Gasthof; SK Sportkomplex; BH Bauhof.). Notice the location of Well No. 1034 in d. 15

Figure 4 Pictures of the weather station at the farm (REF1, left) and at the field site „Bei Lotte“(REF2, right). Images: Reuter. 17

Figure 5 Picture of the field site „Sportkomplex“ with locations of WS1 and WS2. Image: Wendroth. 18

Figure 6 Location of the sampling raster at the field site „Bei Lotte“. Locations of the nests are at positions F2, M3, K9, O11 and E13 of the standard raster. Please note the location of the weatherstation REF2 near the standard raster position D8. 20

Figure 7 Equipment used for field campaigns. ACT-board computer for positioning (a), the mechanical auger (b), the mixing of soil samples (c) and the processing bench (d). Notice the large white freezer in d, which was used to cool the samples immediately after sampling and mixing. Images: Wendroth, Garlinski. 21

Figure 8 Standard error of soil penetration measurements versus depth in dm and number of penetrations performed. 22

Figure 9 The HydroN-Sensor during a recording run in 2000, mounted on top of a lightweight vehicle (Left). Manual measurements of leave chlorophyll content using the HydroN-tester (middle) and of the plant-nitrogen content using a Merck Nitrate test set (right). Images: Reuter, Wendroth. 24

Figure 10 Structural diagram of the ArcInfo procedure to remove „Failure“ elevation values. 27

Figure 11 Different landform elements and their probable water movement and concentrations adapted from Pennock et al. (1987). Black arrows indicate vertical infiltration, empty arrows through flow of water, and dotted arrows surface runoff of water and sediments. 31

Figure 12 Example of a landform classification for the field site „Bei Lotte“. Result of an unfiltered landform classification (a), the same area after the area filtering approach (Threshold was set to 5 cells) (b). The data set was a 10 m by 10 m Digital Elevation Model (DEM) resampled from a 1 m by 1 m Laserscan DEM. 32

Figure 13 Measured daily solar irradiance (SR) versus simulated daily solar irradiance for the time period 1998-2001 for the weather station REF1. 35

Figure 14 Examples of relief parameters for the field site "Bei Lotte" computed from the Laserscan-Digital Elevation Model. Elevation (a), Slope (b), aspect (c), flow direction (d), profile curvature (e) and planform curvature (f). The parameters were computed using the topo.aml and shown as a 3D-View to the south. 43

Figure 15 Examples of relief parameters for the field site "Bei Lotte" computed from the Laserscan-Digital Elevation Model. Flow accumulation (g), basin area (h), watershed area (j), Topographic Wetness Index (k), Sediment Transport Capacity (l) and landscape position (m). The parameters were computed using the topo.aml and shown as a 3D-View to the south. 44

Figure 16 Landform classification for the field site "Bei Lotte" using the landform.aml. (a) aggregated landform units grouped for landform position (Shoulder SH, Backslope BS, Footslope FS, Level LE). (b) detailed landform classification with eleven landforms (Notice that the CSH and the DFS do not occur). View to the south. 46

Figure 17 Pennock's original landform classification algorithm using only major landforms for the field site „Sportkomplex“ (a) and „Bei Lotte“ (b). Results of the extended landform classification algorithm for the field site „Sportkomplex“ (c) and „Bei Lotte“ (d). Differences between both approaches for the field site „Sportkomplex“ (e) and „Bei Lotte“ (f). 48

Figure 18 Example of difference in profile curvature aggregated for a 27 x 27 m raster for the laserscan (LS) - and the topographic map (TK) - DEM at the field site „Bei Lotte“ 49

Figure 19 RMS Percent Slope (top graph) and the intercept of the Root Mean Square Slope (RMSS)(bottom graph) as a function of the DEM resolution for the datasets Laserscan (LS), TK 1:10.000 (TK10), TK 1:25.000 (TK25) and TK 1:100.000 (TK100) 54

Figure 20 Distribution of hand harvest grain yields at the field site „Bei Lotte“ in the years 1999 (a), 2000 (b) and 2001 (c). 59

Figure 21 Semivariogram analysis for processed (proc.) and raw (raw) winter rape combine harvester data in 2002 for the whole field and for single landform elements at the field site „Bei Lotte“. 61

Figure 22 Differences between hand (HH) and combine harvested (CH) grain yield at the field site “Sportkomplex” in the year 2000. Lodging areas are shaded in gray. Additionally, five different fertilizer strategies are given. 63

Figure 23 Differences between hand harvested and combine harvested grain yield in the year 2000 in direction of the harvest tracks (east-west) for the field site “Sportkomplex” (a) and “Bei Lotte” (b) and perpendicular to the direction of the harvest tracks (north-south) for the field site “Sportkomplex” (c) and “Bei Lotte” (d). 65

Figure 24 Semivariogram analysis for hand-harvested yields for the field site “Bei Lotte” for Spring Barley in 1999, Winter Rye in 2000 and Winter Rye in 2001. 66

Figure 25 Relationship between Precipitation Sum for March to July (week 8 to 31) in mm and semivariogram model ranges of hand harvested yield at two field sites “Bei Lotte” and “Sportkomplex” for the years 1999, 2000 and 2001. The semivariogram model ranges of profile curvature (PROF), planform curvature (PLAN), curvature (CURV), flow accumulation (FLACC, slope (SLP) and the topographic wetness index (WI) are shown additionally. 68

Figure 26 Relationship between Precipitation Sum for March to July (week 8 to 31) in mm and semivariogram model ranges of grain yield. The ranges were obtained for the years 1998 to 2001 for combine harvest (CH) data measured at two field sites “Bei Lotte” (BL) and “Sportkomplex” (SK). Specifically indicated data points in 1998 at “Sportkomplex” and in 2000 at “Bei Lotte” were excluded from the Least Square Fit. The semivariogram model ranges of profile curvature (PROF), planform curvature (PLAN), curvature (CURV), flow accumulation (FLACC, slope (SLP) and the topographic wetness index (WI) are shown additionally. 69

Figure 27 Semivariogram analysis for grain yield in tha-1 at the field site „Bei Lotte“ in 1998, seperated for different landform elements (Minimum area 10.000 m2). The top rightmost variogram (FIELD) represents the average field conditions. Other variograms (PSH planar shoulder, PBS planar backslope, CFS convergent foot slope, LCL low catchment level, HCL high catchment level) show semivariances computed at different landform elements. 72

Figure 28 Outliers detected (marked with red squares) by cross validation with (left side, a, c, e) and without (right side, b, d, f) landform element boundaries as faults for the years 1999, 2000 and 2001 at the field site „Bei Lotte“ (n = 192). LF element boundaries are shown with black lines in the left column. 73

Figure 29 Difference between kriged grain yield with and without considering landform boundaries as faults. Values are shown for examples of combine harvested grain yield in 1998 (left part) and hand harvested grain yield in 2001 (right part). 74

Figure 30 Ontogenesis of grain yield (adapted from Damisch, 1971). 75

Figure 31 Correlation between yield components (adapted from Anderl et al., 1981). 76

Figure 32 Relationship between Number of Spikes per m^2 and grain yield in tha^{-1} at the field site “Bei Lotte” for the year 2000, differentiated for Kernel per Spike (KpS): below 33 KpS, between 33 and 36 KpS and above 36 KpS (n=192). 78

Figure 33 Yield classification for the field site „Bei Lotte“. Black cells indicate no stable yield classification in all three years; white painted cells show a consistent yield classification during all three years. 80

Figure 34 Average grain yield classification in percent for landform units (a) and aggregated landform units (b) (Shoulder SH, n = 17, Backslope BS, n = 49, Footslope FS, n = 16, Level LE, n = 110, left) at the field site “Bei Lotte” for the years 1999, 2000 and 2001. 81

Figure 35 Crosscorrelation between grain yield in 1999 and the relief parameter slope at the field sites “Sportkomplex” (SK) and “Bei Lotte” (BL). The dashed line represents the 95% confidence interval. Lag Distance interval 27 m. 87

Figure 36 Selected significant crosscorrelations between grain yield at the field site “Sportkomplex” (SK) for the period 1997-2002. The dashed line represents the 95% confidence interval. Lag Distance interval 27 m. 89

Figure 37 Selected significant crosscorrelations between grain yield at the field site “Sportkomplex” (SK) in 1998 and the relief parameters mean elevation, slope, profile curvature, flow accumulation, topographic wetness index, stream power index, and landscape position. The dashed line represents the 95% confidence interval. Lag Distance interval 27 m. 90

Figure 38 Selected significant crosscorrelations between number of spikes, kernel per spike, TKM, protein content and yield per spike and the relief parameters elevation and slope at the field site "Bei Lotte" (BL) in 1999 and 2000. The dashed line represents the 95% confidence interval. Lag Distance interval 27 m. 91

Figure 39 Selected significant crosscorrelations (CC) between number of spikes, kernel per spike, yield per spike and protein content and soil penetration resistance at different depths at the field site "Bei Lotte" (BL) in 1999 and 2000, Additionally CCs are drawn for the protein content at the field site "Sportkomplex" (SK) in 2000 and 2002 versus penetration resistance and secondly, between grain yield and elevation. The dashed line represents the 95% confidence interval. Lag Distance interval 27 m (54m for SK protein). 92

Figure 40 Selected significant crosscorrelations between grain yield and silt content at three different depths at the field site „Sportkomplex" (SK) and "Bei Lotte" (BL). The dashed line represents the 95% confidence interval. Lag Distance interval 27 m. 93

Figure 41 Selected significant and non significant crosscorrelations (CC) between grain yield and sand content at the field site „Sportkomplex" (SK) and "Bei Lotte" (BL) for the years 1998, 2000 and 2001. Additionally, the CCs between grain yield and clay content in 1998 at "Sportkomplex" are given. The dashed line represents the 95% confidence interval. Lag Distance interval 27 m. 94

Figure 42 Selected significant crosscorrelations between grain yield and yield properties and NDVI at the field site "Bei Lotte" (BL). The dashed line represents the 95% confidence interval. Lag Distance interval 27 m. 95

Figure 43 Selected significant crosscorrelations between grain yield and yield properties and NDVI at the field site "Bei Lotte" (BL). The dashed line represents the 95% confidence interval. Lag Distance interval 27 m. 96

Figure 44 Selected significant crosscorrelations between grain yield in 1997, 1998 and 1999 and NDVI at the field site "Sportkomplex" (SK). The dashed line represents the 95% confidence interval. Lag Distance interval 27 m. 97

Figure 45 Selected significant crosscorrelations between grain yield in 2000 and 2001 and NDVI at the field site "Sportkomplex" (SK). The dashed line represents the 95% confidence interval. Lag Distance interval 27 m. 98

Figure 46 Number of spring barley plants per m^2 at 28.04.1999 and spike numbers per m^2 at harvest (25.07.1999) for the landforms Shoulder (S, n = 17), Backslope (B, n = 49), Footslope (F, n = 16), Level (L, n = 110) and the field (A, n = 192) at the field site "Bei Lotte". Bars show the standard error. 100

Figure 47 Number of winter rye plants per m^2 at the 24.10.2000 and spike numbers per m^2 at harvest (16.08.2001) for the landforms Shoulder (S, n = 17), Backslope (B, n = 49), Footslope (F, n = 16), Level (L, n = 110) and the field (A, n = 192) at the field site "Bei Lotte". Bars show the standard error. 101

Figure 48 Grain yield in tha^{-1} (a) and the yield properties number of spikes m^{-2} (b), yield per spike in g (c), number of kernels per m^2 (d), number of kernels per spike (e) and TKM in $g(1000\ kernels)^{-1}$ (f) for the field site "Bei Lotte" for the landforms Shoulder (S, n = 17), Backslope (B, n = 49), Footslope (F, n = 16), Level (L, n = 110) and the field (A, n = 192). Bars show the standard error. 103

Figure 49 Protein Content in 1999, 2000 and 2001 at the field site "Bei Lotte" for the landforms Shoulder (S, n = 17), Backslope (B, n = 49), Footslope (F, n = 16), Level (L, n = 110) and the field (A, n = 192). Bars show the standard error. 104

Figure 50 Grain yield in tha^{-1} (a) and the yield properties Number of Spikes m^{-2} (b), Yield per Spike in g (c), Number of Kernels per m^2 (d), Number of Kernels per spike (e) and TKM in $g(1000\ kernel)^{-1}$ (f) for the field site "Bei Lotte" for the profile curvature classes 1 (-0.2--0.1, n=12), 2 (-0.1-0.0, n=74), 3 (0-0.1, n=89), 4 (0.1-0.2, n=15), 5(0.3-0.4, n=2) and the field (FI, n = 192). Bars show the standard error. 106

Figure 51 Ear Length (a), Length of the 4.Internode (b), Length of the 3.Internode (c), Length of the 2.Internode (d), Length of the 1.Internode (e) Sum of ear length and the length of the top three internodes (f) for the field site "Bei Lotte" for the landforms Shoulder (S, n = 17), Backslope (B, n = 49), Footslope (F, n = 16), Level (L, n = 110) and the field (A, n = 192). All values given in cm. Bars show the standard error. 108

Figure 52 Ear length (a), Length of the 4.Internode (b), Length of the 3.Internode (c), Length of the 2.Internode (d), Length of the 1.Internode (e) Sum of ear length and the length of the top three internodes (f) for the field site "Bei Lotte" for the profile curvature classes 1 (-0.2-0.1 n=12), 2 (-0.1-0 n=74), 3 (0-0.1 n=89), 4 (0.1-0.2 n=15), 5 (0.3-0.4 n=2) and the field (FI, n = 192). All values given in cm. Error bars show the standard error. 110

Figure 53 Soil information for the field site „Bei Lotte“ according to the German Soil Appraisal (a), for the depth 0-30 cm according to KA4 (AG Boden, 1994)(b), for the depth 30-60 cm (c) and for the depth 60-90 cm (d). Black lines characterize the field borders. 112

Figure 54 Distribution of average N_{min} contents for the total field (FI, n=192, spacing: 27 m) on the 13.08.2001 in the depth 0-30 cm (a), 30-60 cm (b), 60-90 cm(c) and N_{min} content in 0-30 cm depth on the 06.03.2001 (d) on the 08.04.2001 in 0-30 cm depth (e) and on the 08.05.2001 for 0-30 cm depth (f) for 12 selected points versus distance to tramline at the field site „Bei Lotte“ for the year 2001. Distribution of average N_{min} in the five nests (n=120, spacing 5 m) at the 13.08.2001 versus distance to tramlines for 0-30 cm depth (g), 30-60 cm depth (h) and 60-90 cm depth (j). Bars show the standard error. 114

Figure 55 Relationship between the total precipitation in the period from March to July and the semivariogram model range of N_{min} for three depth layers (0-30 cm (30), 30-60 cm (60) and 60-90 cm depth (90)) for the field „Bei Lotte“. For data point 1998 in the depth 0-30 cm outliers were removed for semivariogram processing using cross validation (99% of normal distribution). For data points in 2001 the two largest N_{min} samples (175 kg and 78 kg Nha^{-1}) were removed for semivariogram processing. Additional, the range of N_{min} in 2001 is shown for a dataset combining the standard raster (n=192) and five small scale nests (n=120). 115

Figure 56 Sampling points for silt content at 0-30 cm depth sampled in a 54 m raster (a). The same raster with additional samples spaced at 27 m is shown in (b). The semivariogram for the dataset (a) is shown in (c), for the extended dataset (b) in (d). 118

Figure 57 EUF-P_1-content classes (1 = 0-1 mg(100g soil)$^{-1}$, 2 = 1-2, 3 = 2-3, 4 = 3-4, FI = field average) versus spring soil surface moisture sampling (top row, three different dates) and soil water content from fall sampling in 1999 (bottom row, three different depths). Error bars show the standard error. 120

Figure 58 Distribution of the EUF-P_1 fraction in 1999 for the field site “Bei Lotte”. A) closely related to a small farm yard, B) field entrance C) footslope positions with increased colluvium, D) represents both field entrance as well as increased colluvium. 121

Figure 59 Spatial distribution of soil water content in 0-10 cm depth for the dates 25.03.1999 (a), 29.04.1999 (b), 10.05.1999 (c) and 12.05.2000 (d) for the field site “Bei Lotte”. 122

Figure 60 Soil water content distributions for post-harvest sampling for 0-30 cm, 30-60 cm and 60-90 cm depth for the years 1999, 2000 and 2001 at the field site “Bei Lotte”. Range of values can be seen from Table 41. 123

Figure 61 Significant crosscorrelations between soil surface water content measurements in 1999 and 2000 at the field site “Bei Lotte”. The dashed line represents the 95% confidence interval. 126

Figure 62 Significant crosscorrelations between soil surface water content (SSM) at different dates versus Topographic Wetness Index (TWI), HydroN-Sensor at the 11.Mai 1999, EUF-P_1 in August of 2000, pH in 30-60 cm in 1998, and $K_{(DL)}$ in 30-60 cm in 1998. The dashed line represents the 95% confidence interval. 128

Figure 63 Crosscorrelations between soil water content after harvest (SM) for 0-30 cm depth and relief parameters or HydroN-Sensor. The dashed line represents the 95% confidence interval. 129

Figure 64 Crosscorrelations between residual N_{min} after harvest itself and between N_{min} after harvest versus elevation. The dashed line represents the 95% confidence interval. 130

Figure 65 Significant and nonsignificant crosscorrelations between texture and chemical soil properties (sand, silt and pH) and elevation and NIR. The dashed line represents the 95% confidence interval. 131

Figure 66 Significant Crosscorrelations for penetration resistance in different depths. The dashed line represents the 95% confidence interval. 133

Figure 67 Significant crosscorrelations between penetration resistance in different depths versus relief parameter or versus NDVI. The dashed line represents the 95% confidence interval. 134

Figure 68 Significant Crosscorrelations between EUF-K_1 and EUF-K_2 fractions and the relief parameter elevation and flow accumulation. The dashed line represents the 95% confidence interval. 136

Figure 69 Average profile curvature for the 27 m by 27 m raster at the field site “Bei Lotte”, aggregated from a 10 m by 10 m DEM. 137

Figure 70 Gravimetric soil water content at 0 -10 cm depth (SSM) at the field site “Bei Lotte”, stratified for landform classes (upper part) and curvature classes (lower part). Landform elements are Shoulder (S), Backslope (B), Footslope (F), Level (L) and the field average (A). Profile curvature classes are: -0.2- -0.1 as 1, -0.1-0 as 2, 0-0.1 as 3, 0.1-0.2 as 4, 0.3-0.4 as 5, Field average as FI . Bars show the standard error. 139

Figure 71 Gravimetric soil water content at the field site “Bei Lotte” for the years 1999, 2000 and 2001 for the depth 0-30 cm (left column) and 30-60 cm (right column), stratified for profile curvature. Profile cur-

vature classes: -0.2- -0.1 as 1, -0.1-0 as 2, 0-0.1 as 3, 0.1-0.2 as 4, 0.3-0.4 as 5, Field average as FI. Bars show the standard error. 140
Figure 72 Gravimetric Soil Water content for the years 1999,2000 and 2001 at the field site "Bei Lotte" for the depth 0-30cm (left column) and 30-60cm (right column), separated for landform units. Shoulder (S), Backslope (B), Footslope (F), Level (L) and the field average (A). Bars show the standard error. 141
Figure 73 Residual mineralised soil Nitrogen (N_{min}) after harvest for four subsequent years (1998-2001) for the depths 0-30 cm (left column) and 30-60 cm (right column), stratified for different curvature classes for the field site "Bei Lotte". Profile curvature classes: -0.2- -0.1 as 1, -0.1-0 as 2, 0-0.1 as 3, 0.1-0.2 as 4 and 0.3-0.4 as 5, FI as field average. Bars show the standard error. 143
Figure 74 Residual mineralised soil Nitrogen (N_{min}) after harvest for four subsequent years (1998-2001) for the depths 0-30 cm (left column) and 30-60 cm (right column) at the field site "Bei Lotte", stratified for different landform elements. Shoulder (S), Backslope (B), Footslope (F), Level (L) and field average (A). Bars show the standard error. 144
Figure 75 $P_{(DL)}$ in 0-30 cm and 30-60 cm depth (top row) and EUF-P_1-fractions (left column) and EUF-P_2-fraction (right column) in 0-30 cm depth after harvest for three subsequent years (1999-2001) stratified for profile curvature at the field site "Bei Lotte". Profile curvature classes: -0.2- -0.1 as 1, -0.1-0 as 2, 0-0.1 as 3, 0.1-0.2 as 4 and FI as field average. Bars show the standard error. 146
Figure 76 $P_{(DL)}$ for 0-30 cm (P30) and 30-60 cm (P60) and EUF-P-fraction (P_1 and P_2) for the Years 1999-2001 at the field site "Bei Lotte", separated for different landform units. Shoulder (S), Backslope (B), Footslope (F), Level (L) and field average (A). Bars show the standard error. 147
Figure 77 $K_{(DL)}$ in 0-30 cm and 30-60 cm depth (top row, samples taken in 1999) and EUF-K_1-(left column) and EUF-K_2-(right column) fractions for the Years 1999, 2000, and 2001 at the field site " Bei Lotte, stratified for profile curvature at the field site "Bei Lotte". Profile curvature classes: -0.2- -0.1 as 1, -0.1-0 as 2, 0-0.1 as 3, 0.1-0.2 as 4 and FI as field average. Bars show the standard error. 149
Figure 78 $K_{(DL)}$ in 0-30 cm and 30-60 cm (top row) and EUF-K-fractions (K_1 in the left column and K_2 in the right column) for the years 1999, 2000, and 2001 at the field site "Bei Lotte", separated for different landform units. Shoulder (S), Backslope (B), Footslope (F), Level (L) and field average (A). Bars show the standard error. 150
Figure 79 Texture for 0-30 cm depth (30), 30-60 cm depth (60) and 60-90 cm depth (90) for 64 sampling points at the field site "Bei Lotte", separated for profile curvature classes. Profile curvature classes: -0.2- -0.1 as 1, -0.1-0 as 2, 0-0.1 as 3, 0.1-0.2 as 4 and FI as field average. Bars show the standard error. 152
Figure 80 SOC in 1998, pH in 1998 (98) and 2000 (00) for 0-30 cm depth (30), 30-60 cm depth (60) and 60-90 cm depth (90) for 64 sampling points at the field site "Bei Lotte", separated for different profile curvature classes. (N=15 for pH in 2000). Profile curvature classes: -0.2- -0.1 as 1, -0.1-0 as 2, 0-0.1 as 3, 0.1-0.2 as 4 and FI as field average. Bars show the standard error. 153
Figure 81 Texture for 0-30 cm depth (30), 30-60 cm depth (60) and 60-90 cm depth (90) for 64 sampling points at the field site "Bei Lotte", separated for different landform units. Shoulder (SH), Backslope (BS), Footslope (FS), and Level (LEVEL). Bars show the standard error. 154
Figure 82 SOC in 1998, pH in 1998 (98) and 2000 (00) for 0-30 cm depth, 30-60 cm depth and 60-90 cm depth for 64 sampling points at the field site "Bei Lotte", separated for different landform units. (N=15 for pH in 2000). Shoulder (S), Backslope (B), Footslope (F), Level (L) and field average (A). Bars show the standard error. 155
Figure 83 Three dimensional view of the Topographic Wetness Index (a) and the Topographic Wetness Index with Monte-Carlo-methods (b) at the field site "Sportkomplex" for a DEM resolution of 10 m by10 m. 156
Figure 84 Comparison of Soil Water Content Index versus Topographic Wetness Index (TWI) and Monte-Carlo-(MC) Simulated TWI. Note the decreased standard errors for all five soil water content classes for the MC-TWI. Bars show the standard error. 158
Figure 85 Correlation coefficients between topographic map (TK) based Monte-Carlo (MC) Topographic Wetness Index (TWI) (a); LaserScan (LS) based TWI (b), LS-MC-TWI (c) and TK-TWI (d) and four dates of soil surface moisture measurements. At the x-axis the respective DEM resolution for the LS and the TK 1:10.000 dataset is shown. 159
Figure 86 Optimized Hydromorphic Modification Factor (HMF) in respect to measured dry matter using PEST and HERMES for the field site Sportkomplex in 1998 (a) and 1999 (b). Model runs performed without spatially varying solar radiation. 161
Figure 87 Measured monthly average of daily solar irradiance (SR) at REF1 compared against simulated average monthly daily irradiance at sloping surface. Solid line represents the least squares fit, the dashed line the 1:1 line between measured and simulated solar irradiance. 163

Figure 88 Solar irradiance measured at WS1 and WS2 on June, 12. 2003. Note the shading of WS2 after 5:00 pm with around 150 Wm^{-2} less solar irradiance. 164

Figure 89 Simulated monthly average daily shortwave irradiance at sloping surface for July (a) and November (b) 2002 in Wm^{-2}. The grey lines represent the sampling grid, black lines the contours of the DEM. The backdrop is a colour infrared image of May, 30 1999. 165

Figure 90 Evaluation of model runs using the RSD-Index for spring barley in 1998 (a) and winter rye in 1999 (b) at the field site "Sportkomplex". 168

Figure 91 Example simulation of above ground biomass development for locations WS1 and WS2 in 1998 (spring barley) for model runs assuming spatially constant solar irradiance (P) and solar irradiance influenced by topography (R). Additionally, the dry matter yield obtained at location WS1 and WS2 is shown. 170

Figure 92 Example simulation of above ground biomass development for locations WS1 and WS2 in 1999 (winter rye) for model runs assuming spatially constant solar irradiance (P) and solar irradiance influenced by topographic shading (R). Additionally, the dry matter yield obtained at location WS1 and WS2 is shown. 170

Figure 93 Evaluation of model runs using the RSD-Index for 1999- 2001 at the field site "Bei Lotte". The number 1 shows the location of the depressional area, the number 2 the location of the Backslope area mentioned in the text. 171

Figure 94 Annual precipitation for the time period 1969 -1989 for the precipitation stations Rüseina and Mochau. 196

Figure 95 Precipitation Sum in mm and Average Temperature in degree C for decades in 1998. 196

Figure 96 Precipitation Sum in mm and Average Temperature in degree C for decades in 1999. 197

Figure 97 Precipitation Sum in mm and Average Temperature in degree C for decades in 2000. 197

Figure 98 Precipitation Sum in mm and Average Temperature in degree C for decades in 2001. 198

List of Tables

Table 1 Annual precipitation sum and average temperature in 2 m height for the time period 1998-2001. 16

Table 2 Field Names, Size, Number of sampling points and crop rotation. (Sugar beet (ZR), winter wheat (WW), winter rye (WR), rape seed (WRA), spring barley (SB), triticale (TR), unknown (NA). 19

Table 3 Radial distortion of aerial imagery for field sites in the Luettewitz area for a flying height of 2000 m. 24

Table 4 A list of relief parameters computed using self developed AML-Scripts and the SRAD-Program. 29

Table 5 Classification table for landform elements. 30

Table 6 Example of a text file for a SRAD run. 34

Table 7 An example SRAD parameter file for the year 1998. 36

Table 8 Descriptive Statistics of relief parameters for the field site "Bei Lotte" for a 27 m aggregation. Moran's I and Geary's C were computed based on the 10 m by 10 m DEM. Filter size was set to 200 m. 42

Table 9 Frequency of unclassified and classified landform elements for the field site "Bei Lotte" using a majority filter for the 27 m by 27 m sampling cells. LF classification was based on the Laserscan-DEM with 10 m resolution. 47

Table 10 Statistical parameters for profile curvature aggregated for 27 m x 27 m sampling cells for the Laserscan (LS)- and the topographic map (TK)-DEM at the field site "Bei Lotte". 49

Table 11 Statistical parameters for the relief parameters slope, profile curvature, planform curvature and flow accumulation of generated 10 m resolution DEM. Input data were derived from the datasets Laserscan (LS10), TK 1:10.000 (TK10); TK 1:25.000 (TK25) and TK 1:100.000 (TK100). 51

Table 12 Number of classified landform (LF) units for a 10 m Digital Elevation Model (DEM) for the datasets Laserscan (LS10), TK 1:10.000 (TK10); TK 1:25.000 (TK25) and TK 1:100.000 (TK100) at the field site "Bei Lotte". Classification parameters were 0.1 for planform and profile curvature, 3.0 for slope, the area threshold was set to 5 and the LEVEL condition was set to 500 m2. 51

Table 13 Defined and optimized parameters for the landform (LF)-classification algorithm for a 10 m Digital Elevation Model (DEM) for the datasets Laserscan (LS), TK 1:10.000 (TK10); TK 1:25.000 (TK25) and TK 1:100.000 (TK100) for the agricultural area in the Luettewitz region (approximately 200 ha). The correlation coefficient (r^2) as computed by PEST between the defined and the optimized dataset is given in the last row. 52

Table 14 Semivariogram model parameters of relief parameters for the field site "Bei Lotte" (20ha). Parameters for the first structure are signed 1, for the second signed 2. If a Power model was fitted, the exponent is given under Range 2. For units of relief parameters see Table 4. 57

Table 15 Semivariogram model parameters of relief parameters for agricultural field sites (226ha). Parameters for the first structure are signed 1, for the second signed 2. If a Power model was fitted, the exponent is given under Range 2. For units of the relief parameters see Table 4. 57

Table 16 Descriptive Statistics for hand harvest grain yields for the field site "Bei Lotte" for the entire field and landform units. Values are given in t ha^{-1}. Shoulder (SH), Backslope (BS), Footslope (FS), and Level (LE). 58

Table 17 Statistical parameters for the unprocessed and processed yield for the combine harvest obtained in 2002 at the field site "Bei Lotte". Values are given in tha^{-1}. CV is given in %. 60

Table 18 Differences between field mean grain yield values of hand harvest and combine harvest. In parentheses the CV in % is stated. NA - data not available. 62

Table 19 Example for measurement uncertainty for hand harvest and combine harvest. 64

Table 20 Variogram model parameters obtained from omnidirectional semivariogram fitting of grain yield of combine harvest data at the field sites Sportkomplex (SK), Bei Lotte (BL), Am Teich (AT), Kleinbahn (KB) and Bauhof (BH) for the year 1998. Lag distances were set to 12 m in all cases. (Sph = Spherical semivariogram model, Power - Power semivariogram model). 69

Table 21 Semivariance values in $(tha^{-1})^2$ of hand harvest grain yield at the field site „Bei Lotte" in 1999, 2000 and 2001 for the first four lags (lag distance was set to 27m), stratified for shoulder, backslope, footslope and level landforms. Additionally the values of the whole field is given. (spring barley (SB), winter rye (WR)). 70

Table 22 Statistical parameters for kriging standard deviation for hand harvest (HH) and combine harvest (CH) in the respective year. Kriging was performed using landform boundaries as faults (wf) and without using faults (wof). 74

Table 23 Correlation analysis for different yield components for the year 1999, 2000 and 2001 at the field site „Bei Lotte" for 192 hand harvested plots (0.5 m2). The following abbreviations are used: SB = grain yield of spring barley, TKM = thousand kernel mass; PROT = protein content; KernelN = number of kernels; SpikeN = number of spikes; KpS = kernel per spike; YpS = Yield per Spike; year 1999 = 99; year 2000 = 00; year 2001 = 01. ** 0.01 (2-tailed) significant * 0.05 (2-tailed) significant. 77

Table 24 Semivariogram parameters for different yield components in the years 1999-2001 at the field site "Bei Lotte". TKM = thousand kernel mass in $g(1000\ kernel)^{-1}$; Protein = protein content on %; Kernel = number of kernels; spikeN = number of spikes; KpS = number of kernel per spike; YpS = Yield per Spike; NS-Ratio Nugget-Sill-Ratio. 83

Table 25 Semivariogram parameter of ear length and internode length for spring barley in 1999, winter rye in 2000 and 2001. 84

Table 26 Variogram model parameters for N_{min} in the time period 1998-2001 at the field site "Bei Lotte". Number in parentheses in 1998 show range and sill for an outlier removed dataset by cross validation. 116

Table 27 Semivariogram model parameters for soil organic carbon (SOC) and texture (Sand, Silt, Clay) in three different depths at the field site "Bei Lotte". 117

Table 28 Semivariogram model parameters for $P_{(DL)}$ and EUF-P_1 and EUF-P_2 fractions in the time period 1999-2001 at the field site "Bei Lotte". 119

Table 29 Semivariogram model parameters for $K_{(DL)}$ and EUF-K_1 and EUF-K_2 fractions for 1999-2001 at the field site "Bei Lotte". 121

Table 30 Semivariogram parameter for soil water samples at the field site "Bei Lotte". Additional semivariogram parameters for four soil surface sampling dates at "Sportkomplex" are presented. Semivariogram parameter for the year 2000 are computed without data from the second sampling period and shown in parentheses. 124

Table 31 Significant crosscorrelations between penetration resistance in spring of 2001 in different depth layers and NDVI measurements for the time period 1999 to 2001. A '+' indicates a positive CC at lag distance 0, a '-' a negative CC. 'NA' means no value can be identified as values are below a value of 0.1. 'Sign.' means significant cross correlations above the 95% confidence interval for at least one lag distance around lag 0. 135

Table 32 Correlation coefficients between soil surface moisture at 0-10 cm (SSM) and Topographic Wetness Index (TWI) at four dates and two aggregation methods (point/raster) for the field sites "Sportkomplex" and "Bei Lotte". MC indicates that the results have been obtained using a Monte - Carlo - Simulation approach. 158

Table 33 Average daily irradiances for a summer and a winter period. Values are given in $MJ\ m^{-2}\ day^{-1}$ 166

Table 34 Model evaluation for model runs assuming spatially constant solar irradiance (P) and solar irradiance influenced by topography (R) and observed grain yield (MES) for the years 1998 and 1999 at the field site "Sportkomplex". Statistical parameters of simulated and measured dry matter are given in tha^{-1}. 167

Table 35 Model evaluation for model runs assuming spatially constant solar irradiance (P) and solar irradiance influenced by topography (R) and observed grain yield (MES) for the years 1998 and 1999 at the field site "Sportkomplex". Statistical parameters of simulated and measured dry matter are given in tha^{-1}. 172

Table 36 List of Abbreviations. 193

Table 37 Depth profiles of the TK4845 from Härtel (1931) and own investigations. 195

Table 38 Average monthly precipitation in mm from Härtel(1931) and own calculations from data of the meterological service of the GDR. 195

Table 39 Descriptive parameters of Spring Soil Moisture in $g(100g\ soil)^{-1}$ for the field site "Bei Lotte". 198

Table 40 Descriptive parameters of Spring N_{min} in $kgha^{-1}$ at the field site "Bei Lotte". 199

Table 41 Descriptive parameters for soil surface water content (SSM) and soil water content (SM) in $g(100g\ soil)^{-1}$ for the field site „Bei Lotte". (Values given in parentheses for soil water content in 2000 show results from the first sampling date). 200

Table 42 Descriptive parameters for soil organic carbon (SOC) and soil texture in $g(100g\ soil)^{-1}$ for the field site „Bei Lotte". 200

Table 43 Descriptive parameters for phosphorus in $mg(100g\ soil)^{-1}$, potassium in $mg(100g\ soil)^{-1}$, pH, EUF-P_1 in $mg(100g\ soil)^{-1}$, EUF-P_2 in $mg(100g\ soil)^{-1}$, EUF-K_1 in $mg(100g\ soil)^{-1}$ and EUF-K_2 in $mg(100g\ soil)^{-1}$ for the field site „Bei Lotte". 201

Table 44 Descriptive parameters for residual N_{min} in $kgha^{-1}$ for the field site „Bei Lotte" for the time period 1998-2001. 201

Table 45 Significant cross correlations between different variables at the field site „Bei Lotte". Additionally some signifcant cross correlations for the field site „Sportkomplex" are shown which are specially signed with SK. The abbrevitions used are described below the table. Length indicates the number of significant lag classes (lag distance 27 m, except for the SK protein content, here 54 m). 232

Table 46 Abbreviations used in Table 45 239

Table 47 N_{min} Content in $kgha^{-1}$ for Profile curvature in1/100m for 1998-2001 (Mean = Average, STD = Standard deviation, N = number of samples, CV = Coefficient of Variation, 030 0-30cm depth layer, 360 30-60cm depth layer, 690 60-90 cm depth layer). All samples taken at the field site „Bei Lotte". 241

Table 48 N_{min} Content in $kgha^{-1}$ for Landforms for 1998-2001 (Mean = Average, STD = Standard deviation, N = number of samples, CV = Coefficient of Variation, 030 0-30cm depth layer, 360 30-60cm depth layer, 690 60-90 cm depth layer). All samples from the field site „Bei Lotte". 242

Table 49 Nutrient Content for potassium (K_{DL}) and phosphorus (P_{DL}) using the DL method and the EUF-method (Fractions are denoted with subscripts 1 and 2) seperated for Profile curvature in 1/100m for 1998-2001 (Mean = Average, STD = Standard deviation, N = number of samples, CV = Coefficient of Variation, 030 0-30cm depth layer, 360 30-60cm depth layer). All samples from the field site „Bei Lotte". 242

Table 50 Nutrient Content for potassium (K_{DL}) and phosphorus (P_{DL}) using the DL method and the EUF-method (Fractions are denoted with subscripts 1 and 2) seperated for Landforms for 1998-2001 (Mean = Average, STD = Standard deviation, N = number of samples, CV = Coefficient of Variation, 030 0-30cm depth layer, 360 30-60cm depth layer, 690 60-90 cm depth layer). All samples from the field site „Bei Lotte". 244

Table 51 Soil surface moisture content (SSM) in $g(100g\ soil)^{-1}$ for Landforms for 1999 and 2000 (Mean = Averag, STD = Standard deviation, N = number of samples, CV = Coefficient of Variation, Sampling depth 0-10cm). All samples from the field site „Bei Lotte". 245

Table 52 Soil moisture (SM) content in $g(100g\ soil)^{-1}$ for Landforms for 1999-2001 (Mean = Average, STD = Standard deviation, N = number of samples, CV = Coefficient of Variation, 030 0-30cm depth layer, 360 30-60cm depth layer, 690 60-90 cm depth layer). All samples from the field site „Bei Lotte". 245

Table 53 Soil surface moisture content (SSM) in $g(100g\ soil)^{-1}$ for Profile curvature in1/100m for 1999 and 2000 (Mean = Average, STD = Standard deviation, N = number of samples, CV = Coefficient of Variation, sampling depth 0-10cm). All samples from the field site „Bei Lotte". 246

Table 54 Soil moisture content (SM) in $g(100g\ soil)^{-1}$ seperated for Profile curvature in1/100m for 1999 and 2000 (Mean = Average, STD = Standard deviation, N = number of samples, CV = Coefficient of Variation, 030 0-30cm depth layer, 360 30-60cm depth layer, 690 60-90 cm depth layer). All samples from the field site „Bei Lotte". 246

Table 55 Texture, Corg and pH for profile curvature in1/100m (Mean = Average, STD = Standard deviation, N = number of samples, CV = Coefficient of Variation, 030 0-30cm depth layer, 360 30-60cm depth layer, 690 60-90 cm depth layer). All samples from the field site „Bei Lotte". 247

Table 56 Texture, Corg and pH for landform elements for 1998-2001 (Mean = Average, STD = Standard deviation, N = number of samples, CV = Coefficient of Variation, 030 0-30cm depth layer, 360 30-60cm depth layer, 690 60-90 cm depth layer). All samples from the field site „Bei Lotte". 248

Table 57 Correlation between SOC in 0-30cm and 30-60cm and EUF-P content and $P_{(DL)}$ content. All samples from the field site „Bei Lotte" (n= 64). 249

Summary

Farmers currently consider yield and soil variability to optimize their production under mainly economic considerations using the technology of precision farming. Researchers at the same time try to understand the variations of crop yield and yield patterns for subsequent years, as well as soil properties within individual fields. One example of the latter are homogeneous soil conditions, where one would expect a similar crop yield throughout the field. However, distinct grain yield variability is often observed even though management and soil conditions are relatively homogenous.

The main objective of this work is to quantify the spatial and temporal variability and dynamics of grain yield components and soil properties and their interactions. The investigation area is a loess landscape in Saxony - Germany. The relief can be described as hummocky with plateau like areas interspaced throughout the whole area. Therefore, special attention has been drawn to the incoperation of relief information into the analysis performed.

Relief analysis tools have been developed, which allow the quantification of a landscape in terms of relief parameters. Based on these relief parameters, a landform classification algorithm was developed within the Geographic Information System ArcInfo, which allows to partition a landscape into eleven landforms based on a Digital Elevation Model. Examples are shoulder, backslope, footslope and level areas. The influence of DEM generalization on landform classification was quantified, showing that the occurrence of footslope and shoulder positions decreased significantly with increasing topographic map scale. A nonlinear parameter optimization approach was applied to overcome these problems.

The variability of grain yield components and soil properties across several years was analysed using semivariogram and crosscorrelograms. A decrease in grain yield semivariogram range with increasing precipitation could be observed. Different semivariogram parameters were identified for landform units. Several significant crosscorrelations could be identified.

At each classified landform unit a characteristic yield development could be observed. Different developments throughout the years were based on soil and meteorological conditions, as well as management history. Similar results were drawn for soil properties.

To understand the observed grain yield and soil property differences, the HERMES model was used to simulate the nonlinear processes of biomass development and nitrogen dynamics. The solar irradiance influenced by topographic shading and the topographic wetness index were found helpful for explaining crop yield and soil properties variability.

KEYWORDS: Relief Analysis, Landform Classification, Soil Properties, Grain Yield, Spatial, Temporal, Variability, Solar Radiation, Topographic Shading, Modelling, Geostatistics

Zusammenfassung

Landwirte beschäftigen sich mit der räumlichen und zeitlichen Variabilität von Erträgen und Bodeneigenschaften um ihre Produktionsverfahren unter ökonomischen Gesichtspunkten zu optimieren. Im Rahmen der Anwendung des teilflächenspezifischen Managements besitzen sie dazu entsprechende Steuerungsinstrumente. Zugleich versuchen Wissenschaftler die räumliche Variabilität der Ertragsentwicklung innerhalb von Ackerflächen zu verstehen. Ein Beispiel dafür sind homogene Böden, wo ein gleichmäßiger Ertrag erwartet werden würde. Trotz gleichmäßiger Bewirtschaftung weisen diese aber Ertragsunterschiede auf.

Das Ziel dieser Arbeit ist die Quantifizierung der räumlichen und zeitlichen Variabilität von Ertragskomponenten und Bodeneigenschaften und ihren Interaktionen. Das Untersuchungsgebiet ist eine Lößlandschaft in Sachsen. Aufgrund der hügeligen Landschaftsgestalt wurde bei der Analyse der Daten das Relief besonders berücksichtigt.

Verschiedene Skripte zur Quantifizierung der Reliefeigenschaften wurden im Geographischen Informationssystem ArcInfo entwickelt. Basierend auf diesen Parametern wurde ein Algorithmus zur Landschaftsklassifikation entwickelt. Die Landschaftsklassifikation erlaubt eine Segmentierung der Landschaft in bis zu elf Klassen basierend auf einem Digitalen Geländemodel. Mit zunehmender Generalisation der Eingangsdaten zeigte sich eine Abnahme der Anzahl von Oberhang- und Hangfusspositionen. Ein erster Ansatz, Einflüsse der Generalisierung zu verringern, wurde mittels nichtlinearer Optimierung aufgezeigt.

Die Variabilität von Pflanzenparametern und Bodeneigenschaften wurde mit Semivariogrammen und Crosskorrelogrammen untersucht. Es konnte eine Verringerung der Semivariogrammreichweite der Erträge mit zunehmendem Niederschlag nachgewiesen werden. Zudem konnten auf unterschiedlichen Landschaftseinheiten differenzierte räumliche Gültigkeitsbereiche für Erträge ermittelt werden.

Eine charakteristische Ertragsentwicklung konnte auf jeder Landschaftseinheit quantifiziert werden. Diese Differenzen basieren auf den Boden- und Witterungseigenschaften, und dem Management des jeweiligen Jahres. Ähnliche Ergebnisse können für die Bodeneigenschaften aufgezeigt werden.

Um die nichtlinearen Prozesse des Pflanzenwachstums und der Bodeneigenschaften zu verstehen, wurde das Biomasse- und Stickstoff Modell HERMES eingesetzt. Die Sonneneinstrahlung mit topographischer Abschattung und der topographische Bodenfeuchteindex wurden in Hinsicht auf die Erklärung der Ertrags- und Bodenvariabilität positiv bewertet.

KEYWORDS: Relief Analyse, Landschaftsklassifikation, Bodeneigenschaften, Ertragskomponenten, Räumliche, Zeitliche, Variabilität, Einstrahlungsmodellierung, Geostatistik

CHAPTER 1 *Introduction*

For centuries farmers have been managing agricultural fields based on their experience gained from farm management and daily observations. Typically, plants with higher growing requirements are planted on soils richer in nutrients, while crops with lower requirements are planted on fields not as fertile. As the demands of agriculture increased, the requisition on soils were such that the use of fertilizers and other methods to improve fertility has become more common.

However, with the advance of technology over the last 50 years, such as mechanization and automation has increased farming efficiency and has lead to a decreasing number of people farming, although the amount of land in production over the time has increased (1900 - 38.2% of the population; 1950 – 24.3%; 2000 – 2.5%, (Maul and Richter, 2001)). The technical development during that time period is quite impressive; however, the accompanying scientific knowledge has not always kept pace.

Much of the understanding of agricultural crop production was developed on the regional level during the 1950's through the 1970's. During this time period, researchers measured and analyzed the spatial distribution of certain parameters, in order to derive regional soil, climate and phenological maps. Various models have followed capable of predicting agricultural production and allowing the researchers new insight in the complexity of agricultural production (Penning deVries and van Laar, 1982, Ivanov et al. 1980). The amount of literature and the variety of complex models for the decision support of farmers demonstrates that much of the research during this time period was directed to the regional scale.

During the last decade as technical development was very fast, pronounced problems became obvious. Farmers were now able to vary crop and soil management in agricultural fields using a technology called PRECISION FARMING (PF). Farmers were soon concerned, that precision farming equipment was available to them but advice on what to do with it was not forth coming (Blackmore et al., 1999). Tools and methods needed to be developed, because a different spatial scale - field scale - was now under investigation as opposed to the regional scale. The experience gained at the regional scale could not and still can not be transferred on a one to one relationship to a smaller scale - the field level. An example of this would be that if in two regions due to meteorological and soil properties, ripeness of winter wheat differ by 10 days. Does that mean that soil and meteorological differences are also the triggers for differences in ripeness in a given field? Researchers and farmers applying PF are now in a similar position as the researchers back in the 1960s working on regional scale soil maps

(Lieberoth et al., 1985) or even earlier investigators of the relationships between crop grain yield and soil information (Roth, 1956, Baumann, 1949).

For the investigations of agricultural production at the field scale level, either a deductive or an inductive research approach is possible. A deductive approach requires a broad series of parameters, which are obtained and analysed with respect to their ability to explain crop production inside an agricultural field site. In most cases those results should be transferable to similar fields in the investigated landscape, provided that the fields under investigation are representative of the original site. The inductive approach starts with a small set of parameters. If further refinements are required, additional parameters will be added.

Related to this rather abstract view of research is a more practical question. Which parameters influence agricultural crop production, at what times and what method is required to sample at the field scale in a given landscape under research as well as production conditions? These types of questions are not only related to research conducted under the vision of agronomy or soil science, but to the application of precision farming as well. Similar research questions concerning the field scale level can also be found in the discipline of landscape ecology (Wu and Hobbs, 2002).

Until today, the main goal for a farmer has been to have optimal production to maximize economical results. Before introducing site specific farming, farmers managed fields typically as one single unit, applying similar management practices across different textural classes, landform units and soil nutrient contents. Managed in this manner, grain yields varied within the fields. Those differences were a result of the crop yield potential varying across a field, which showed the plant dependency in terms of soil nutrients, soil moisture, soil texture and meteorological conditions. The application of a single uniform rate of nitrogen (N) within a field with different crop yield potentials, for example, often results in areas with an over or under supply of nutrients, which in turn results in differences in economical outcomes.

Research focused on understanding crop yield variability is not only aimed at maximizing the economical outcomes, it is also an attempt to understand the underlying processes. Researchers investigate, analyse and interpret hydrological, pedological, meteorological, biological, geomorphologic and agricultural management data at the field scale level in terms of economical as well as ecological outcomes. As shown above, relationships can not be transferred easily from larger to smaller scales; therefore, methods, processes and parameters influencing agricultural production must be identified and evaluated for each field, each farmer, each agricultural zoning like the „Lommatscher Pflege“ and each ecological zoning like tropical/subtropical areas.

Having drawn the larger picture now, one of the major research questions in agronomy and soil science should be addressed. Under heterogeneous soil and field conditions the observed grain yield variability is mainly attributed to the underlying soil properties. In contrast, for homogeneous soil conditions a homogeneous distribution of yields is expected across a field. However, yield variability is often observed even if management (i.e. fertilizer application) and soil conditions (texture) are relatively homogeneous. If it is not solely the soil properties that cause the yield to vary, which other processes or parameters are contributing to the yield variability?

The questions that will be discussed in the following chapters are related to the field scale in a specific landscape. Beginning with a presentation of spatial and temporal variability of crop yield and soil parameters under special consideration of relief information (section Chapter „Literature Review“ on page 4), the objectives of the undertaken research will be provided (Chapter „Objective“ on page 11). A description of the site condition will be given in Chapter „Site Description“ on page 13, followed by the applied methods in Chapter „Methods“ on page 17. Further, the role of relief information will be discussed (Chapter 6 on page 41) not only in terms of crop yield variability (Chapter 7 on page 58), but also for soil, nutrient and moisture parameters (Chapter 8 on page 111). Lastly some advantages of implementing relief information for explaining grain yield variability using model simulation for two different fields will be evaluated (Chapter 9 on page 156). Finally, some concluding remarks will be made in Chapter 10 on page 173.

CHAPTER 2 *Literature Review*

Relief analysis is a tool to analyse a landscape based on a Digital Elevation Model (DEM). One of the simplest parameters might be the elevation itself, also included is the slope or the exposition of a given point in a landscape. Moore et al. (1991) state that the spatial distribution of topographic attributes can often be used as an indirect measure of the spatial variability of hydrological, geomorphologic and biological processes. The advantage compared to other information such as soil parameters or biomass production estimation is based in the relatively simple and fast techniques to model processes in large areas and complex spatial patterns of environmental systems as seen by Moore et al. (1993b).

Relief parameters can be computed based on algorithms implemented into different GIS packages (ArcInfo, GRASS, Idrisi). Several standalone packages are also available to researchers (TAPES (Gallant and Wilson, 1996), TOPOG (CSIRO, 2002), SARA (Breuer, 2001), Landform Element Complex – algorithm by Pennock and Corre (2001), Landlord (Florinski et al., 2002)). For a general overview of relief attributes please refer to Zevenbergen and Thorne (1987) or Gallant and Wilson (1996).

One secondary relief parameter related to topography is the solar irradiance. Geiger et al. (1961) state: "if the sun is shining, slopes of different inclination and orientation receive different amounts of radiant energy,...where rainfall is in plentiful supply, but solar energy is somewhat lacking, southern slopes provide a better habitat for most plants. However, where there is plenty or solar energy, but a shortage of water, the shady northern slopes support more luxurious vegetation"

A series of models exists to quantify the effect of shading on incoming solar radiation (Schaab, 2000). In general, incoming solar radiation models (ISRM) can be classified:

- the spatial scale of the simulation at a plant stand scale (i.e. Dubsky et al., 1999) or at a landscape scale (Moore et al., 1993b),
- complexity based on the number of input parameters (Digital Elevation Model, latitude, longitude, T_{max}, T_{min}, Cloudiness, Albedo, and other),
- the computed output parameters (potential solar radiation, diffuse solar radiation, direct solar radiation, duration of sunshine, soil and air temperature),
- incorporation into a Geographical Information System (GIS) for data transfer via direct or indirect methods (GRASS, ArcInfo, Idrisi, None) and
- use of computing resources (fast/slow, precise).

Moore et al. (1993b) identified occurrences of three eucalyptus species based on the minimum temperature in the coldest month and the annual net radiation at a landscape level. Similarly at the field scale level, Schaab (2000) used their model output to upscale biogenic emissions for two forest sites. Anthoni et al. (2000) investigated an open canopy juniper sagebrush landscape. Shading resulted in temperature differences of 31 °C for soil surface measurements and 2 °C for plant leaves; however, ISRM results were between 20 Wm^{-2} for longwave radiation and 10 Wm^{-2} for shortwave radiation and could not help to explain the energy budget difference of 200 - 250 Wm^{-2}.

Wessolek et al. (1992) used a photosynthesis model including only slope and exposition to account for the spatial variability of incoming solar radiation for a point model at the field scale. However, the effects of topographic shading were not included in their model. DeWit (1965) in Wessolek et al. (1992) showed that the maximum photosynthesis rates obtained under clear sky conditions and optimum water availability doubled compared to diffuse solar radiation conditions. This highlights the importance of solar radiation modelling for agriculture purposes. As a given point in a field is shaded (receives only diffuse solar radiation), the potential gross photosynthesis rate is lowered compared to un-shaded conditions (direct beam solar radiation). This rate can also be influenced by other parameters, especially the actual evapotranspiration, capillary rise and precipitation. Therefore, the actual biomass production can behave even opposite to the computed gross photosynthesis rate (see Geigers statement above). At the plant scale, Dubsky et al. (1999) used ISRM to characterize crop stands to characterize space and time variability of solar radiation insight of plant development, biomass balance and gas exchange.

Beresneva and Popova (1983) investigated vegetative elements at a landscape scale for northern and southern slopes. Differences were observed in blue grass between 10 to 14 days for germination, formation of generative shoots, the amount of dry matter and the death of the vegetation. The authors attributed that to the microclimatic variability, primarily to the amount of solar radiation received. Ratnam and Goudreddy (1977) attributed the differences due to light saturation of only the upper part of the foliage at high light intensities, while lower foliage may be photosynthetically inert during cloudy days. Bonner (1962) stated that the efficiency of light utilization is higher at low sunlight than at full sunlight. This presents two major considerations: the direction of sowing and the ability of solar radiation to penetrate plant stands (Kler, 1987) as well as plant properties created by plant breaders (Wernecke et al., 1999). For a more in-depth discussion of the relation between incoming solar irradiance and topography see Rich et al. (2002). Therefore, one objective has been to test if ISRM are helpful tools to understand differences in crop yield development observed within field using modelling techniques.

Another relief parameter relevant for this work is landforms or relief units. Each of these contains certain characteristic physical, chemical, and biological processes and parameters (see Dehn et al., 2001). Milne (1936) was one of the first scientists, who recognized the catena principle of soil formation in a hilly terrain (Ruhe, 1960). Several authors followed on that concept (Ruhe, 1956, Conacher and Dalrymple, 1977), which eventually is the basis for all landform classifications systems found today, however, in the catena concept the third dimension is missing. Hugget (1975) presented a soil landscape system (not only a landform system), which overcomes the two-dimensional character and simulated whole systems behaviour. Following on that, Willgose et al. (1991b) have recently developed their SIBERIA model, to understand evolution of landforms over geomorphologic time scales.

Before the introduction of DEM, manual delineation of landforms was performed using field survey or by interpretation of stereo aerial photographs. McBratney et al. (1992) showed that surveyors delineate a complex landscape according to their personal bias. On the other hand, Burrough et al. (2000) stated that a global set of rules found at a national or international level, does not provide the most sensible divisions at the local level; therefore, different types of landform classification algorithms were developed.

Pennock et al. (1987) used values published by Young (1972) to classify seven/nine three-dimensional landform elements. The relief parameters profile curvature, planform curvature, slope and catchment size were used for classification. A general disadvantage is the omitting of planar landforms at all landscape positions. A second disadvantage is that this classification is only valid for a DEM cell size of 10 m by 10 m.

In a recent paper by Park et al. (2001) surface curvature and a terrain characterization index were used to delineate a five landform element model. The terrain characterization index is based on the multiplication of surface curvature and the logarithm of the upslope contributing area. Park et al. (2001) concluded, that the values of their four criteria are rather arbitrary and may need several trials to achieve a reasonable approximation of landform delineation. Another approach was developed by MacMillan and Pettapiece (1997) using the percentage of a location in relation to the minimum and maximum in a given landscape, to assign landform elements.

Discontinuous landform classification as shown in the two examples above separates a landscape into definite facets with "hard" boundaries, whereas a continuous classification assigns the strength, or degree of membership of each cell in each of n defined classes (MacMillan et al., 2000) either based on expert judgment (Brown et al., 1998) or statistical analysis (Burrough et al., 2000, Irvin et al., 1997).

An example for a continuous classification is the work by MacMillan et al. (2000), who performed a fuzzy classification, however work by Burrough et al. (2000) used expert knowledge to identify landform facets via a semantic import (SI) model (Dehn et al., 2001). Ten different relief parameters were incooperated to classify up to 15 different landforms using a rulebase containing 20 rules. MacMillan et al. (2000) stated, that these 15 classes contain an undesirable level of complexity and simple fragmentation. Therefore, similar to work done before by MacMillan and Pettapiece (1997) without the fuzzy approach, the 15 classes were aggregated to four classes. Results were found to be comparable to Pennock's classification.

The general limitation of the approaches above is based on the human expert knowledge. MacMillan et al. (2000) states, that statistical classifications based on ordination of numerical data in n-dimensional feature space (Burrough et al., 2000) can identify the optimum number of natural classes for a given landscape. The drawback of this approach is that classes and class definitions will never be the same for any two sites and are optimized for a given landscape. Still, an advantage is that classifications performed base on statistical evidence.

Researchers have been frustrated by computational issues related to the scale of investigation. MacMillan et al. (2000) described the three major problems for an automated landform classification algorithm for identifying different types of landscapes using DEMs as follow:

- Selecting and computing an appropriate suite of terrain derivatives from DEM data
- Identifying an appropriate number of meaningful different landform classes and describing their salient or defining characteristics
- Selecting and applying a classification procedure capable of using the available terrain derivatives to produce the required classes.

Throughout several investigations a similar set of relief parameters were found, which were processed with a variety of methods. Three main questions need to be addressed:

- some methods are able to handle different scales (i.e. continuous models), whereas other may not be operable,
- continuous models suffer from the arbitrariness of the criteria used for their classification process
- a lack of understanding of process-response relationships between terrain attributes and other parameters such as soil properties and vegetation (Park et al., 2001).

Pennock et al. (2001) mention that there is quite a number of papers using topography to delineate practical, agronomical meaningful field segments, which contain distinct plant growth controlling parameters (see Stevenson and van Kessel, 1996). Nevertheless, the yield response upon soil properties will be different from year to year due to different weather conditions. During a wet spring footslope positions may exhibit an elevated level of soil water content, leading to air deficiency, and therefore a deceleration of plant growth. In contrast during a drought at the same landscape position biomass will be produced, whereas plants at other positions suffer from moisture deficits. The general process becomes even more complex if different radiation regimes and different soil types are considered, which in turn lead to different evapotranspiration rates.

In general two broad research approaches exist for soil properties and yield components: the first relating properties to discontinuous relief classification (i.e. Shoulder (SH), Backslope (BS), Footslope (FS), Level (LE) positions) with the majority of papers published in that area and second papers relating properties to continuous relief parameters like elevation or slope.

First, the relationships between relief parameter for continuous and discontinuous classification and their relevance for yield components will be discussed here.

Researchers found an increase in grain yield with increasing curvature character of the landscape, as moving from shoulder to footslope position (Miller et al., 1988, Simmons et al.,

1989, Fiez et al., 1994, Jowkin and Schoenau, 1998, Bakhsh et al., 2000, Manning et al., 2001b, Pennock et al., 2001). During the years, where no such relation occurred (see Stevenson and van Kessel, 1996, Ciha, 1982), authors often state that soil water content lead to the differences – either due to air deficiency in spring or due to logging of the plants before harvest.

Higher test weights were found at convex positions than on concave positions by Ciha (1982) and Mulla et al. (1992). Vachaud and Chen (2002) found culm height related to grain yield for different landforms. Ciha (1982) stated, that the tallest plants revealed the highest grain yields. Haque and Lupwayi (2002) and Jowkin and Schoenau (1998) found an increase in straw / stover from shoulder to footslope positions, closely related to an increase in grain yield. This relationship between landscape and amount of stover is even valid for the results of Stevenson and van Kessel (1996), although if there was a higher yield at SH positions.

Ciha (1982) state, that Thousand Kernel Mass (TKM) and number of spikes $\cdot$ m^{-2} may be the most important yield components along slope position. Unfortunately, they presented only the aggregated results for TKM (0.2 g difference) and spikes $\cdot$ m^{-2} (100 spikes m^{-2}) across all grain varieties, which show both an increase from convex to mid to concave positions. Even larger differences for TKM were found by Simmons et al. (1989) with up to 1.2 g difference in 1984 and 1.8 g in 1985. Similar results were found for number of spikes $\cdot$ m^{-2} in 1990 and 1991 by Fiez et al. (1994), who reported differences in spikes $\cdot$ m^{-2} between landforms of around 100.

Grain quality gains increased attention by producers due to economical constrains; therefore, producers try to optimize grain protein content across a landscape. Fiez et al. (1994) mentioned that grain protein content tend to be lowest at north faced backslope positions, and highest at shoulder positions. Fiez et al. (1994) attributed these differences to late season moisture stress due to reduced A-horizon depth at Shoulder positions. Manning et al. (2001b) observed the lowest protein content at Footslope positions due to a greater biomass production together with a dilution effect. Grain protein content by Jowkin and Schoenau (1998) showed no statistical differences between landforms in contrast to the observations above. Still, differences in grain nitrogen in % were found for Shoulder and Footslope position of 3.55 % and 4.4 % for spring tissue sampling and 1.9% and 1.95% at harvest, respectively. Jowkin and Schoenau (1998) attributed those differences in spring to the plant availability of soil Nitrogen.

The use of continuous relief parameters together with yield and yield components for analysis are quite sparse.

Li et al. (2001) determined cotton lint yield in relation to elevation under irrigated conditions and found lower yielding areas at the top positions of their transect. Yang et al. (1998) analysed data for wheat yield, slope and aspect for five different fields and found a negative correlation between grain yield and elevation. Another observation found was that grain yields were highest at gentle slopes, and lowest at steep slopes for four fields with grain yield differences between 0.4 and 2.7 t ha^{-1}. As well, the aspect was found to have a low influence on grain yield with differences between 0.1 and 0.6 t ha^{-1} for northern and southern slopes. The authors used first and third order regression analysis with the parameters elevation, slope and aspect and could predict yield variability for whole fields and for specific field sub units. Yang et al. (1998) found out that third order regression performed better, because the effects of interactions among the topographic attributes were considered. Timlin et al. (1998) used spectral analysis with the relief parameter elevation, curvature and grain yield. They found no dominant component of variance of grain yield and curvature at any particular frequency; however, both variables showed a strong coherency of frequencies. Timlin et al. (1998) mentioned that grain yields were higher in normal to dry years at concave positions, whereas no difference in grain yield was observed in wet years. Kravchenko et al. (2000) used multifractal analysis to specify the relation between slope and crop yield. The authors concluded that higher yields were observed with lower elevation values and lower slopes, as well as that a wider range of grain yields were found at steep slope locations.

The role of topography was not only evaluated in the case of yield and yield properties. Similar research for soil properties with continuous and discontinuous classification has been conducted. One disadvantage for continuous classification is the assumption of a linear change of soil properties with relief parameters. A disadvantage of discontinuous models is the assumption, that the occurring substantial variation will be attributable to differences between sub regions, which are largely uniform with respect to terrain properties (Lark, 1999). Both approaches contain certain advantages and disadvantages: one major drawback of both approaches is that they do not account for nonlinear behaviour of landscape development.

Brubaker et al. (1993) investigated the relationship between soil properties and landscape position. Silt content, pH-value, $CaCo_3$-, Ca-, Mg content increased downslope with a marked drop at FS positions. In contrary, Clay content, Soil Organic Matter (SOM), Cation Exchange Capacity (CEC) and K content decreased with downslope character. The results presented by Brubaker et al. (1993) indicate the highest values of SOM, Clay content, CEC and K content at ridge (in their case interfluve) positions. This may be unexpected since one might have expected to find lower values of OM at that position due to erosion events. The question remains open, why shoulder positions exhibit similar conditions as FS positions.

Manning et al. (2001a) investigated the soil nitrogen and soil moisture distribution in an undulating glacial till landscape using Pennock's landform element complexes (LEC). They found LEC to be useful in capturing gross variability at a manageable landscape level. Increasing convergent character of the landscape led to an increase in volumetric soil water content. Absolute differences between LEC were smallest during times, when soil approached field capacity. Moreover, Manning et al. (2001a) reported, that the absolute differences of soil moisture between LEC decreased during the growing season. Additional, the observed higher nitrate content in the lower LEC may imply a higher mineralization, which was not considered to be the case with season long estimates of N-mineralization (Manning et al., 2001a). Further investigations included Phosphorus, Potassium, and Sulphur, whereby an increasing content was found with increasing curvature. However, differences ($\propto = 0.2$) were hardly significant.

A manipulative experiment was performed by Pennock and Corre (2001) to identify the relationship between different fertilizer rates and N_20 emissions for a field site. The N_20 emission rates were found to be lowest at shoulder positions, whereas backslope positions showed 5-6 times higher rates as compared to shoulder positions across all fertilizer treatments. Rates 18 times higher were observed at level LEC compared to shoulder positions due to better internal drainage and topographically influenced soil moisture conditions (Pennock and Corre, 2001). A low N fertilizer level results in similar N_2O emissions from all LEC; however, applications of larger amounts of N fertilizer led to low emissions at shoulder positions and high emissions at level positions.

Examples for investigations using continuous classification are similar to the continuous yield-relief relationships quite rare.

Sinai et al. (1981) investigated the effect of curvature on moisture distribution in a loess loam. They observed a wide spatial variation at their 70 by 70 m field site with 12 % slope in the Negev desert. Generally, soil water content increased with depth and was strongly correlated with curvature (the second derivative of the surface) with an R^2=0.9. Sinai et al. (1981) stated, that the high correlation might be due to lateral movement and accumulation of water and not due to overland flow. Their investigations were performed under conditions, where the evapotranspiration is higher than under middle European conditions; therefore, the transferability of these results needs to be closely investigated.

Moore et al. (1993a) investigated soil texture, phosphorus, Soil Organic Carbon (SOC) and pH and their relation to the spatial distribution of topographic attributes. The authors used classical stepwise regression to predict several soil parameters. Moore et al. (1993a) identified the relief parameters slope and topographic wetness index explaining up to half of the variability found.

Florinski et al. (2002) used regression analysis to predict soil moisture, solumn thickness, P, depth to $CaCO_3$, and OM from different relief parameters. Florinski et al. (2002) stated that four different regression equations with different coefficients and sets of independent variables were obtained for four different dates. The authors concluded that this is further evidence of the temporal variability in soil topography relationships.

Last, two novel approaches for analysing soil variability should be mentioned; however, such methods were not applied in this work. Willgose et al. (1991a) developed the coupled Channel Network Growth and Hillslope Evolution Model SIBERIA, which might help to understand landscape variability as well as landscape evolution in the future. Another approach is the application of wavelets to study variations in scale and magnitudes, which would have been lost by a low or high pass filtering method (Lark and Webster, 1999).

Generally, no "average" relief attributes can be given to estimate soil properties or yield components. At different scales, as well as in different landscapes, the relief parameter used to estimate soil and plant properties with a variety of methods differed strongly. Adapted regional approaches are needed to describe crop yield and soil variability.

CHAPTER 3

Objective

1. develop and evaluate a relief analysis tool,
2. quantify the spatial and temporal variability of crop yield components for a hummocky loess landscape,
3. quantify the spatial and temporal variability of soil properties for a hummocky loess landscape,
4. quantify the spatial and temporal dynamics of crop yield components and soil properties for the profile curvature classes and the landform element classification,
5. include incoming solar radiation models and the topographic wetness index into an agricultural crop yield model. Evaluate advantages / disadvantages of using such approach for explaining crop yield variability.

CHAPTER 4

Site Description

Investigations were performed at the research farm of the Südzucker company in Luettewitz-Dreißig in south eastern Germany. The farm is located in the agricultural region „Lommatscher Pflege", near Doebeln, approximately 50 km South, South-West of Dresden (51°7'N, 13°14'W; see Figure 1 for a characteristic image). The region is part of a landscape which is described as the middle Saxony loess plateau. A sharp boundary distinguishes the area to the north as the northern Saxony lowlands; whereas, a slow incline in elevation occurs to the south continuing to the Erzgebirge mountains.

Figure 1. View to the Southeast at the field site „Sportkomplex". Image: Reuter.

The relief can be described as a hummocky with plateau like areas interspaced throughout the area. For this landscape, valleys with up to 30 m elevation differences are typical (Härtel, 1931). The Mapping Units of the medium scale soil map (MMK) (Neuhof et al., 1999) characterised the relief in the landscape as a level slope with moderate slope areas. The mean elevation above sea level taken from the topographic map (TK) TK4845 (Scale 1:10.000) is 213 m with a minimum of 117 m and a maximum height of 297 m. The average slope is 4.2° with a maximum slope of 64° found in the area. Slope length analysis revealed a mean L-factor of 1.29; whereas the mean S-factor reached 0.961 (own calculations).

Härtel (1931) identified old palaeozoic sediments underlying the investigation area. The major geological sediments are clayey shale, shist, diabase, tuff, greywacke and conglomerate with strongly disturbed deposits (Härtel, 1931). The old sediments are overlaid by diluvial deposits of loess with an approximate thickness of 25 - 30 m. Härtel (1931) stated that the depth range of this layer decreases from north to south for the geological map 4845 and masks older surface formations. Based on the blow out of fine material in the bald moraine and sand areas, the material has a porous structure with a considerable stability and a calcareous content between 10-14% (Härtel, 1931). However, the top 1 m of the soil profile is depleted of calcareous material and described as silt-loam. Depth profiles up to 6 m reveal a increase in sand content with depth (Table 37 in Appendix Site Description).

The MMK classification (Neuhof et al., 1999) describes the area as partly influenced by groundwater. The major soil unit in the farm area is described as a Luvisol. Among the agricultural land, the areal distribution according to the German soil appraisal (GSA) is L1 (9.7%), L2 (4.2%), L2Lö (1.6%), L3Lö (26.9%), L4Lö (52.9%), L5Lö (3.1%), sL4Lö (1.2%) and 0.4% farm yard and building areas (Figure 3c). The soil fertility is rated between 55 and 79 (100 represents the best soil found for agricultural production under German conditions). The arable soils are rated with 67 (9% of the total area distribution), 70 (25%), 71 (13%) and 76 (21%). When deep chiselling, the farmer of the Südzucker estate noticed stone particles of the underlying shale at the depth of 0.45 m at the field site Gasthof (see Figure 3a).

As mentioned above, the soil is influenced by the depth of the ground water table. Härtel (1931) mentioned that the ground water table depth (DGW) was found between 3 m below soil surface in the alluvial valleys and up to 20 - 35 m at the highest plateaus. Figure 2 shows some of the groundwater conditions at TK4845, with well No. 878 representing a valley condition with differences between minimum and maximum of 0.4 m; however, the fields under investigation are better represented by well No. 1034 (see Figure 3d for the location and Figure 2 for fluctuation of the DGW). This well is in close proximity to the investigation area and shows a much shallower DGW with a mean of 3.96 m. The minimum was 2.71 m and the maximum was 4.93m. These higher differences in DGW were attributed to the geological conditions of the plateaus which contain small ground water storage capacity (Härtel, 1931).

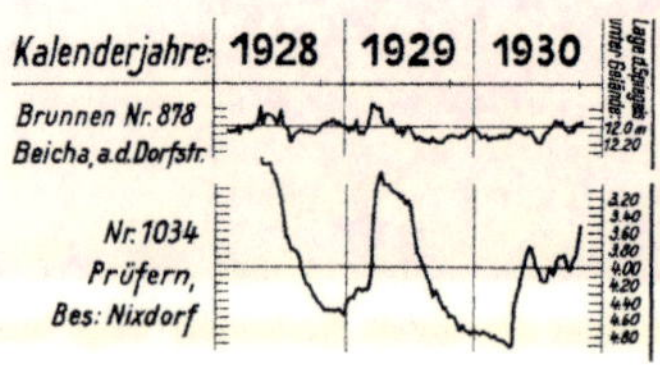

Figure 2. Depth of ground water table (measured biweekly) for the time period 1928-1930 as published by Härtel (1931). See Figure 3d for well location No. 1034.

The meteorological conditions in the investigation area were described in old records with a mean precipitation of 687 mm and a mean air temperature of 8.2°C (200 m above sea level (asl)) for the time period 1866-1925 (Härtel, 1931). Precipitation records for the period 1969-1989 show an annual precipitation sum of 659 mm for the Rüseina precipitation station (51°10', 13°25', 213 m asl) and 578 mm for the Mochau precipitation station (51°14',13°19', 122 m asl). Both stations are in close proximity to the investigation area. The differences of 80 mm of annual precipitation for the two stations approximately 6 km apart from each other, seem to be attributed to the protected conditions of the Mochau station in a valley approximately 90 m below the more exposed position of the Rüseina station at a plateau position. For the annual precipitation in the „Lommatscher Pflege“ of both stations see Appendix Site. The annual average conditions during the time of the investigations are shown in Table 1. Precipitation and temperatures for 10-day periods (decades) for 1999-2001 are also found in Table 38 in Appendix Site Description.

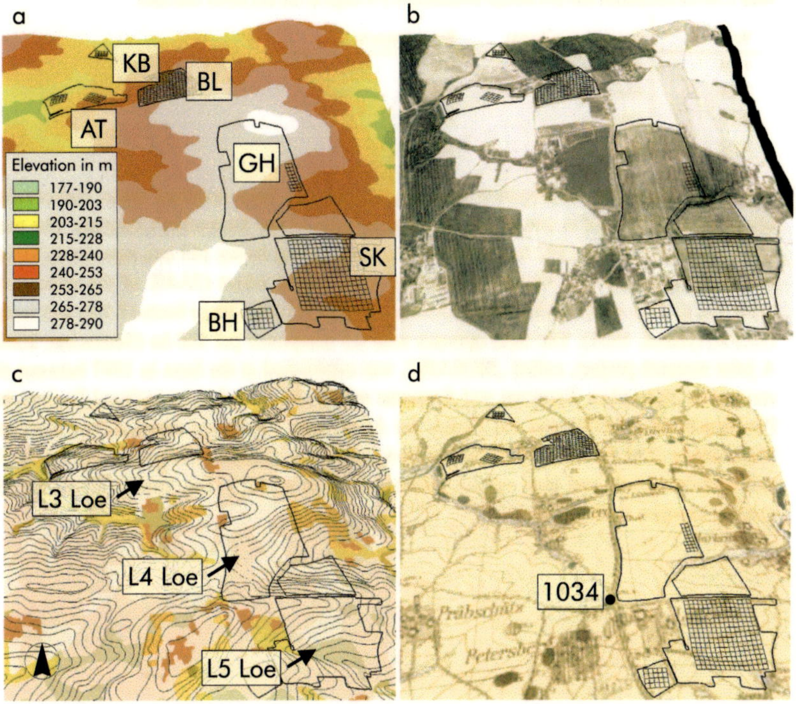

Figure 3. Height in m above sea level (a), satellite image of the investigation area (b), soil types of the german soil appraisal (c), and the geological map (d). Field borders with sampling grids are shown with black lines. Letters in (a) give the field names (AT Am Teich; KB Kleinbahn; BL Bei Lotte; GH Gasthof; SK Sportkomplex; BH Bauhof.). Notice the location of Well No. 1034 in d.

The year 1999 was characterized by elevated temperature in March and April and low precipitation throughout the year to November leading to the further development of drought stress (Löpmeier, 1999). The year 2000 showed a similar beginning with 8-10 K higher temperatures than usual in the third decade of April. The vegetation had to survive the very dry conditions of the following months in the year 2000. A wet July led to a decline in grain yield due to lodging (Löpmeier, 2000). In contrast, the year 2001 was characterized with elevated precipitation with delayed fieldwork in spring. Temperatures in May appeared to be above normal, followed by low temperatures and rainy conditions in June, while in July, thunderstorms followed extremely hot days (Löpmeier, 2001). The frost-free periods for each of the years were March 26 to Oct. 15 for 1999, March 28 to Dec. 18 for 2000, and March 28 to Nov. 09 for 2001. Previous observations for the time period of 1866-1930 documents the period April 26 to Oct. 16 as frost free days (Härtel, 1931).

Table 1. Annual precipitation sum and average temperature in 2 m height for the time period 1998-2001.

Year	Precipitation in mm	Temperature in °C
1998	691.0	8.44
1999	651.4	9.29
2000	556.9	9.58
2001	662.2	8.49

The farm, where the investigation sites are located, has been operated under conservation tillage since 1992, and is one of the 16 research farms of the Südzucker Company. The main crops are winter wheat (192 ha), winter rye (23 ha), spring barley (83 ha), rape seed (31 ha), and sugar beets (38 ha). Historical accounts show winter wheat (33.4% of the agricultural area), rye (6.2%), oat (15.5%), barley (7.5%), potatoes (14.2%), sugar beets (7.9%) and forage (15.3%) (Härtel, 1931). The farm operates with 0.5 manpower per 100 ha (Südzucker, 2000). A joint research project, called „MOSAIK“, was established at the farm in 1997 between the companies Südzucker, Agrocom and Amazone and the Institute for Agricultural Engineering Potsdam-Bornim (ATB) and the Center for Agriculture Landscape and Landuse Research Müncheberg (ZALF). Aims of the project are to identify the spatial relation between crop yield and underlying soil and land surface processes and to adapt a crop growth model for site specific nitrogen fertilization.

CHAPTER 5 *Methods*

5.1. Weather data

Weather conditions were monitored automatically by two weather stations located approximately 3 km apart. The first weather station (REF1, Lambrecht-company, 51°7'33",13°,13'43", 278 m asl) is located in the village of Luettewitz around 50 m south of the farm house (Figure 4). At this station, air temperature in °C, wind speed in ms^{-1}, global solar irradiance (SR) in Jcm^{-2}, relative humidity in %, precipitation in mm and soil temperature in ° C at three depths (5, 10 and 20 cm) were measured at 10 minutes intervals. Data were stored locally on a computer and compiled to hourly and daily values for further analysis, respectively. This weather station gives the advantage of a quick access and fast support, on-site weather conditions were measured using a second weather station (REF2, Thiess-company, 51°08'30", 13°13'16", 256 m asl) established on the field site „Bei Lotte"(Figure 4). Similar to the first station data for precipitation in mm, air temperature in °C, global solar irradiance in Jcm^{-2}, soil temperature in °C at 20 cm depth, wind speed in ms^{-1} and wind direction in degrees were recorded at 10 minute intervals. Precipitation was measured at both locations at 1 m height above ground using a heatable „Hellmann" rain gauge with a precision of recording 0.01 mm rainfall. At both locations air temperature was recorded in 2 m height inside a ventilated protection hut (Ventilation not implemented in the „Bei Lotte" station before March 2002).

Figure 4. Pictures of the weather station at the farm (REF1, left) and at the field site „Bei Lotte"(REF2, right). Images: Reuter.

Additionally in 2002, two logging devices measuring global solar irradiance (WS1, WS2 Thiess-company) were installed for field evaluation of solar irradiance (SR) influenced by topographic shading. The locations of WS1 (51°7'37", 13°14'8", 271 m asl) and WS2 (51°7'27", 13°14'00", 255 m asl) are shown in Figure 5. Before installing the logging devices at the fieldsite „Sportkomplex", WS1 and WS2 were placed at the same location as REF2 for a period of 14 days. Investigations showed differences between the amounts of solar irradiance observed between WS1 and WS2 as well as REF2. The differences of daily solar irradiance sums between two sensors should be smaller than 10% of the daily sum (Adolf Thiess GmbH &CO.KG, 1999). To minimize these differences a least squares fit was performed on hourly SR values to correct values to reference SR values of REF2. The equations used were for WS1 as $SR1_c = 1.0836 \times SR1 - 2.838$ ($R^2 = 0.9883$) with SR1 representing the measured SR at location WS1 and for WS2 with $SR2_c = 1.0788 \times SR2 + 4.4297$ ($R^2 = 0.9882$) with SR2 representing the measured SR at location WS2. Corrected values with negative solar irradiance were excluded from further analysis.

Figure 5. Picture of the field site „Sportkomplex" with locations of WS1 and WS2. Image: Wendroth.

5.2. Soil data

5.2.1. Sampling Design

At the 19 ha field site „Bei Lotte" the main sampling grid layout consists of 192 sampling points, evenly spaced at a separation distance of 27 m (Figure 6). Moreover, additional sampling points located with shorter separation distances to determine spatial covariance at short distances. Five crosses in 1998/1999 with a distance of 1 m as well as five nests with a distance of 5 m distance in 2000/2001 were laid out, to examine small scale variability of mineralized Nitrogen (see Figure 6 for location of the nests).

Additional sampling positions existed on the fields „Kleinbahn" (KB), „Bauhof" (BH), „Sportkomplex" (SK), „Gasthof" (GH) and „Am Teich" (AT). Information about size, area, number of sampling points and crop rotation are given in Table 2. Respective raster layouts were designed on 5 by 5 grid with a sampling distance of 27 m. Locations are shown in the site description in Figure 3 on page 15.

Table 2. Field Names, Size, Number of sampling points and crop rotation. (Sugar beet (ZR), winter wheat (WW), winter rye (WR), rape seed (WRA), spring barley (SB), triticale (TR), unknown (NA).

Field	Kleinbahn (KB)	Bauhof (BH)	Am Teich (AT1) (AT2)	Gasthof (GH)	Bei Lotte (BL)	Sportkomplex (SK)
Area in ha	4.13	2.9	22.34	45.64	19.13	29.63
No- of Points	16	20	25 24	21	192	225/64
Seperation distance in m	27	27	27	27	27	28.8/54
1994	ZR	NA	ZR	NA	ZR	NA
1995	NA	NA	WW	WW	WW	NA
1996	NA	WW	TR	ZR	TR	WW
1997	WW	TR	PE	WW	NA	Tr
1998	TR	SB	WW	SG	WW	SB
1999	SB	WR	SB	ZR	SG	WR
2000	WR	WW	ZR	WW	WR	WW
2001	WR	WRA	NA	NA	WR	WRA
2002	WRA	WW	NA	NA	WRA	WW

5.2.2. Global Positioning System

The Global Positioning System (GPS) consists of 24 satellites with between five and eight satellites visible from any point on the earth. GPS provides specially coded satellite signals that can be processed in a GPS receiver, enabling the receiver to compute position, velocity and time. The signals are specially encrypted due to the military nature of the signal. By using additional references sources (Base Stations with known positions, Satellites without encryption) a system called Differential GPS (DGPS) allows for the determination of the position on earth with horizontal accuracy lower than one meter.

All positions at the field sites were located using a Landstar-DGPS-receiver linked to an ACT-board computer (Agrocom-Company) (Figure 7 A). In cases when an ACT was not useable or available a Pentop equipped with a DGPS receiver (AgLeader GPS2100) was used.

5.2.3. Sampling procedure

Soil samples were taken using three different methods - based on the amount of labour and the accessibility of the field site. During spring sampling, samples were taken by hand, whereas during the post-harvest sampling period, machine augering as seen in Figure 7 was deployed.

Gravimetric water content was sampled in the springs of 1998-2000 to obtain the spatial water content pattern of the entire field. Samples from the soil surface were taken in two replicates in the 0-10 cm layer using a hand auger of 1 cm diameter. The samples were stored in sealed plastic cups (GREINER company), which could withstand temperatures up to 130°C. Sampling for determination of unsaturated hydraulic properties were obtained using 300 cm^3 metal cylinders. These were taken using the Eijkelkamp soil sample Model C kit. Cylinders with a volume of 250 cm^3 were used for measuring soil bulk density.

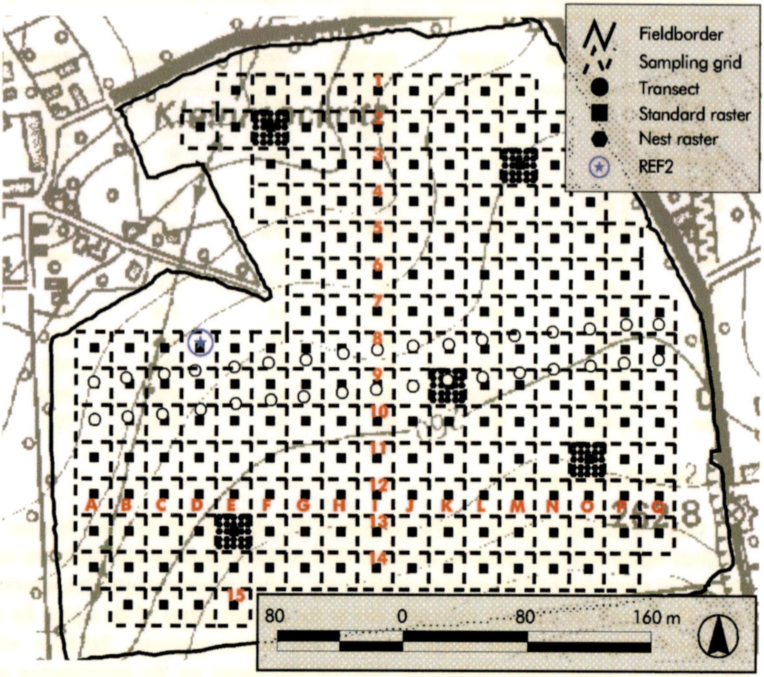

Figure 6. Location of the sampling raster at the field site „Bei Lotte". Locations of the nests are at positions F2, M3, K9, O11 and E13 of the standard raster. Please note the location of the weatherstation REF2 near the standard raster position D8.

In spring samples of mineralized nitrogen (N_{min}) were taken using cone-shape Pürckhauer soil augers together with an electrically powered breaker „Hilti TE805". This way, the labour efficiency was high, while large disturbances of the field vegetation could be avoided. Soil samples were stratified in three different depths: 0 - 30 cm, 30 - 60 cm and 60 - 90 cm, mixed by hand before being placed in plastic bags and placed in a cooling container. Depending on the crop species three (cereals) or two (rape seed) sampling dates were performed prior to fertilization during spring time.

During the post-harvest sampling campaign large soil sampling equipment was used (Figure 7). A buggy based soil auger (2 cm diameter) took 4 - 5 samples down to 90 cm depth and divided them automatically in three different depths (Figure 7 B). The samples were mixed using an electrically powered Mixer (Moulinex-company) (Figure 7 C) and cooled in a deep freezer directly in the field. Throughout all campaigns each sample was identified using a unique number, which related the sample to the depth and the location within the field.

Figure 7. Equipment used for field campaigns. ACT-board computer for positioning (a), the mechanical auger (b), the mixing of soil samples (c) and the processing bench (d). Notice the large white freezer in d, which was used to cool the samples immediately after sampling and mixing. Images: Wendroth, Garlinski.

5.2.4. Soil Analysis

Gravimetric water content was determined from parallel soil samples, containing each around 15 g of soil material. The gravimetric water content in g(100 g)$^{-1}$ (SM) was determined using:

$$SM = \frac{W_w - W_d}{W_d} \qquad \text{(EQ 1)}$$

with W_w as the wet soil in g and W_d as the soil in g after drying at 105°C for a period of 24 hours.

Bulk density (BD in gcm^{-3}) was determined by weighing a sample (W_d in g), which was dried at 105 ° C up to 24 h, and cooled in a desiccator. The bulk density was then calculated by:

$$BD = \frac{W_d}{V} \qquad \text{(EQ 2)}$$

with V as the Volume in cm^3 of the soil sample.

Soil samples were analysed in the central laboratory facility of the ZALF Müncheberg for the following components Phosphorus (P), Potassium (K) (Thun et al., 1991), pH (DIN 10390), mineralized nitrogen (N_{min}) (Thun et al., 1991), mineralized Sulphur (S_{min}) (DIN 11048) and total organic carbon (SOC) (DIN 10694). Textural analysis of soil samples was performed at the ZALF (Institute for Soil Landscape Research) using the sieving method for sand fractions and pipette analysis (DIN 19638, Part 1 & 2, DIN 10693) for silt and clay fractions. Each sample was analysed in three replicates.

Additionally soil properties were determined at the „Bodengesundheitsdienst“ laboratory of the Südzucker AG. The Electro-Ultra-Filtration method (EUF) was used to extract different components using 200 Volt and a time period of 30 minutes for the first fraction at 20-25°C (Signed with a subscript 1 in the later chapters) and 400 Volt and a time of 30 minutes for the second fraction at 80°C (Signed with a subscript 2). The components extracted were NO_3, N_{org}, P, K, B, Mg, S, Ca and K. For a more detailed description of the EUF-method refer to Nemeth (1982).

Penetration resistance was measured using an Eijkelkamp Penetrologger 06.15.01 (Eijkelkamp Agriresearch Equipment, 1999) at field capacity - thereby assuming that soil moisture conditions were similar at each location sampled as mechanical resistance to penetration depends on texture, soil water content, and the bulk density. In general six penetrations were made at a distance of 2 m apart around a given point in a circle down to a depth of 80 cm. The numbers of penetrations were determined from preliminary investigations by plotting the standard error against the number of penetrations as seen in Figure 8. The formula for the standard error is given later. The resolution of the penetration force was 1 N and the depth resolution was 1 cm. The top at the probing rod had a surface of 1 cm^2 (cones were replaced if

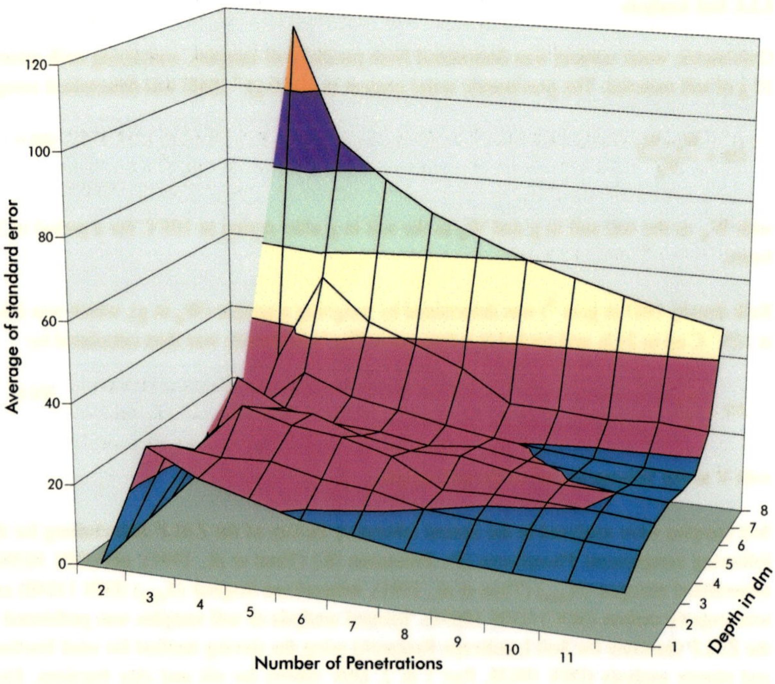

Figure 8. Standard error of soil penetration measurements versus depth in dm and number of penetrations performed.

the surface area differed by more than 10 %) and an angle of 60 degrees. Data obtained from a 10 cm depth increment was averaged.

5.2.5. Sensor based sampling

Airborne Colour-Infra-Red (CIR) photos were taken using an approach described by Jürschik and Flemming-Fischer (1995). A sports aircraft (Cessna 172) was used together with a standard reflex camera (Film material: KODAK-Ektachrome) and a GPS-receiver to obtain relatively low cost images. The film was developed, scanned and georeferenced to the Gauss-Krueger-coordinate system using a set of different map sources. Field borders obtained by DGPS-receivers after harvest, and topographic maps were sources to rectify the images either using the software systems Erdas Imagine, or Topol using a second order polynomial. A set of 15 to 20 ground control points were used to georeference the images to the field borders as well as topographic maps (Scale 1:10.000), given a total Root-Mean-Square-Error smaller than 5 m. Two aspects were not considered during the process of image processing. The radial distortion of aerial imagery due to the underlying elevation was not corrected. Shifts in the spectral frequencies could not be corrected because no ground truth was obtained using a spectrometer. At the time of image processing, software to incorporate available digital elevation models was not considered for the rectification process. The error due to radial distortion (Kraus, 1997) is specified in Table 3. An Arcview - Script to compute the radial distortion - based on a DEM and a vector cover containing the field borders - was developed and is provided at the website http://arcscripts.esri.com/ due to size limitations.

Jürschik and Flemming-Fischer (1995) stated, that the quality of the obtained low cost imagery is sufficient for most applications, even if the standard of accuracy of planned missions is not reached. For the rectified images, a normalized difference vegetation index (NDVI, Rouse et al., 1974) was computed. The spectral bands infrared (IR, 700-1300 nm) and the visible red (VISR, 600-700 nm) yielded the NDVI according to:

$$\mathrm{NDVI} = \frac{\mathrm{NIR} - \mathrm{VISR}}{\mathrm{NIR} + \mathrm{VISR}} \qquad \text{(EQ 3)}$$

NDVI values were averaged, whereby the averaging area differed between the years, due to the developing experience. First an area of 27 by 27 m - representing the grid structure was used, whereas later a round buffer of 10 m around the sampling points was used for data analysis. Additionally, a near infrared image was obtained by videography (NIR).

Ground based plant sensor signals were sampled using two sensors from the company Hydro Agri. The vehicle-mounted HydroN-sensor system (Figure 9 left) recorded the spectral reflectance of leaf surfaces in four different directions for a total area of around 50 m^2 (Hydro, 2000). Based on that value a crop and variety specific nitrogen fertilizer recommendation was given using internal functions based on field experiments (Hydro, 2000). The output of the sensor is the recommended local amount of fertilizer to be applied and a Biomass Index (BI). This index is relevant to characterizing areas with a low plant density (120-140 plants m^{-2},

Agricon Precision Farming Company, 2000), which will receive less fertilizer. If this biomass index were ignored, erroneously high nitrogen recommendations were obtained for areas with a low plant density. Data recorded by the sensor were brought into ArcInfo and averaged to the given raster cells. The HydroN-sensor was calibrated to the local conditions using data from the HydroN-tester. Please note, that if the word „N-Sensor", or „HydroN-Sensor" is used throughout the thesis, actually the N-status of the plants is meant.

Table 3. Radial distortion of aerial imagery for field sites in the Luettewitz area for a flying height of 2000 m.

Field	Field Diagonal in m	Range of DEM Height in m	Radial Distortion in m
Am Teich	787.58	35.64	7.0
Bauhof	230.65	4.68	0.3
Bei Lotte	601.23	23.74	3.6
Gasthof	1075.04	31.61	8.5
Kleinbahn	388.02	12.61	1.2
Plantage	542.84	27.50	3.7
Sportkomplex	894.42	25.00	5.6

The second sensor was the „HydroN-tester" (Figure 9 middle), which measures the chlorophyll content of the plant leaves using an optical system (HYDRO, 2002). Before each nitrogen-sampling campaign the HydroN tester was calibrated against a given standard. A representative site-specific value was obtained from 30 measurements. This mean value was then corrected using a crop variety specific factor (HYDRO, 2002). Additionally, plant nitrogen was determined using a Merck Nitrate test set (Figure 9 right).

Lastly, the soil sensors Veris 3100 as well as the EM38 were used to measure electrical conductivities of the soil.

Figure 9. The HydroN-Sensor during a recording run in 2000, mounted on top of a lightweight vehicle (Left). Manual measurements of leave chlorophyll content using the HydroN-tester (middle) and of the plant-nitrogen content using a Merck Nitrate test set (right). Images: Reuter, Wendroth.

5.3. Crop Yield Sampling and Data Processing

Reliable yield data are one of the most important parameters to be obtained in precision farming, mainly because the harvested yield is the factor on which every strategy or management decision must be evaluated. Based on that demand special care was taken to ensure yield data recording. Two different approaches were applied. The first one was hand harvest, whereas the second approach used was a farm-based combine harvester with an automated yield monitoring system.

5.3.1. Hand Harvesting

Manual yield sampling was performed with a gas driven hedge-clipper at 192 locations at the field site „Bei Lotte“ in 1999, 2000, and 2001 and at 64 locations at the field site „Sportkomplex“ in 2000 two or three days prior to combine harvest. This was performed in order to (I) receive a dataset independent from the combine harvester and (II) to determine aboveground dry matter production and related yield components. Two plots of 0.25 m^2 were sampled for a total sampling area of 0.5 m^2. Plants were cut the nominal distance of 5 cm above ground. The harvested material was air-dried, and analysed for number of spikes, length of internodes, and length of spike. For this work the fourth internode is defined as the distance between the ear base and the next node downwards, the third internode between this node and the next node downwards, and so on. Such approach need to be taken, because the harvest method did not allow to determine the first internode precisely. Afterwards, the material was threshed using a laboratory thresher (KP100 - ZVM Nordhausen). Grains were cleaned from dust, straw particles and weed seeds using an air separator Mini Petkas (RÖNER-company). Dry mass of kernels in g m^{-2} (DM), the total number of kernels (N_{Kernel}) using the Contador E (PFEUFFER-company), thousand kernel weight in g(1000 kernel)$^{-1}$ (TKM), kernel size fractions using a SORTIMAT K (PFEUFFER-company) and plant nitrogen in % (ISO 13878), which was estimated using an LECO-CNS2000 Analyser. The protein content was calculated by multiplying the plant nitrogen in % times 6.5. The number of kernels per spike (K_{Spike}) was computed using:

$$K_{Spike} = \frac{DM \bullet 1000}{TKM \bullet N_{Spike}} = \frac{N_{Kernel}}{N_{Spike}} \qquad \text{(EQ 4)}$$

The second part of the equation was used for the data processing in 1999 and 2001, whereas for 2000, the first part had to be used due to missing values. The error associated using the first part of the equation can be specified with a deviation up to 1600 kernel m^{-2} for spring barley in 1999. This leads to differences in K_{Spike} up to value of 4 at individual sampling locations.

5.3.2. Combine Harvesting

In the combine harvester (CLAAS DO 218 Mega, 270 hP), the yield monitoring system CEBIS recorded the geographical position using DGPS, grain moisture content, grain yield, cutter bar length, and several other information. The yield was automatically corrected for the influence of field slope on yield estimation inside the combine harvester. Besides the automatic records, every grain load dumped to the trailer was recorded manually and later compared to the results of the weighbridge. Grain moisture using a HE-50 (PFEUFFER-company) device and the hectolitre weight were recorded by taking a sub sample from the trailer in double replicates.

Problems existed in 1999 with the automatic recording for the field „Bei Lotte" due to a software problem; therefore, only hand harvest data was available for 1999. For 2000 both datasets exist; however, combine harvest data could only be used after severe assembler programming by Dipl.-Ing. M. Heisig of ATB Potsdam-Bornim to read a PCMCIA-card at which a hardware defect occurred. In 2001 an internal module of the harvester did not work properly. Only data from the hand harvest was available. For the field site „Sportkomplex" data were recorded with a combine harvester between 1997 and 2001. Problems existed in 2001, due to the internal module failure.

The recorded data were transferred via an PCMCIA-Memory card to a personal computer and processed according to a method described by Jürschik et al. (1999) and an internal report of the project MOSAIK (Heisig et al., 2001). In this procedure the following steps are performed:

- Project data from World Geodetic Reference System 84 (WGS84) into the Gauss Krueger Reference System
- Mark and exclude from further analysis all points which contain yield = 0 or where the DGPS was not in differential mode
- Calculate Mean (MW) and standard deviation (SD)
- Exclude from further analysis all points which are outside of MW +/- 2 SD if yield data of the headland are excluded, used values of MW +/- 3 SD if headland areas are included
- Evaluate and exclude from further analysis all points where grain moisture content based on expert knowledge exceed certain limits (see Heisig et al., 2001). Thereby, measured grain moisture of every trailer was compared against the averaged moisture readings from the combine.
- Convert grain yield (Yie_{std}) to desired moisture content ($Moist_{std}$ i.e. 15%) by using the yield values (Yie_{harv}) and grain moisture in % ($Moist_{meas}$) from the harvester:

$$Yie_{std} = Yie_{harv} \bullet \left(\frac{100 - Moist_{meas}}{100 - (100 - Moist_{std})}\right) \qquad \text{(EQ 5)}$$

- The corrected yield was aggregated for each cell and mean and standard deviation are calculated.

5.4. Topographic Analysis

5.4.1. Data Sets

Two different sources of Digital Elevation Models (DEM) were obtained for the investigated sites in Luettewitz. The first dataset consists of an airborne generated laser-scan (LS, also known as LIDAR) with a spatial resolution on 1 m by 1 m and a vertical accuracy better than 0.15 m. The five single tiles were georeferenced and merged using Erdas Imagine. 3D-Views with a 10 times exaggeration showed, that certain points inside agricultural fields contained erroneous information. A filter was used to detect such anomalies and to eliminate these values. The maximum and the minimum values were computed for a window of 3 by 3 cells. If the empirical threshold of 30 (i.e., if the elevation difference between two points was larger than 30 m) was exceeded, the value at that location was removed and marked as unknown. These unknown pixels are also called Nodata-values. The mean elevation values for a 3 by 3 cell window were computed for the temporary grid. These values were then used to replace the Nodata values. For a flowchart of the applied ArcInfo Grid Commands see Figure 10.

For computational purposes Laserscan DEM-data were resampled to a coarser resolution of 10 m by 10 m. Thereby the amount of harddisk storage (100 MB to 1 MB), memory usage and computing time for the area of 200 ha could be reduced significantly.

The second data source was topographic map sheets (TK). DEMs were digitized manually from contour lines and break lines at 1:10.000 (TK10), 1:25.000 (TK25) and 1:100.000 (TK100) map sources using ArcInfo. Again raster DEMs with a resolution of 10 m were created. The accuracy of TK-contour lines is known to be better than 80 cm compared against static DGPS measurements (Grenzdörfer and Gebbers, 2001).

Simple data checks during processing were performed to evaluate the correct allocation of elevation height to the contour lines: (I) using the hill shade command, and (II) looking at the data in a 3D perspective. Further quality assessments of the different DEM's were achieved using visual assessment of slope, aspect, and profile curvature. Failures in data processing were clearly visible and were removed.

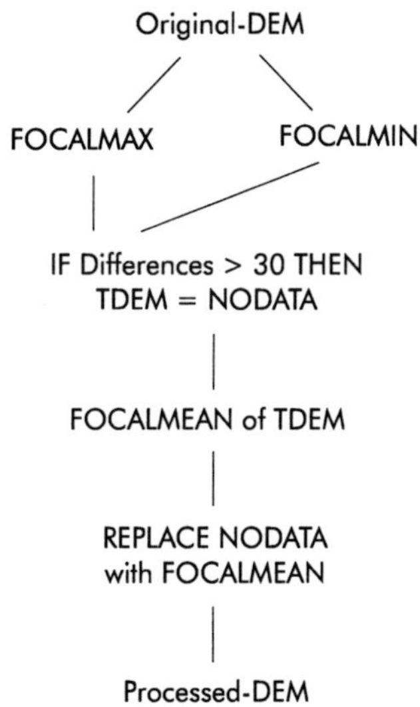

Figure 10. Structural diagram of the ArcInfo procedure to remove „Failure" elevation values.

The resolution of the TK generated DEM was evaluated using the Root Mean Square Slope Criterion (RMSS, Hutchinson, 1996). An algorithm was developed to create DEMs with varying resolution using the TOPOGRID command and to compute the RMSS using ArcInfo. The AML-script is given in Appendix AML.

5.4.2. Methods for topographic analysis

Primary topographic attributes were calculated from directional derivatives of a topographic surface. They include for example slope, aspect, profile and plan curvature, flow path length and are computed using a second-order finite difference scheme or by fitting a bivariate interpolation function (Wilson and Gallant, 2001).

Secondary topographic attributes (i.e. sediment transport capacity, topographic wetness index) are computed from two or more primary attributes and offer an opportunity to describe patterns as a function of landscape processes. An example may be the Topographic Wetness Index, which quantifies the role of topography for redistributing water in the landscape. The TWI assumes steady-state conditions and spatially invariant conditions for infiltration and transmissivity as well as that sub surface flow follows surface morphology.

Primary and secondary topographic attributes were computed for the DEM's using the ArcInfo GRID module. All primary and secondary variables, which can be computed with the developed tools provided in Appendix AML, are summarized in Table 4. The parameters 1-15, based on available ArcInfo commands, were grouped into the *topo* AML-script, which allows to compute a comprehensive relief analysis with one model call (i.e. &r topo <DEM> {streamflow threshold} {streamcover}).

The secondary relief parameters (Number 13-15) are computed according to Moore et al. (1993) for the topographic wetness index (TWI):

$$TWI = \ln\left(\frac{A_s}{\tan\beta}\right) \qquad \text{(EQ 6)}$$

with A_S the specific catchment area in m^2 and β the slope angle in degrees. The stream power index (SPI) was computed:

$$SPI = A_s \bullet \tan\beta \qquad \text{(EQ 7)}$$

and the sediment transport capacity (STC):

$$STC = \left(\frac{A_s}{22 \cdot 13}\right)^{0 \cdot 6} \bullet \left(\frac{\sin\beta}{0 \cdot 0896}\right)^{1 \cdot 3} \qquad \text{(EQ 8)}$$

For the computations of secondary relief parameter, A_S was set to half the cell size, if a Nodata value was observed, β was set to 0.001 if a zero value was observed.

The parameters 16 - 19 are described in more detail later. The parameter 20-26 (Wilson and Gallant, 2001) were implemented in the *elevres* AML-script to compute relief analysis for a given window size with the moving window technique. The command call is: &r elevres <DEM> {cell size}.

The parameter 27 in Table 4 is based on a simple Monte-Carlo simulation approach to account for uncertainties/inaccuracies of the DEM. It computes the TWI n times by adding a given

Table 4. A list of relief parameters computed using self developed AML-Scripts and the SRAD-Program.

No.	Attribute	Unit	Values	Characterization Example	AML-Script
1	Height (M)	M	NA	Height above a specific datum	Topo.aml
2	Slope (SLP)	°	0..90	Gradient, Runoff Rate	Topo.aml
3	Plan curvature (PLAN)	1/100m	-30..30	Contour Curvature, Convergent/Divergent flow	Topo.aml
4	Profile curvature (PROF)	1/100m	-30..30	Slope Profile Curvature, Flow Acceleration	Topo.aml
5	Aspect (ASP)	°	0..360	Slope azimuth, solar irradiance	Topo.aml
6	Flowaccumulation (flacc)	Cells	NA	Accumulated flow to each cell	Topo.aml
7	Flowdirection(FLDIR)	ORDINAL	0-64	Direction of Flow	Topo.aml
8	Strnet(STRN)	Boolean	0/1	Stream network	Topo.aml
9	Basin(BAS)	No.	NA	Unique basin number	Topo.aml
10	Watershed (WSHD)	No.	NA	Unique watershed	Topo.aml
11	RDG	Boolean	0/1	Ridges	Topo.aml
12	PCTG1	%	0..100	Position in landscape based on basin	Topo.aml
13	STC	Unitless	NA	Sediment Transport Capacity	Topo.aml
14	TWI	Unitless	NA	Topographic Wetness Index	Topo.aml
15	SPI	Unitless	NA	Stream power index	Topo.aml
16	SRAD	Wm^{-2}	0..200	Characterize incoming long- and shortwave solar irradiance at sloping surface	SRAD
17	LF	ORDINAL (11)	1-11	raw landforms	Landform.aml
18	LFR	ORDINAL (4)	1-4	aggregated raw landforms	Landform.aml
19	LFC	ORDINAL (11)	1-11	Filtered landforms	Landform.aml
20	MEAN	m	NA	Mean height for filter	Elevres.aml
21	SD	m	NA	SD for filter	Elevres.aml
22	DIFF	m	NA	Range for Filter	Elevres.aml
23	DEV	m	NA	Deviation for filter	Elevres.aml
24	PCTG	%	0..100	Position in landscape	Elevres.aml
25	MIN	m	NA	Min height for filter	Elevres.aml
26	MAX	m	NA	Max height for filter	Elevres.aml
27	MWI	Unitless	NA	Topographic Wetness Index using Monte-Carlo-Simulation	montewi.aml

probability distribution (STD) to the original DEM, and delivers the mean TWI of all model runs. The AML is terminated when (I) the number of given iterations (N) is reached or (II) the solution converges between two iterations smaller than a given threshold value (break). The threshold is computed by dividing the STD by N. The command call is: &r montewi <INPUTDEM> <OUTPUTDEM> <standard deviation> <number of iterations> {break}.

The slope in degrees was computed using the steepest downhill slope method (D8), aspect in degrees as line of steepest descent, and curvature values as the second derivative of the height. For profile curvature this is the direction of the flowline of a cell, whereas plan curvature is the direction perpendicular to that direction. The values are given as 1/100 meter.

5.4.3. Methods for landform analysis

Primary and secondary attributes were used to classify the DEM's into different landforms. A method by Pennock et al. (1987) and Pennock et al. (1994) was implemented to allow an automatic classification. For a distribution of landform elements see Figure 11. In the original paper the slope, profile curvature, plan curvature and the watershed size were used to classify eight different landforms. A limitation existed in this classification, due to the fact that only convex or concave landforms are classified. However, in Pennock et al. (1987) the recommendation is given to use a criterion of +/- 0.116 1/100 m of profile curvature (Young, 1972) to separate planar areas from convex/concave areas. This criterion was added and used to identify three additional relief units (planar landforms) - yielding a total of 11 units (Table 5). The criterion of +/-0.1 1/100 m profile curvature was taken as granted for a DEM resolution of 10 m by 10 m. Pennock et al. (1987) found the extended classification not useful in their study without giving any reasons. Preliminary results for the field site „Bei Lotte“ showed, that certain relief units were biased after the original (without planar) classification, i.e., for the shoulder positions were 16 positions classified as convex and only 2 for divergent, compared to a distribution of 2 for Convergent Shoulder (CSH) and 15 for Planar Shoulder (PSH) positions for the extended (including planar) classification.

Table 5. Classification table for landform elements.

Landform Elements	Abbrev.	Slope in°	Profile Curvature in 1 / 100 m		Plan Curvature in 1 / 100 m		Watershed area in m^2
Divergent SHoulder	DSH	>0	>0.1		>0.1		NA
Planar SHoulder	PSH	>0	>0.1		<=0.1	>=-0.1	NA
Convergent SHoulder	CSH	>0	>0.1		<-0.1		NA
Divergent BackSlope	DBS	>3.0	>=-0.1	<=0.1	>0.1		NA
Planar BackSlope	PBS	>3.0	>=-0.1	<=0.1	<=0.1	>=-0.1	NA
Convergent BackSlope	CBS	>3.0	>)-0.1	<=0.1	<-0.1		NA
Divergent FootSlope	DFS	>0	<-0.1		>0.1		NA
Planar FootSlope	PFS	>0	<-0.1		<=0.1	>=-0.1	NA
Convergent FootSlope	CFS	>0	<-0.1		<-0.1		NA
Low Catchment Level	LCL	<3.0	>-0.1	<0.1	NA		<=500
High Catchement Level	HCL	<3.0	>-0.1	<0.1	NA		>500

DEMs provided by different data sources may contain certain errors from different origin - in maps due to cartographic errors or generalization, in a laser scan DEM due to positioning errors or false values caused by backscattering (Huising and Gomes Pereira, 1998). In a terrain analysis or landform analysis as described above, errors become accentuated, mainly because the planform curvature is the second derivative of the height. As a result, „misclassified“ areas

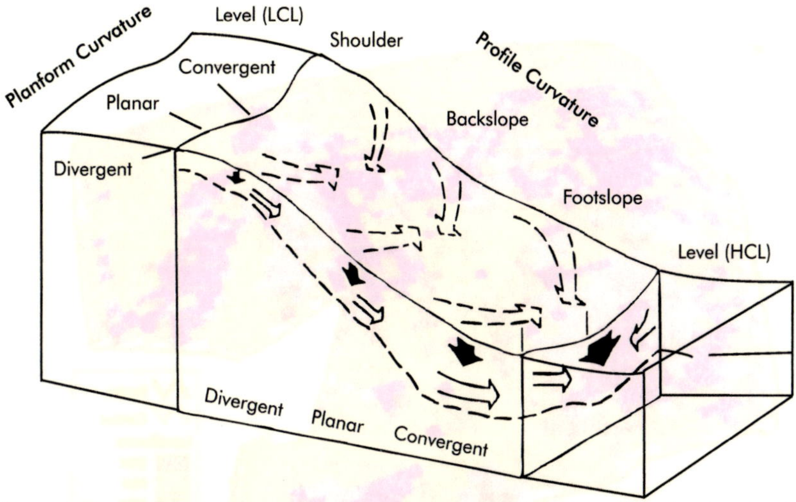

Figure 11. Different landform elements and their probable water movement and concentrations adapted from Pennock et al. (1987). Black arrows indicate vertical infiltration, empty arrows through flow of water, and dotted arrows surface runoff of water and sediments.

can appear in the results. „Misclassified" pixels represent either (I) true micro-topographic landform elements that differ strongly from their surrounding positions, or (II) „misclassified" landforms due to errors in the DEM. Both results increase the difficulties to understand landform relationships connected to other processes. A classification, which minimizes „misclassified" areas was implemented in the GIS ArcInfo based on the work of Pennock et al. (1994) and extended. Five steps were implemented:

- Perform a preliminary landform classification
- Group the classification results into the main relief positions (Shoulder, Backslope, Footslope, Level.)
- Check if four adjacent cells are in the same relief position. If this is the case, no further classification occurs. Otherwise the next two steps are performed.
- First, a clustering is performed to aggregate areas of similar relief positions. Secondly, if one of the adjacent cells of a given cell meets the minimum size criterion (5 cells), the value of that cell is used for the cell in question. Multiple iterations of that step are performed until no further reclassification is necessary. However, if all four cells meet that criterion, the question remains open, which value will be assigned.
- The last step classifies cells, which did not meet the minimum size criterion, and have no assigned value. The modal class of the eight cells surrounding it, was assigned. However, the question remains open, which value might be assigned, if a tie between 2 classes occurs.

The procedure described above was used as a basis to develop the *landform* AML-script (see Appendix AML). The major difference between the published work by Pennock et al. (1994) and the program presented here is, that in all steps instead of the relief positions (SH, BS, FS, LE) the classified landforms were used. Additionally, step four and five used multiple iterations with step five using increasing window sizes to classify cells. Certain cells exist where the modal class of a 3 by 3 window does not allow to determine a landform unit. In this

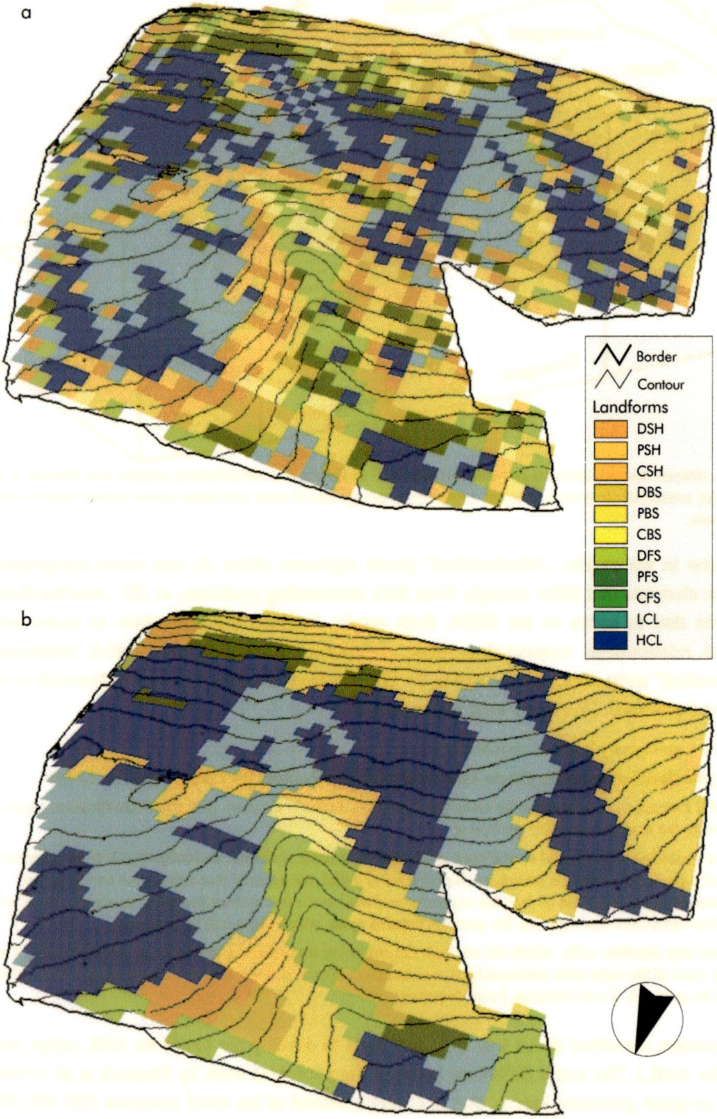

Figure 12. Example of a landform classification for the field site „Bei Lotte". Result of an unfiltered landform classification (a), the same area after the area filtering approach (Threshold was set to 5 cells) (b). The data set was a 10 m by 10 m Digital Elevation Model (DEM) resampled from a 1 m by 1 m Laserscan DEM.

case, an iterative process with increasing window sizes is initiated until a modal value can be determined or a certain threshold depending on computing efficiency (window size is 11 by 11) is met. The arbitrary threshold of 11 was set to the number of multiple iterations in step 5, because no performance gain in classification results could be observed with the inclusion of additional steps. Only two to five single cells remained unclassified after the threshold of 11 for preliminary experiments, which did not satisfy larger computing times. Figure 12 shows the results of relief classification before and after the iterative process. The command call is: &r landform <InputDEM base name> <OUTDEM> {method (eleven or eight LF} {threshold for area filter in number of cells} {threshold profile curvature} {threshold planform curvature} {threshold slope} {threshold for delineation of low and high catchment level areas} {all/ original (extended classification as described in the text or LEC by Pennock} {graphic off/on}. Default values will be assumed if no parameters are provided for a DEM with a 10 m by 10 m Resolution.

5.5. Solar Irradiance Modelling

The distribution of solar irradiance inside an agricultural landscape was calculated using two different approaches. The first used the computationally fast model SRAD, which can be used to calculate the direct and indirect radiation budget on a monthly basis. Secondly the computational time consuming procedure solarflux.aml by Rich et al. (2002), which characterizes the direct irradiance budget for a given day and gives an estimate of the hours of direct sunshine.

5.5.1. SRAD

A short description of the SRAD-model is given, for a more in-depth program description the reader is referred to Wilson and Gallant (2001).

The potential solar radiation is influenced by the following parameters (Wilson and Gallant, 2001): latitude, slope, aspect, topographic shading, time of the year (julian day), sunshine hours, cloudiness.

These parameters influence the short and long wave solar radiation, and are implemented in SRAD. Additionally, air temperatures are calculated and corrected for elevation, slope/aspect and for vegetation via the Leaf Area Index (LAI). Air temperature and potential solar irradiance are then used to compute incoming long wave irradiance. Soil surface temperature and the fraction of the sky visible due to topographic shading are input parameters for the determination of the outgoing long wave radiation. In order to calculate the monthly SR a simulation run was performed every 10 days. For one given day the time interval of computations was set to 12 minutes. Thereby the sky was divided into 16 sections (sky increments).

The model SRAD needs several input parameters, which can be divided into four different categories: 1) elevation data, 2) atmospheric data (concerning solar irradiance) 3) vegetation data and 4) soil and air temperature data. Height data were derived as a floating grid from the laser scan DEM using the FLOATGRID command in GRID for a cell size of 6 m basically due to time and memory constraints; however since SRAD has some limitations with NODATA values, data of the TK-DEM were used to fill the surface around the LS-DEM. For a particular month, an input data file of 6000 columns by 4000 rows was created. The computation time of 20 minutes on a SUN Ultra Enterprise resulted in an approximately 50 MB file.

Table 6. Example of a text file for a SRAD run.

s7.bin	Name of the output file
Y	Specify binary(y)/ ASCII(n) file output
ls015ws.flt	Input DEM as floating grid file exported from ArcInfo
0	Specify grid cell size, if using 0 the parameters will be read from the float file header
5	Specify that DEM is a ARCINFO *.flt file
16	Number of sections the sky is divided into
4	Determine Solar Constant (4 = 1534 W/m2). Results have to be multiplied by 8.64 to convert to MJ cm^2 day^{-1}
1	Lumped transmittance (1) or component approach(2)
bl98.srad	Radiation Parameter File
Y	Specify Temperature Output(Y/N)
07 1 08 1 10	Time period (Start month, Day, End month, Day, Interval of Days in between

The solar radiation and the temperature coefficients for SRAD were parameterized using data from REF1. Using the total global solar irradiance of the weather station described in 5.1 on page 17, the following parameters were calculated as daily values and averaged to monthly means. Due to the enormous weight of the measured solar irradiance on the SRAD parameters, measured SR was first evaluated against simulated maximum clear sky SR (Figure 13). The simulated clear sky SR was estimated using an algorithm described by Penning deVries and van Laar (1982) (Appendix AML), thereby taking the theoretical duration of sunshine based on the geographic position and the mean photosynthetic active irradiance into account. The majority of the measured SR values are below maximum values except for the winter periods, where an overestimation occurred. One possible reason could be the build up of thin ice layers on top of the pyranometer in calm clear sky nights. This can lead to an increase in SR for the first few hours after sunrise up to 50% (Adolf Thiess GmbH &CO.KG, 1999). This reason is supported by the number of 110 calm nights (wind speed minimum below 1 ms^{-1} for the time period 4:00 am to 10:00 am) of a total of 186 days, which showed measured SR exceeding simulated clear sky SR.

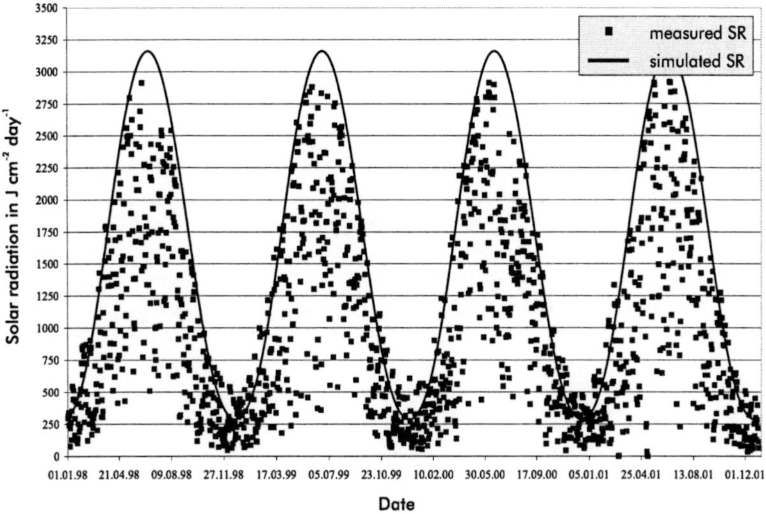

Figure 13. Measured daily solar irradiance (SR) versus simulated daily solar irradiance for the time period 1998-2001 for the weather station REF1.

Further SRAD parameters were computed based on Wilson and Gallant (2001) and McKenney et al. (1999). The Circumsolar coefficient (CIRC, no units) is the fraction of diffuse radiation originating near the solar disk, and is calculated based on:

$$CIRC = \frac{R_{meas} \bullet 0 \cdot 36}{I} \qquad \text{(EQ 9)}$$

where R_{meas} is the incoming solar irradiance in $Jcm^{-2}day^{-1}$, 0.36 a constant to convert Jcm^{-2} in Wm^{-2} and I the solar constant in Wm^{-2} (1354 Wm^{-2} see Wilson and Gallant, 2001). The sunshine fraction (sf) as ratio of the actual sunshine hours (sund in h) to the theoretical hours of sunshine (DL in h) at that latitude is defined as:

$$sf = \frac{sund}{DL} \qquad \text{(EQ 10)}$$

The atmospheric transmittance (AT) describes the fraction of solar irradiance transmitted by the atmosphere. Idso (1969) for example used monthly clear day irradiance divided by the solar constant to determine values of AT. Typical values range between 0.6 and 0.7, but can be as high as up to 0.8 under clear sky mountain conditions (Gates, 1980). The AT was estimated on a daily basis with R_{calc} as the calculated clear sky irradiance in J cm^{-2} day^{-1}:

$$AT = \frac{R_{meas}}{R_{calc}} \qquad \text{(EQ 11)}$$

The cloud transmittance (α) describes the ratio of measured irradiance to possible clear sky irradiance during cloudy periods. The coefficient was estimated for cloudy days (i.e., sf > 0):

$$\alpha = \frac{R_{meas} \cdot (1 - sf)}{R_{calc} \cdot sf} \quad \text{(EQ 12)}$$

Data were calculated based on the measured daily solar irradiance and averaged to monthly values. Data for monthly soil surface temperatures and daily temperatures in 2 m height were averaged from temperatures in 5 cm depth and 2 m height, respectively.

The remaining parameters described are related to the vegetation. The albedo characterizes the fraction of sunlight, which is reflected by the surface. A value of 0.15 for the winter season and a maximum albedo of 0.25 for the summer season was used.

Monthly elevation lapse rates as well as LAI were directly taken from Wilson and Gallant (2001). As an example the parameter file with the specific monthly parameters used for the year 1998 is shown below.The computed monthly irradiance values were imported back into ArcInfo as grids (see sradaggreg.aml, tapesarc.aml in Appendix AML) and average values for every raster cell were determined (rastergrd2.aml- see Appendix AML).

Table 7. An example SRAD parameter file for the year 1998.

1.138 51.143	Latitude min. and max
0.08 0.14 0.24 0.34 0.49 0.47 0.43 0.43 0.26 0.13 0.08 0.07	CIRC (for month 1 to 12)
0.15 0.15 0.16 0.19 0.21 0.21 0.21 0.21 0.17 0.15 0.15 0.15	Albedo
0.16 0.20 0.29 0.45 0.39 0.55 0.47 0.39 0.42 0.34 0.38 0.21	Cloud Transmittance
0.66 0.56 0.52 0.47 0.56 0.46 0.43 0.55 0.46 0.43 0.47 0.71	Sunshine fraction
9.10 10.80 13.10 16.20 21.30 25.90 26.90 25.50 19.10 14.90 8.90 9.40	Max. Air Temperature
-8.60 -12.00 -1.80 4.40 7.10 9.90 10.00 9.50 8.40 2.80 -8.30 -9.40	Min. Air Temperature
2.42 4.35 3.88 9.51 13.58 16.15 16.51 16.69 13.12 8.30 1.09 0.51	Avg . Surface Temperature
6.50 6.50 6.50 6.50 6.50 6.50 6.50 6.50 6.50 6.50 6.50 6.50	Avg. Air Temp Lapse Rate
7.30 7.30 7.30 7.30 7.30 7.30 7.30 7.30 7.30 7.30 7.30 7.30	Min. Air Temp Lapse Rate
6.00 6.00 6.00 6.00 6.00 6.00 6.00 6.00 6.00 6.00 6.00 6.00	Max. Air Temp Lapse Rate
1	Number of LAI profiles
2.25 2.25 2.25 2.25 2.25 2.25 2.25 2.25 2.25 2.25 2.25 2.25	Monthly LAI
10.0 0.96 0.00008 278	Max.LAI, Surface Emissivity, Transmissivity lapse rate, Elevation of reference weather station
0.81 0.74 0.63 0.57 0.64 0.56 0.55 0.64 0.57 0.54 0.62 0.93	AT

5.5.2. SOLARFLUX

The duration of direct sunshine was calculated by using the procedure solarflux.aml (Rich et al., 2002). The procedure solarflux requires a DEM and the day of the year (also known as Julian day, for which Schaab (2000) recommended using three specific days (winter solstice, December 22, summer solstice, June 21, and the spring equinox, March 21). Other parameters were start time, end time, increment of time steps, and the transmissivity of the atmosphere. As start and end times 4.00 and 22.00 were specified, the time steps were spaced similar to SRAD 12 minutes apart and the transmissivity of the atmosphere was set to a value of 0.6. Results were produced directly inside ArcInfo, and mean values averaged for every raster cell were determined (rastergrd2.aml).

5.6. Crop growth, nitrogen and soil water modelling

The crop growth and nitrogen model HERMES (Kersebaum, 1995, Kersebaum and Beblik, 2001) was adapted to calculate the water balance, net-nitrogen mineralisation, denitrification, nitrate transport, crop growth, yield and nitrogen uptake. The model consists of sub-modules for water balance, nitrogen transport and transformations and crop growth. Crop growth reduction by water and nitrogen stress is considered. The model operates on a daily time step using daily weather data for precipitation, temperature, vapour pressure deficit and global irradiance from REF1. Basic soil data (texture, soil organic matter, stone content, wetness, groundwater level), start values (mineral nitrogen, water content and grain yield in the previous year) and management data (sowing, harvest, date and amount of fertilization and irrigation) were provided as initial values. Model runs were performed for each of 225 standard grid points at the field site „Sportkomplex“ for 1998 and 1999 and for each of the 192 standard grid points in the years 1999, 2000 and 2001 for the field site „Bei Lotte“.

To account for topographic shading of solar irradiance a solar radiation correction factor was determined for each location. First, simulated solar irradiance was averaged for the given raster cell size and a given month n ($SISSR_n$). Secondly, the simulated solar irradiance values at the location of REF1 for a given month were determined ($REF1_n$). To obtain this value, 2 m elevation were added to the DEM grid cell of REF1 prior to SRAD-simulation to represent the conditions observed by the pyranometer. The solar radiation correction factor (SRCF) was computed by (see also Kersebaum, 2002):

$$SRCF = \frac{SISSR_n}{REF1_n} \qquad \text{(EQ 13)}$$

By applying those procedures, the measured SR was corrected based on simulated SR-values, which were for each of the simulated points influenced by topographic shading. Additionally, a possible shading of the weather station will be corrected as well. Values below one indicate less SR than observed at $REF1_n$, whereas values above one indicate larger SR than $REF1_n$.

To account for the influence of flow pattern upon soil water caused by topography, the TWI as well as the TWI generated by Monte Carlo simulation (MontWI.aml, see Appendix AML) was computed and averaged for each cell. The median of all TWI values of a given field was subtracted from each TWI value and the difference divided by 2. This step accounts for changes in TWI due to DEM resolution. If TWI values were below zero, values were set to zero. Values above three were set three. These bounded values of the TWI were used as a hydromorphic modification factor (HMF) in the HERMES model.

Model results (x_i) were standardized (signed with subscript SC) with an extended Z-transformation according to Wendroth et al. (2001):

$$x_{SC} = \frac{x_i - (\bar{x} - 2\sigma)}{4\sigma} \quad \text{(EQ 14)}$$

with respect to mean ($\bar{x}$) and standard deviation (σ) in order to evaluate the effect of shading in the simulation of irradiance on a relative basis. The resulting RSD-index for the year 1998 was computed with the measured grain yield in the year 1998 (1998MES), the model run including spatially variable SR (1998R) and the model run with spatially constant SR (1998P) as:

$$RSD_{98} = ABS(1998MES_{SC} - 1998P_{SC}) - ABS(1998MES_{SC} - 1998R_{SC}) \quad \text{(EQ 15)}$$

Additionally, the logarithm of the averaged squared differences between measured and simulated results were provided (Ln(δ)).

5.7. Non-linear optimization techniques

The parameter optimization program PEST (Doherty, 2002) was used to optimize variables to „minimize an objective function compromised of the sum of the weighted squared deviations between certain model outcomes (i.e. simulated dry matter grain yield) and their corresponding field-measured counterparts“. PEST is written in Fortran and is available as freeware for UNIX, DOS and Windows computing environments. It uses a Gauss-Marquardt-Levenberg algorithm, which uses the „least square principal to define a straight line to three parameter-observation pairs and to take the derivative as the slope of the line“ (Doherty, 2002). Any software, which provides some input and output files can be combined with this tool to allow for non-linear optimization. Such were performed for the model HERMES, and the landform classification algorithm. Boundary parameter, input parameter, start values, and increments were chosen, if not known beforehand, in a iterative process, evaluating optimized parameters against the model results.

5.8. Statistical Methods

The coefficient of variation (CV) is determined using s as the standard deviation and $\bar{\chi}$ as the arithmetic mean:

$$CV = \frac{s \cdot 100}{\bar{x}} \quad \text{(EQ 16)}$$

The standard error of the arithmetic mean s_e is defined as:

$$s_e = \frac{s}{\sqrt{n}} \qquad \text{(EQ 17)}$$

with n as the number of samples. The Nalimov-test (see Lozan and Kausch, 1998) was used to test for outliers (x_i) in datasets using:

$$r^{o} = \frac{|x_i - \bar{x}|}{s}\sqrt{\frac{n}{n-1}} \qquad \text{(EQ 18)}$$

Significant differences between soil properties and yield components for relief parameters were determined using SPSS by applying the following procedures: Kolmogorov-Smirnov(KS)-test for normal distributions, followed by test of Error-variances (Levene, α = 0.05). Groups contained different numbers of samples, therefore, if similar variances and normal distributions were observed, Scheffe tests were applied, by different variances the Games-Howell-test was used. If data showed no normal distributions, the KS test with α = 0.2(Manning, 2001a,b) was applied. However, results by statistical analysis are beyond the scope of this work and will not be presented.

Geostatistical analysis was performed using ISATIS software (Bleines et al., 2000). Due to missing import filters for ArcInfo Coverages at the time of analysis a conversion program between ArcInfo and ISATIS was developed (Appendix AML). Experimental semivariograms describe the spatial covariance of a variable with itself as a function of distance(lag distance). Semivariograms were computed as a function $\gamma(h)$ of the distance h for N pairs of values of A_i at a location x_i:

$$\gamma(h) = \frac{1}{2N(h)}\sum_{i=1}^{N(h)}[A_i(x_i) - A_i(x_i + h)]^2 \qquad \text{(EQ 19)}$$

Correlation neglects the spatial distribution of its properties. Therefore, to take into account the measurement locations and to describe the relationship between two variables in time or space, a cross correlogram was computed as a function $r_c(h)$ of a property A_i and another property B_i at locations x_i and x_i +h (Nielsen and Wendroth, 2003) with cov denoting the covariance and var the variance:

$$r_c(h) = \frac{\mathrm{cov}[A_i(x_i)B_i(x_i + h)]}{\sqrt{\mathrm{var}[A_i(x_i)]}\sqrt{\mathrm{var}[B_i(x_i + h)]}} \qquad \text{(EQ 20)}$$

The sill is the value, at which a computed semivariogram reaches a limit - the total semivariance of the dataset. The distance up to this limit is defined as the range. The value a semivariogram reaches at the lag distance zero is defined as the nugget. The Nugget to Sill ratio (NS-ratio) was computed by fitting a semivariogram model to the dataset and dividing the nugget semivariance by the sill semivariance (Cambardella et al., 1994). For nested semivariograms the sill semivariances was computed as the sum of the nugget and the semivariances of all structures. If a linear semivariogram was fitted, the sill and range was derived from the intersection of the fitted function with the semivariance of the data. Moran I and Geary C (Goodchild, 1986), which describe the spatial autocorrelation or similarity between objects, were computed using ArcInfo Grid commands.

CHAPTER 6

Results and Discussion for relief parameters

6.1. General Description of Relief Parameters

The results of the relief analysis using the topo.aml script given in the Appendix AML are presented as an overview in Figure 14 and Figure 15. These figures represent different relief parameters, starting with the elevation (Figure 14a) . The first derivative of the elevation is the slope, which describes the steepest descent between the center raster cell and all neighbouring cells (Figure 14b). The direction of this maximum difference can be used to determine the aspect (Figure 14c). The second derivatives of the elevation are the planform curvature and profile curvature, which describe the shape of the landscape perpendicular to the direction of the slope, and along the direction of the slope, respectively (Figure 14e, f). Further relief parameters that are computed directly of the elevation are the flow direction (Figure 14d) and flow accumulation (Figure 15g). If both these parameters are determined, basin areas and watershed areas can be determined (Figure 15h, j).

Secondary relief parameters are based on a combination of primary relief parameters. The topographic wetness index (Figure 15k), the sediment transport capacity or the landscape positions serve as examples (Figure 15l, m).

The descriptive parameters of the relief analysis, averaged to the 27 m by 27 m raster, are shown in Table 8. Special attention should be drawn to the highly uneven distribution for certain parameters (e.g. flow accumulation); therefore, the mean represents a biased value. The high values of CV values for some relief parameters are quite erroneous, as the mean value is close to zero or even negative (Krebs, 1998).

The spatial autocorrelation based on Moran's I and Geary's C showed similar, regionalized values for all height parameters (M, Max, Mean, Min, SD). The values of Moran's I approaching 1 and Geary's C approaching 0 are similar to values shown by Lee and Marion (1994) for 920 DEM's in the continental US.

Results of the second derivative of the height (PROF, PLAN, CURV) showed dissimilar, checkerboard patterns for Moran's I and Geary's C. The autocorrelation is less than that of the height relief parameter. A significantly larger autocorrelation was found for the second derivative as compared with those of the first derivative (ASP, SLP) and other relief parameters (TWI, FLACC, LFC).

Table 8. Descriptive Statistics of relief parameters for the field site "Bei Lotte" for a 27 m aggregation. Moran's I and Geary's C were computed based on the 10 m by 10 m DEM. Filter size was set to 200 m.

Relief Parameter	N	Min	Max	Mean	CV	Moran I	Geary C
Aspect (ASP)	192	4.07	351.27	255.02	39.65	0.75	0.24
Curvature (CUR)	192	-0.77	0.29	-0.01	-1411.08	-0.26	1.26
Deviation (DEV)	192	-0.70	0.52	-0.01	-4242.61	0.88	0.10
Flow accumulation (FLACC)	192	0.11	469.00	24.01	218.17	0.31	0.68
Flow direction (FLDIR)	192	14.67	112.00	45.29	45.56	0.75	0.24
Landform elements classified (LFC)	192	1	11	8.21	37.79	0.79	0.20
Height (M)	192	247.82	267.54	258.69	1.59	0.99	0.00
Maximum Height for filter (MAX)	192	258.02	285.18	269.83	2.48	0.99	0.00
Average Height for filter (MEAN)	192	249.74	267.14	258.75	1.50	0.99	0.00
Maximum Height for filter (MIN)	192	239.45	258.29	248.94	1.76	0.99	0.00
Position in the landscape (PCTG1)	192	1.48	97.11	47.80	46.50	0.89	0.08
Planform curvature (PLAN)	192	-0.46	0.14	0.00	-4434.90	-0.09	1.08
Profile curvature (PROF)	192	-0.19	0.32	0.01	975.05	-0.23	1.23
Standard deviation for filter (SD)	192	2.50	5.54	4.31	15.54	0.99	0.00
Slope (SLP)	192	0.50	5.96	2.78	40.03	0.60	0.39
Topographic Wetness Index (TWI)	192	7.56	12.87	9.65	10.88	0.45	0.54

The landform classification process, as described in 5.4 on page 26, was applied to the above described DEM. Results are presented as a 3D-view in Figure 16. The lower part of the Figure 16 (b) represents the detailed classification of eleven landform units. Notice that two classes, CSH and DFS are not represented in the figure. A more aggregated image is shown in the top portion of Figure 16 (a). The view is presented from northeast to southwest. The field is classified for the most part as level (blue), with one valley area directing to the front. This is classified as a Footslope (FS) position (green). Some shoulder positions (red) are visible around the FS positions. The second largest area is covered by Backslope landform positions (yellow), which appear in the rightmost upper part of the image.

Generally, the sequence of a geomorphologic landform catena would be Level (LE) - Shoulder (SH) - Backslope (BS) - Footslope (FS) - Level. Especially, the change from SH-BS-FS is not always visible in the Figure 16. This is mainly an artifact of the algorithm due to the classification and filtering process. In the unfiltered LF classification 10 m by 10 m DEM results, such sequences are more clearly visible (see Figure 17). These LF positions are strongly scattered and will therefore be eliminated by the iterative process. This is a disadvantage of the chosen raster based approach. With a tin based classification, linear zones along a valley border might be represented more adequately (see Ivanov et al., 2003).

Initially, an unsupervised classification algorithm using the ISODATA command in ArcInfo as found by Irvin et al. (1997) was implemented using different subsets of the relief parameters elevation, slope, profile / plan curvature, TWI and incident solar radiation. Irvin et al. (1997)

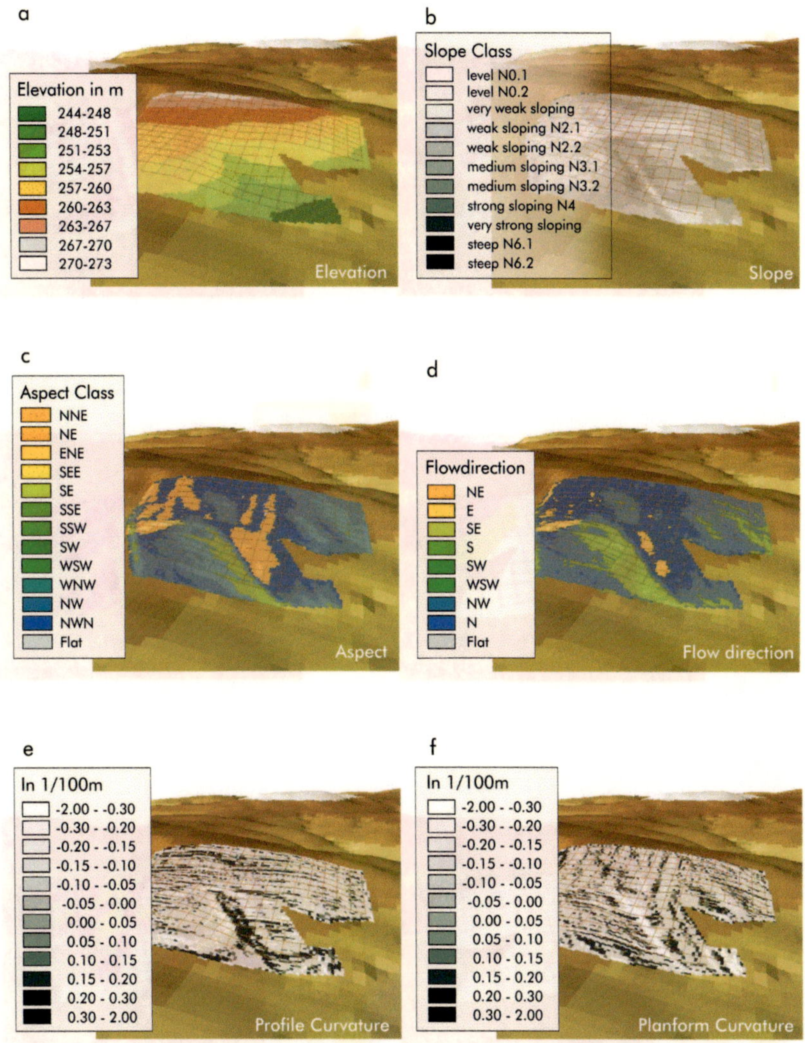

Figure 14. Examples of relief parameters for the field site "Bei Lotte" computed from the Laserscan-Digital Elevation Model. Elevation (a), Slope (b), aspect (c), flow direction (d), profile curvature (e) and planform curvature (f). The parameters were computed using the topo.aml and shown as a 3D-View to the south.

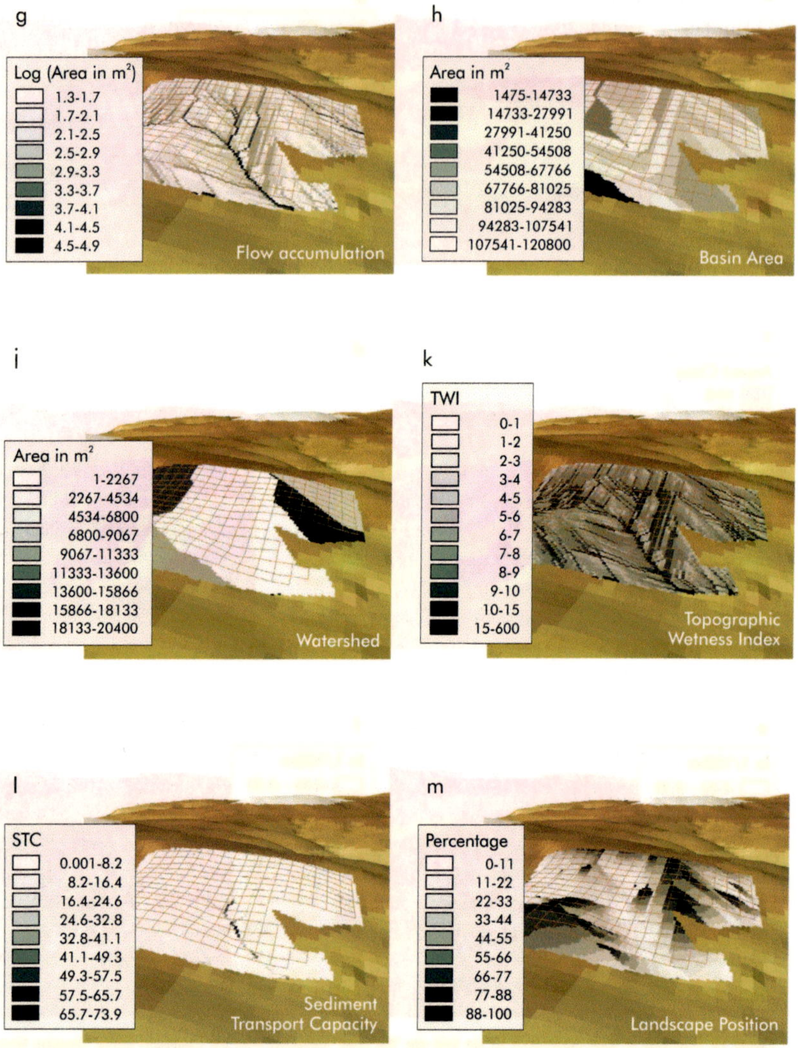

Figure 15. Examples of relief parameters for the field site "Bei Lotte" computed from the Laserscan-Digital Elevation Model. Flow accumulation (g), basin area (h), watershed area (j), Topographic Wetness Index (k), Sediment Transport Capacity (l) and landscape position (m). The parameters were computed using the topo.aml and shown as a 3D-View to the south.

compared their results against manual delineations and found similar results, however more detail was included in the automatic approaches. Results obtained from the unsupervised classification algorithm for the Luettewitz data set led to a low number of classes, which were found difficult to interpret. Still, additional information about each landform had to be determined.

6.2. Comparison of the Original and Extended Landform algorithm

The landform classification algorithm presented extends the original approach (Pennock et al., 1987, 1994) by adding planar LF classes. Additionally, the shape of the LF (e.g., Convergent FS instead of FS alone) is conserved in the classification process. The change in the methodology is evaluated against the results of the original coding in the following part.

This evaluation is shown as an example in Figure 17 for the field sites "Bei Lotte" (left) and "Sportkomplex" (right). The 400 ha 10 m by 10 m DEM, including the investigated sites, was classified using the landform.aml with an area threshold of 5 pixel, a value of 0.1 for planform curvature and profile curvature, a value of 3.0 for slope and a threshold of 500 m^2 to differentiate Level Landforms. The original classification results in so called Landform Element Complexes (LEC, Pennock and Corre, 2001), which include only Shoulder, Backslope, Footslope and Level. To compare both methods, the extended data sets were reclassified from the extended (including planar landforms) 11 LF's to the four LECs. Figure 17 a+b represents the original classification, whereas Figure 17 c+d shows the reclassified extended classification, and Figure 17 e+f shows the differences between both classification approaches.

For the field site "Bei Lotte" a larger zone of SH positions is found in the northern part of the field, surrounding the depressional area running SW-NE. Additional FS and SH positions are classified at the southern end of this depression, which are not as distinct in the extended classification (Figure 17d). These differences are visible in Figure 17f, representing an area of 280 m^2 for the SH and LF positions and an area of 250 m^2 at the southern end of the field. Several extended structures in east-west direction are visible in the original classification (Figure 17b), which are not represented by the extended classification. Finally, the differences between classifications are found at the field borders: It should be noted that classification was performed in each case for a larger area, therefore the clipped area contains only the boundary effects of the field site and no boundary effects of the DEM itself.

Similar results can be found for the field site "Sportkomplex". Again, the original classification resulted in an image with more pronounced heterogeneity. Especially the major west-east running depressional area (FS), which is well represented. Certain structures (see the small SH-BS position visible at the northern side of the depressional area) are better represented at the original classification (own investigations), than with the extended classification. Some linear east-west running LF's classified using the original classification, which are not visible

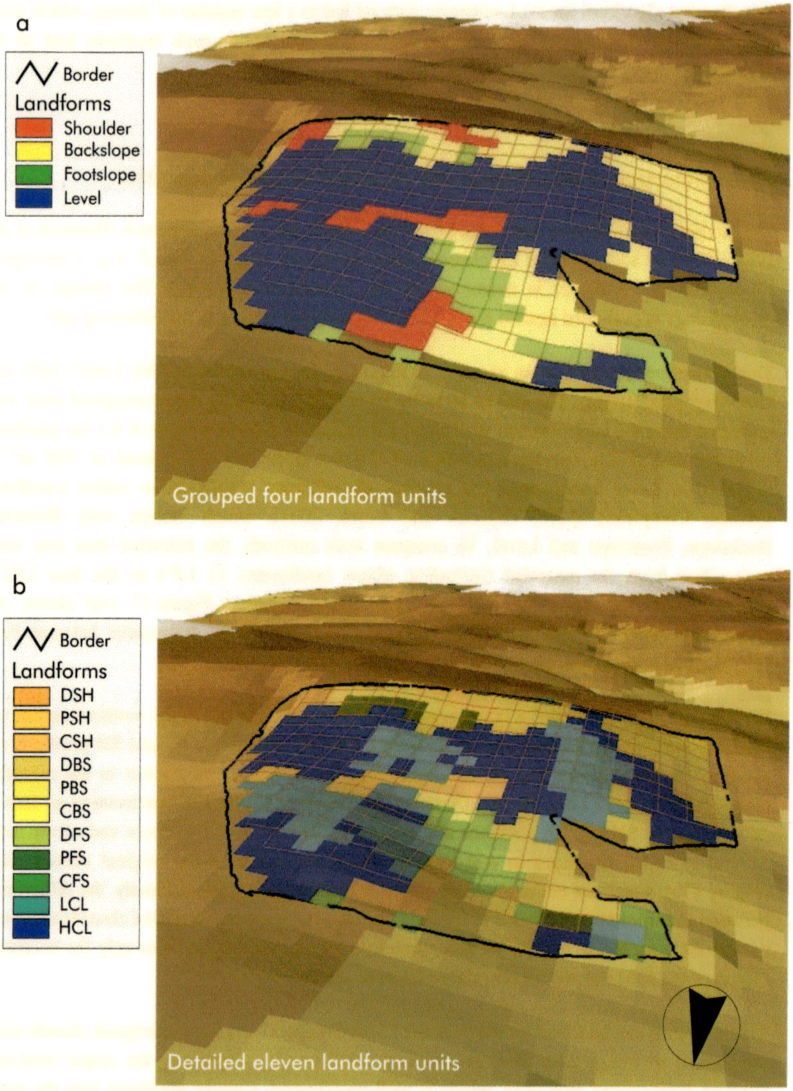

Figure 16. Landform classification for the field site "Bei Lotte" using the landform.aml. (a) aggregated landform units grouped for landform position (Shoulder SH, Backslope BS, Footslope FS, Level LE). (b) detailed landform classification with eleven landforms (Notice that the CSH and the DFS do not occur). View to the south.

in Figure 17c, and could not be validated with the gained field experience. Similar to results for "Bei Lotte" differences are found at the field borders as well as single scattered LF's.

Generally, differences between different LF classifications can be seen for the original and the extended approach: (I) associated with the field border, (II) related to certain linear features, (III) around the borders of certain SH and FS positions and (IV) single LF elements distributed throughout the field sites. The original approach has the advantage of classifying more consistent areas for SH and FS positions (see SH and FS at field site "Bei Lotte"). One major disadvantage for the original procedure is the missing underlying shape of the landform, which is provided by the extended classification. Additionally, the linear shaped contours of landform features observed by the original classification disappear in the extended classification.

Generally, two factors influence the LF-classification results: the algorithm itself and the resolution and accuracy of the provided dataset. This effect should be evaluated for the LF classification, which contributes considerably to the resulting landforms. As outlined in 5.4 on page 26, an iterative classification process is applied based on an area threshold. This is performed to remove certain small area LF pixels, which are results of a local micro topography or of failures in the DEM and increase the difficulty to interpret the data set.

Two different approaches using a majority aggregation for the 27 m x 27 m investigation raster are shown in Table 9, one without the area threshold (raw data- please see Figure 12 on page 32) and the other using an area threshold (Filtered data see Figure 16). The number of SH positions decreases from 26 to 20 as well as for FS positions from 20 to 16 using the filtering approach, which in turn leads to an increase in the number of Level positions (96 to 110). Even if we loose some information about landforms at SH and FS positions using the area threshold approach, it is certainly an improved way to aggregate LF rather than just using the raw data for two reasons: first, the data might hardly be useable due to the highly scattered appearance (see Figure 12 on page 32), and second the results shown here, depend strongly on the cell size of the aggregation cover.

Table 9. Frequency of unclassified and classified landform elements for the field site "Bei Lotte" using a majority filter for the 27 m by 27 m sampling cells. LF classification was based on the Laserscan-DEM with 10 m resolution.

Landform	DSH	PSH	DBS	PBS	CBS	PFS	CFS	LCL	HCL
Filtered	2	15	3	43	3	7	9	47	63
Raw	11	17	8	36	4	16	4	44	52

6.3. Quality of relief parameters obtained from different data sources

The process of landform classification is mainly based on the quality of the underlying relief parameter. The differences for the relief parameter profile curvature will be shown as an example for data sets with different quality in Figure 18. The data presented is a subpart of the

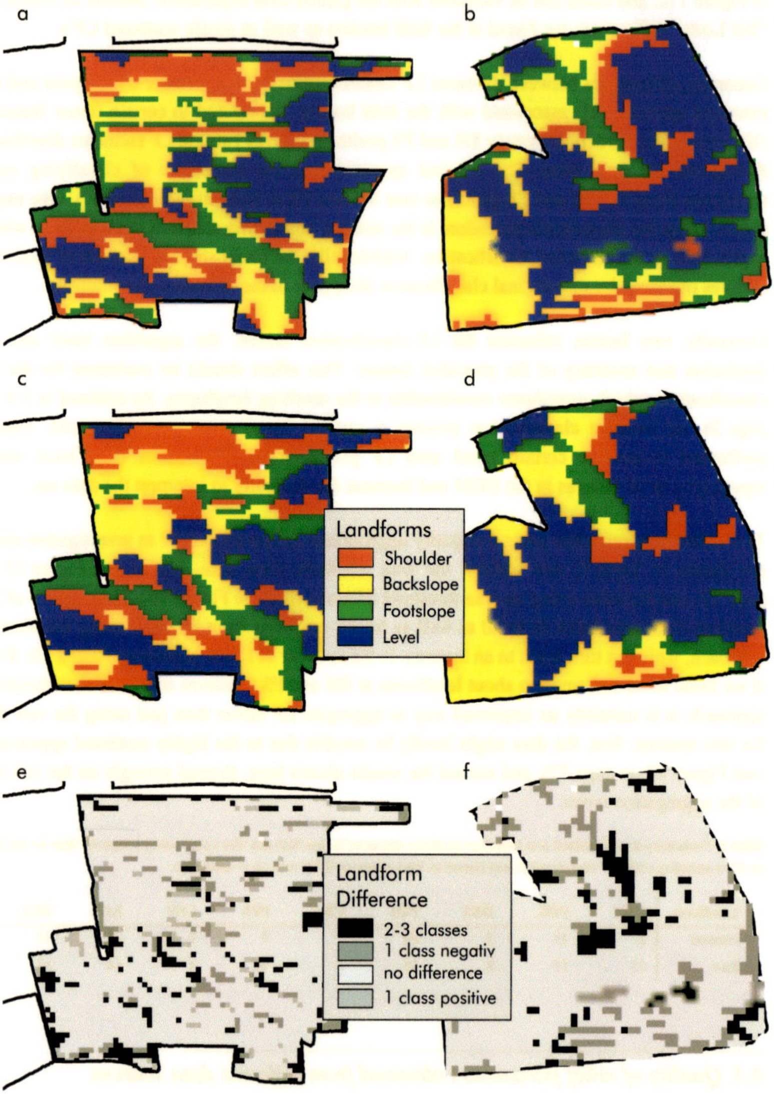

Figure 17. Pennock's original landform classification algorithm using only major landforms for the field site „Sportkomplex" (a) and „Bei Lotte" (b). Results of the extended landform classification algorithm for the field site „Sportkomplex" (c) and „Bei Lotte" (d). Differences between both approaches for the field site „Sportkomplex" (e) and „Bei Lotte" (f).

field site "Bei Lotte" using the LS-DEM and the TK-based DEM (see Table 11 for statistical parameters).

Profile curvature was computed using the ArcInfo *curvature* command, which utilizes the equations derived by Zevenbergen and Thorne (1987). These were ranked second regarding quality from four different methods investigated by Florinski (1998). Profile Curvatures for the TK-DEM are in most cases closer to zero and show less variation than LS-DEM profile curvatures from the visible observation, even though the statistical parameters in Table 10 do not reflect this.

Table 10. Statistical parameters for profile curvature aggregated for 27 m x 27 m sampling cells for the Laserscan (LS)- and the topographic map (TK)-DEM at the field site "Bei Lotte".

DEM-TYPE	STD	Mean	Min	Max	Median
LS-DEM	0.1115	-0.0059	-0.4320	0.5200	-0.0011
TK-DEM	0.1185	-0.0002	-0.4659	0.8102	-0.0009

Results from the TK-DEM are „smoother" as less variation occurred. Further investigations showed that the apparently lower variation led to a disappearance of small-scale heterogeneity of the landscape. To quantify the generalization process or „loss" of variability, the datasets TK10, TK25, TK100 and a DEM generated from contourlines of the LS with a vertical resolution of 0.1 m for the agricultural area (LS10) were examined. The statistical parameters

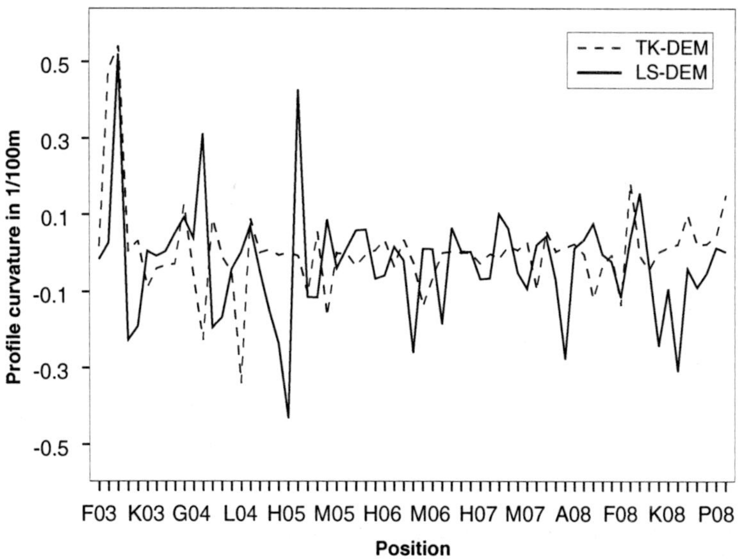

Figure 18. Example of difference in profile curvature aggregated for a 27 x 27 m raster for the laserscan (LS) - and the topographic map (TK) - DEM at the field site „Bei Lotte"

from each dataset for the relief parameters are important for the described landform classification and are reported in Table 11.

In general, with an increasing spatial resolution of the data sets, the median slope and maximum slope decrease (except the TK25). This is especially important as the slope is used to separate BS landforms. The profile and planform curvature show a decrease in minimum and maximum values in the order of the Datasets TK10 > TK25 > TK100. The loss in resolution, especially of the small scale topographic features, are visible in the high resolution data sets (LS10, TK10). From a statistical point of view the TK10 DEM and the LS10 DEM yield very similar values for minimum, maximum and median; however, LF classification results discussed below still show some important differences.

Finally, DEM accuracy was evaluated based on flow accumulation, which provides a representation of the flow pattern. According to Holmes et al. (2000) compound relief parameters, which are affected from a large number of cells on a DEM, are most drastically influenced by elevation data errors.

As the data sources of the generated DEM gains precision, a more detailed flow network is constructed. With lower precision, an increase in the median values of flow accumulation would be expected, which is visible in the order of LS10 > TK10 > TK25. This agrees with Gallant and Hutchinson (1997), who showed that with increasing spacing of elevation samples, fine scale features are lost and the surfaces became more generalized. In general slope gradients decreased and flow accumulation increased (see also Thieken et al., 1999).

The differences in statistical parameters in Table 11 show for certain parameters only small changes. These differences are smaller than results by Wilson et al. (1998) who tested the intensity and patterns of RTK-GPS collection on elevation and topographic attributes. Differences of up to 25 % in slope, 38 % in Specific Catchment Area (SCA) and 22 % in Topographic Wetness Index (TWI) were found (Wilson et al., 1998). Nevertheless, even with the low differences found, the influence of the quality loss in input DEM on the LF classification process is being tremendous. The results of the LF classification of different datasets are presented in Table 12 for the field site „Bei Lotte".

The analysis showed that with decreasing resolution (e.g. TK10 < TK25 < TK100) a more generalized image was produced. Detailed information about specific landforms is lost, for example a loss of 62.5% of all SH position between the LS10 and the TK10 and 87.5% from LS10 to TK25. A similar pattern can be observed for FS positions, where losses occur in both the TK10 (18%) and the TK25 (35%) as compared to the LS10 dataset. Both these LF positions characterize specific positions in any given landscape with specific geomorphologic properties (e.g. plant development, erosion processes). The „lost" LF areas were mainly classified either as BS or as LEVEL areas. The BS positions increased by 31% in the TK10 and by 42% in the TK25 analyses, when compared to the LS10 dataset, whereas, for the TK100 a reduction for 39% is found. The total number of Level LF's for LS10 and TK10 reached similar values. Still the number of LCL cells was lower for the LS10 dataset. The TK25 and

TK100 datasets showed an increase of 10% and 52% for LEVEL areas compared to the LS10 dataset.

Table 11. Statistical parameters for the relief parameters slope, profile curvature, planform curvature and flow accumulation of generated 10 m resolution DEM. Input data were derived from the datasets Laserscan (LS10), TK 1:10.000 (TK10); TK 1:25.000 (TK25) and TK 1:100.000 (TK100).

LS10	Slope	Profile Curvature	Planform Curvature	Flow Accumulation
Median	2.71	0.01	0.01	9.17
Minimum	0.50	-0.19	-0.46	0.11
Maximum	5.96	0.32	0.14	469.00
TK10				
Median	2.67	0.00	0.01	13.24
Minimum	0.53	-0.29	-0.36	0.50
Maximum	4.66	0.26	0.17	565.00
TK25				
Median	2.63	0.00	0.01	14.50
Minimum	0.89	-0.17	-0.20	0.50
Maximum	7.30	0.22	0.08	453.67
TK100				
Median	2.14	-0.01	0.01	9.00
Minimum	0.09	-0.08	-0.05	0.50
Maximum	4.54	0.02	0.05	92.50

The purpose of this investigation is not to find any scale differences. For such an approach different cell sizes would have to be computed; instead, the aim has been to specify the loose of precision in the determination of landforms by using a less precise dataset. Especially as the "lost" SH and FS positions are environmentally and economically important, care should be taken to acquire the dataset with the highest quality for landform analysis. Still, LS datasets presented in this analysis are out of reach for most researchers and farmers due to cost of collecting the high level of precision.

Table 12. Number of classified landform (LF) units for a 10 m Digital Elevation Model (DEM) for the datasets Laserscan (LS10), TK 1:10.000 (TK10); TK 1:25.000 (TK25) and TK 1:100.000 (TK100) at the field site "Bei Lotte". Classification parameters were 0.1 for planform and profile curvature, 3.0 for slope, the area threshold was set to 5 and the LEVEL condition was set to 500 m^2.

DEM/LF	DSH	PSH	DBS	PBS	CBS	PFS	CFS	LCL	HCL
LS10	1	15	4	42	3	8	9	47	63
TK10		6	8	56		11	3	51	57
TK25		2		54	5	10	1	42	78
TK100				30				79	83

In the previous section the disappearance of certain LF elements is discussed if classification parameters remain stable over different quality datasets with the same resolution. To overcome the disappearance of LF elements, parameters in the landform classification process such as profile curvature or DEM resolution would need to be optimized.

The nonlinear optimization program PEST was used together with the landform.aml to optimize planform and profile curvature, slope and the threshold value. DEM resolution was not varied. The optimization shown in Table 13 is based solely on the number of grid cells of certain LF units observed for the agricultural area of the LS-DEM under defined LF classification conditions (Parameters are provided in Table 13 under heading LS). These results are compared against the number of grid cells of certain LF units by varying LF classification conditions. The number of LF units was determined for DSH, PSH, PFS and CFS from the LS-DEM for the agricultural area. These LF's were chosen carefully since these are locations of environmental and ecological importance. Input datasets were DEM's with 10 m resolution generated from the map sources TK1:10.000, TK 1:25.000 and TK 1:100.000. The ranges for curvatures were allowed to vary between 0.001 and 0.4 and for slope between 0.001 and 5. The threshold value was allowed to vary between 1 and 8. The boundary optimization values for the TK25 had to be set to a lower slope boundary of 0.5 and an upper threshold of 5 due to unreasonable optimization results.

Table 13. Defined and optimized parameters for the landform (LF)-classification algorithm for a 10 m Digital Elevation Model (DEM) for the datasets Laserscan (LS), TK 1:10.000 (TK10); TK 1:25.000 (TK25) and TK 1:100.000 (TK100) for the agricultural area in the Luettewitz region (approximately 200 ha). The correlation coefficient (r^2) as computed by PEST between the defined and the optimized dataset is given in the last row.

Type	LS (defined)	TK10 (optimized)	TK25 (optimized)	TK100 (optimized)
Profile curvature	0.1	0.0900042	0.065799	0.0291868
Planform curvature	0.1	0.164316	0.126976	0.04117573
Slope	3	2.19853	2.57722	0.604286
Threshold	5	2.53277	5	1.55838
Number of Cells per LF	LS	TK10	TK25	TK100
DSH	917	744	804	929
PSH	1572	1359	1368	1345
PFS	870	1126	1166	1240
CFS	804	986	954	946
r^2		0.74	0.77	0.73

Results in Table 13 are an example of scale variant processes in DEM resolution. The major classification parameter profile curvature decreases logarithmic from LS -> TK10 -> TK25 -> TK100. For the parameter planform curvature again a logarithmic decrease from TK10->TK25-> TK100 can be observed. Profile curvature as variable to classify the landscape position (i.e. SH, BS) showed always smaller values than used for the classification of the LS. The parameter planform curvature (which describes the convergent / divergent character of each landscape position) showed in contrast to profile curvature slightly larger values for the TK10 and the TK25 dataset compared to the value used for the original LS classification.

The parameter slope and the threshold value show a decrease with the datasets LS10->TK10 ->TK100, with the exception of the TK25 dataset, which exceeds values optimized for the TK10 dataset.

By using the nonlinear optimization technique parameters for the classification of landform elements can be estimated for a range of DEM datasets. Therefore, if one set of classification values for a specific landscape is determined for a high resolution dataset like the LS dataset, it can be applied for larger areas where only coarser DEM information is available. It should be noted that the optimization technique applied was only focused in the optimization of number of cells, other approaches would include the spatial location of the classified LF as well as made use of the occurrence of certain soil types, soil properties at specific landforms and base the optimization results on that particular technique.

6.4. Optimum resolution for different data sets

The evaluation of an "optimum" resolution is one step in the process of the DEM generation. Gallant and Hutchinson (1996) recommended the use of the root mean square slope (RMSS) of a DEM. Flaws exists within this concept in general. A landscape may on the first hand contain different parts of landforms in the spatial distribution (e.g., steep carved valleys within large plateaus), which cannot be aggregated to an optimum resolution. Second, different relief parameters like profile curvature and elevation need different resolution parameters. This was further supported by Gallant and Hutchinson (1997), who later stated that topographic attributes are scale-dependent.

The RMSS was computed using the rsme.aml to provide a complete assessment of the DEM data sets (see Appendix AML). Results are shown in Figure 19 for the RMSS against an increasing DEM cell size. In general, the RMSS decreases slowly for the dataset TK10 (values for TK25 in brackets) from a value of 22 (19) down to a value of 1 (0.9) with increasing resolution. A similar decrease is visible for the TK100 dataset; however, the highest values reach approximately 8 and decrease down to a value of 0.9. The LS dataset (please notice that only the agricultural areas were considered here) show a linear decrease, starting with 5.9 and decreasing to a value of 0.6 at the largest resolution tested. The optimum resolution for a given dataset from Figure 19 appears to be difficult to determine; therefore, the intercept of a three point linear fit to the RMSS was performed (Figure 19). Based on a significant decrease shown on this graph, the following conclusions were made: For the TK10 dataset the resolution should be less than 16 m, for the TK25 dataset less than 22.6 m, and for the TK100 dataset less than 32 m. The LS dataset does not provide a distinct decrease for any given resolution as seen from the TK datasets. The optimum resolution should be less than 5.7 m as a clear visible decrease in RMSS occurs at that resolution. Secondly, the optimum resolution could also be based on a value less than 16 m, because below this resolution, the RMSS of the LS dataset exhibits its last decrease. Results obtained for different datasets are in agreement with Gallant and Hutchinson (1996); however, the results for a TK25 shown here are above the optimum resolution of 15 m reported by Wilson and Gallant (2001) for a 1:24.000 DEM.

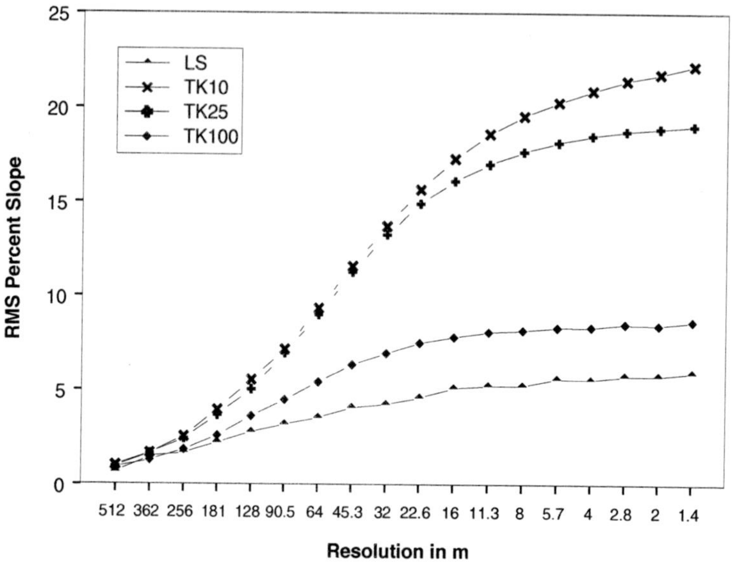

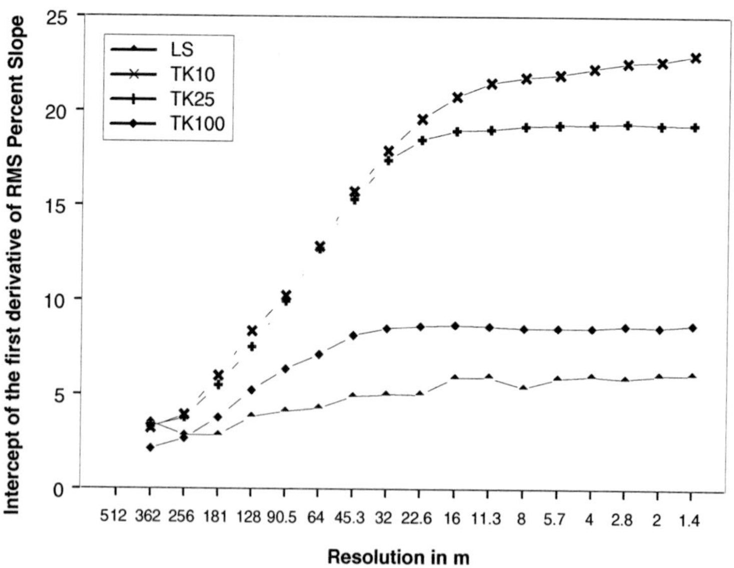

Figure 19. RMS Percent Slope (top graph) and the intercept of the Root Mean Square Slope (RMSS)(bottom graph) as a function of the DEM resolution for the datasets Laserscan (LS), TK 1:10.000 (TK10), TK 1:25.000 (TK25) and TK 1:100.000 (TK100)

Still questionable is the use of the relief parameter slope for the "optimum cell size" as it neglects the different behaviour of other terrain attributes (see also next section). Wolock and McCabe (2000) investigated the aggregation size of DEM on related terrain attributes for slope, TWI and flow accumulation area (SCA). Slope values were found to decrease with increasing cell sizes; whereas, TWI and SCA showed an increase. Also the general concept remains questionable, as a landscape may contain different parts of relief in the spatial distribution (such as the deep carved valley in the Luettewitz region in contrast to the flat plateaus in between), which can't be aggregated to an optimum resolution. Gallant and Hutchinson (1997) state that in regard to scale dependencies in DEM, topographic attributes are scale dependent and different grid resolutions are needed for different applications. In that context, results by Bindlish and Barros (1996) look promising. In their work, a fractal interpolation scheme is used to aggregate a landscape to coarser resolutions. Bindlish and Barros (1996) state that especially for high level aggregations this method preserves the spatial structure of elevation.

6.5. Semivariogram analysis of relief parameters

Knowledge about the spatial variability of relief parameters is important, as the topography is one of the influencing factors in landscape development (Willgose et al., 1991b). Examples are relief parameters (i.e. slope, landforms), which influence soil properties (Moore et al., 1993a) or yield parameters (Kravchenko et al., 2000). In order to avoid possible problems of loosing high frequency information due to under sampling of certain relief parameters and to characterize a site adequately, information about the spatial extend of a certain spatial attribute are critical to management decisions.

Similar to an analysis about spatial variability, a temporal analysis could determine differences between relief parameters with respect to time. Thereby processes influencing the soil surface could be investigated, i.e. amount of soil erosion or changes in glacier mass (see Willgose et al., 1991b). Due to missing elevation datasets from different dates this could not be performed.

Semivariogram analysis of relief parameters was performed for the 20 ha field site "Bei Lotte" (BL, see Table 14). This field site presents only a subsection of the landscape; therefore, the total agricultural area covered by the Laserscan (ALS) was used to characterize the spatial variability for a larger part of the investigated landscape (Table 15).

Height (m) and related parameters (min, mean, max) show a semivariogram range of 200-220 m for the field site "Bei Lotte"; whereas, the ALS data presented a range of approximately 440 m. The fitted power semivariograms indicated the presence of a drift in the elevation values, which is apparent in Figure 3. These range parameters are smaller than values by Bian and Walsh (1993) with a nested semivariogram, the first providing a range up to 70 m, flattening up to 140 m and a second linear semivariogram model behind that. Still, all elevation range values are smaller than the linear fit for a USGS DEM by Holmes et al. (2000), with N-S and E-W directional semivariograms conforming up to 1200 m.

The first derivative of the height, i.e. the slope, showed a smaller semivariogram range of 165 m, and aspect 158 m for BL. The ranges for slope and aspect found for the ALS dataset are extended to 286 m for slope and 222 m for aspect. As slope and aspect for the BL do not show any drift occurrence, a reason for the larger fitted ranges might be the larger area used for the semivariogram calculation. This would lead to the conclusion that the field site "Bei Lotte" shows less spatial dependencies than the total area. Ranges observed are above values by Bian and Walsh (1993) with a range of 70-75 m for slope and 30 m for aspect. These values were interpreted as resulting from landscape-specific formation (Bian and Walsh, 1993).

The second derivatives of the height show much smaller ranges. The planform curvature approaches a range of 37 m. This means, that the distribution of ridges and valleys perpendicular to the profile curvature were related up to 40 m on BL, which is half of the length of the approximated valley profile at that field (see Figure 16). A similar observation can be seen on the ALS dataset with 44 m. For profile curvature (i.e. in direction of the steepest descent) a strong spatial range is observed up to 44 m; however, for the ALS dataset the range is extended to 63 m. The increase must be interpreted in that sense, that the LF distribution at the field site BL is more compacted in terms of slope length compared to the total ALS dataset (see Figure 17 for the spatial extend of landforms between two fieldsites).

The TWI shows a total range of 160 m; however, the first structure of the fitted nested semivariogram model shows a range of 25 m. In contrast, the TWI range found for the ALS extends to 74 m. The different structures found for the secondary relief parameter TWI are attributed to the combination of slope and flow accumulation. The least range of both variables is represented as a first semivariogram structure by the TWI. This is the case for the field "Bei Lotte", where one single structure determined the flow pattern (slope 165 m and flow accumulation 27 m). For the ALS a range of 74 m for TWI was determined; however, slope and flow accumulation differed. A reason for this difference is the scale variance of the process, i.e. several structures with different extend are represented by the ALS than one observed at BL.

As relief is one of the components influencing landscape character, soil development and a number of other parameters, the spatial variability of relief parameters need to be considered in the development of a sampling design. One example might be the development of a sampling design for soil water content using the Topographic Wetness Index. Another example might be considerations for sampling points examining grain yield components. The lesson learned is: If, for example, the sampling distance of 60 m was chosen, sampling points could have been located accidentally only on the shoulder landforms at the field site „Bei Lotte", and the footslope positions would be missed, resulting in biased values of the sampling properties.

Generally, nugget sill ratios show a variety of spatial dependencies. However, as different semivariograms are used to obtain a value of total semivariance, care must be taken by comparing NS ratios. The planform curvature, which shows a strong spatial dependency of 0.07, shows a lower spatial dependency than the relief parameter exposition with 0.05. This does not imply that exposition contains a stronger spatial dependency, as the range of the exposition is larger due to the nested semivariogram functions. For types of semivariogram models please see Bleines et al. (2000).

Table 14. Semivariogram model parameters of relief parameters for the field site "Bei Lotte" (20ha). Parameters for the first structure are signed 1, for the second signed 2. If a Power model was fitted, the exponent is given under Range 2. For units of relief parameters see Table 4.

Type	Nugget	Range 1	Sill 1	Model 1	Range 2	Sill 2	Model 2	N/S-Rat.
ASP	799.9853	31.98m	7276.6190	Spherical	158.73m	9203.8824	Spherical	0.05
CUR	0.0049	41.83m	0.0381	Spherical				0.11
FLACC	5323.9584	27.22m	1235.7965	Spherical				0.81
FLDIR	90.6159	56.27m	129.8576	Spherical	190.89m	445.0870	Spherical	0.14
LFC	0.6532	66.77m	6.3340	Spherical	243.35m	2.3457	Cubic	0.07
M	0.1000	220.00m	19.6824	Power	1.8500			
MAX	0.5668	200.68m	45.2878	Power	2.0000			
MEAN	NA	203.97m	15.2216	Power	2.0000			
MIN	NA	222.00m	20.0376	Power	1.5000			
PCTG1	NA	184.93m	670.7177	Spherical				
PLAN	0.0011	37.28m	0.0150	Spherical				0.07
PROF	0.0028	44.31m	0.0105	Spherical				0.21
SD	0.0031	220.00m	0.5876	Power	1.7000			
SLP	NA	165.24m	1.4978	Spherical				
TWI	0.6000	25.42m	0.9482	Spherical	136.35m	0.7873	Cubic	0.26

Table 15. Semivariogram model parameters of relief parameters for agricultural field sites (226ha). Parameters for the first structure are signed 1, for the second signed 2. If a Power model was fitted, the exponent is given under Range 2. For units of the relief parameters see Table 4.

Type	Nugget	Range 1	Sill 1	Model 1	Range 2	Sill 2	Model 2	N/S-Rat.
ASP	475.4094	221.51m	7379.2698	Exponential				0.06
CUR	0.0163	64.79m	0.0298	Exponential				0.35
FLACC	3407.1851	32.69m	1916.9808	Exponential	201.34m	1084.8529	Spherical	0.53
FLDIR	57.4673	143.02m	644.0089	Exponential	800.00m	1063.7153	Cubic	0.03
LFC	1.1984	178.49m	8.5549	Exponential				0.12
M		440.00m	67.9605	Power	1.4000			
MEAN	1.0000	443.38m	47.7511	Power	1.5000			
MIN		443.38m	71.8938	Power	1.5000			
PCTG1		327.75m	658.7550	Exponential				
PLAN		41.34m	0.0170	Exponential				
PROF	0.0049	63.51m	0.0098	Exponential				0.33
SD		406.72m	1.4371	Cubic				
SLP		286.55m	2.0976	Exponential				
TWI	0.5690	74.03m	1.5614	Exponential				0.27

CHAPTER 7

Results and Discussion for grain yield components

7.1. Statistical and geostatistical grain yield analysis

In this section the spatial and temporal variability of crop yields and yield components using different methods will be discussed. Basic statistics for the hand harvested yields for the years 1999, 2000, and 2001 at the field site "Bei Lotte" are shown in Table 16 and Figure 20. Additionally, the yields are presented with respect to landform units. Mean yield of Spring Barley in 1999 was 6.7 tha^{-1}, whereas the winter rye in 2000 and 2001 yielded 11 t ha^{-1} and 10.4 tha^{-1}, respectively. The yield for spring barley had a standard deviation (SD) of 0.90. In comparison the yield SD for 2000 almost doubles this values with 1.55 during a dry season. During the wet season of 2001 these values are lowered to 1.34. The SD for the whole field was lowest in 1999, and highest in 2000. These results agree with the results by Porter et al. (1998), who found a higher SD of grain yield under drier conditions. Generally, coefficient of variation (CV) of grain yield values found at the field site "Bei Lotte" are similar to those reported in the literature (see Eghball and Varvel, 1997, Porter et al., 1998, Yang et al., 1998, Bakhsh et al., 2000). The CVs of grain yield for Shoulder (SH) and backslope (BS) positions showed higher values in all three years compared to footslope (FS) and level (LE) positions. At FS and Level LF positions a similar CV was observed in 1999 and 2001, whereas for the dry season of 2000 the CV is slightly elevated. Pennock et al. (2001) showed that with an increasing fertilizer application a lowering of the CV of grain yields occurred. Such decrease can not be observed from the CV of spring barley in 1999 with less fertilizer application and winter rye in 2000, where the opposite is the case for the CV.

Table 16. Descriptive Statistics for hand harvest grain yields for the field site "Bei Lotte" for the entire field and landform units. Values are given in t ha^{-1}. Shoulder (SH), Backslope (BS), Footslope (FS), and Level (LE).

	YEAR	Field	SH	BS	FS	LE
Spring barley yield	1999	6.64	6.55	6.50	6.57	6.73
Winter rye yield	2000	10.99	10.36	10.97	10.88	11.12
Winter Rye yield	2001	10.38	11.10	10.00	10.02	10.49
SD	1999	0.90	0.85	1.01	0.85	0.87
SD	2000	1.55	1.91	1.66	1.52	1.44
SD	2001	1.34	1.62	1.50	1.09	1.18
CV in %	1999	13.6	13.0	15.6	12.9	12.9
CV in %	2000	14.1	18.4	15.1	13.9	13.0
CV in %	2001	13.0	15.0	15.2	11.3	11.3
Number of plots		192	17	49	16	110

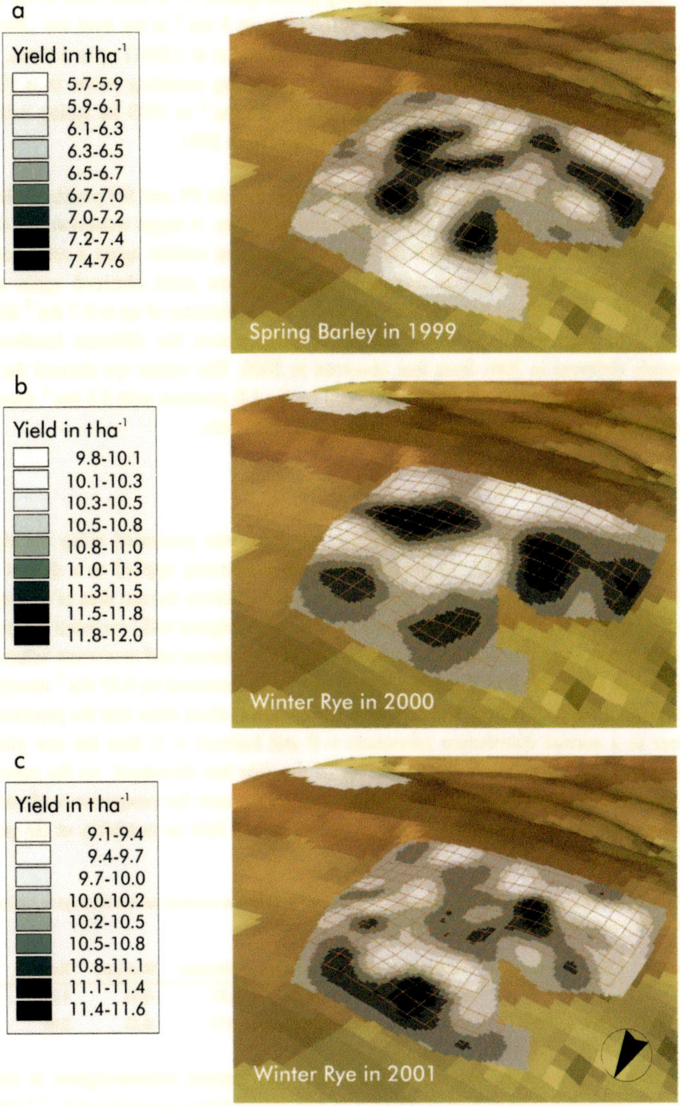

Figure 20. Distribution of hand harvest grain yields at the field site „Bei Lotte" in the years 1999 (a), 2000 (b) and 2001 (c).

Stevenson et al. (2001) reported that grain yields varied up to 1 tha^{-1} in a distance of up to 50 m. Differences between individual sampling points spaced 27 m apart were in the majority up to 1 tha^{-1}, however some extreme cases differed up to 5 tha^{-1} at the field site „Bei Lotte", therefore exceeding the differences reported by Stevenson et al. (2001) up to 5 times. The SD of the differences in grain yield between two neighbouring sampling points (n = 153 pairs, because field border areas were not included) were 1.18 tha^{-1} in 1999 for spring barley, 1.97 tha^{-1} for winter rye in 2000 and 1.83 t ha^{-1} for winter rye in 2001.

Grain yields are highest for spring barley at the Level LF. At FS and SH positions yield were observed approximately 0.15 tha^{-1} lower than level positions. A larger reduction was observed at BS position (0.23 tha^{-1}). In 2000 the winter rye harvest exhibit again a reduction at BS positions of around 0.2 tha^{-1} as compared to the highest yield obtained again at Level positions. However, FS and SH positions showed a larger decrease of up to 0.7 tha^{-1} difference in yield. The distribution of winter rye grain yield across the different landforms was completely different in 2001 from that observed in 2000. The winter rye showed the highest yields at SH positions with 11.1 tha^{-1}, followed by the LE position with 0.5 tha^{-1} lower yield. At the BS and FS positions 10 tha^{-1} were harvested (Table 16).

7.1.1. Grain yield processing

Grain yield data collected from the combine harvester was processed using the algorithm outlined in 5.3.2 on page 25. In general, any data processing applied to a dataset usually influences the statistical parameters. An example is the combine harvest data of the year 2002 at the field site "Bei Lotte": Table 17 shows that from the original 6457 recorded points around 2059 data points were removed due to various reasons. Minimum and maximum values show a smaller range for grain yield; the mean of the grain yield increased by 0.85 tha^{-1} mostly due to the removal of the zero values. The skewness and kurtosis values show that the processed yield is closer to a normal distribution (skewness = 0 and kurtosis = 3) than the raw yield data. Variance and CV indicate that the variability of the data has decreased. As the mean of the processed data is larger than the raw data, SD can not be used for comparisons of variability. Instead the CV has to be used, which showed a decrease of 16% in variability of the processed data (Table 17).

Table 17. Statistical parameters for the unprocessed and processed yield for the combine harvest obtained in 2002 at the field site "Bei Lotte". Values are given in tha^{-1}. CV is given in %.

Type	N	Minimum	Maximum	Mean	Std.Dev.	Variance	Skewness	Kurtosis	CV
raw	6457	0	8.5	3.74	1.37	1.87	-0.94	4.25	36.63
processed	4398	2.2	6.6	4.59	0.87	0.75	-0.32	2.64	18.95

The influence of the data processing upon the experimental semivariogram is shown in Figure 21. The entire field nugget semivariance was reduced from approximately 1.2 $(tha^{-1})^2$ to slightly less than 0.25 $(tha^{-1})^2$(Figure 21) and the total semivariance of the datasets decreased from 1.5 $(tha^{-1})^2$ to $(0.5\ tha^{-1})^2$. The range of the semivariogram for the whole field dataset extended by one lag class (12 m). Furthermore, a specific behaviour for LF elements can be observed. For example, the unprocessed Level LF semivariance increased from a value of 1.0 $(tha^{-1})^2$ up to 1.4 $(tha^{-1})^2$ for the first three lag classes and then decreases and approaches a value of 0.8 $(tha^{-1})^2$ for the remaining lag classes. In contrast, unprocessed BS semivariance increased from a value of 1.3 up to 2.5 over 5 lags and decreased afterwards. The semivariance based upon processed data still show the increase in the BS-LF up to 3 lag distances. Generally, the semivariance at the BS position approached the highest values for all LF-yield combinations and semivariances at Level LF extended their range similar to the whole field for one lag (Figure 21).

The data processing led to a larger Nugget to Sill ration and caused a decrease in the total variance. Therefore, the local variance related to combine and yield recording uncertainties was removed (reduction in nugget effect). The visible hole effects at certain LF are removed and showed a semivariogram shape similar to the whole field dataset (Figure 21).

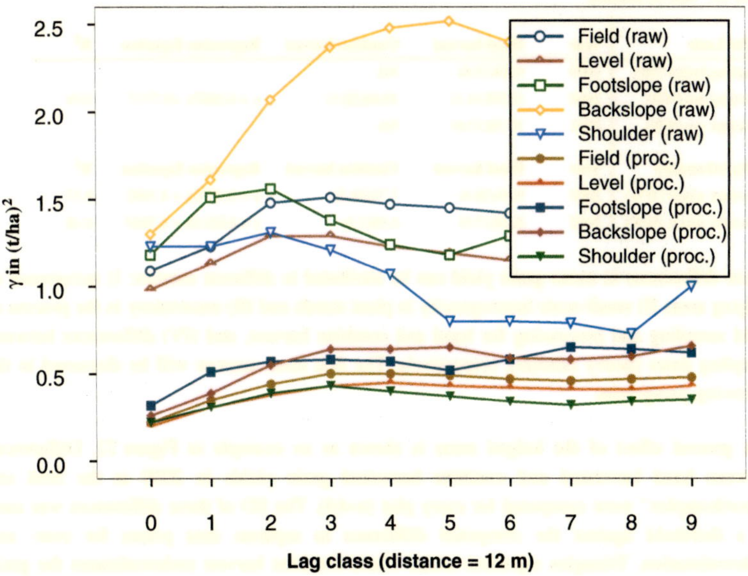

Figure 21. Semivariogram analysis for processed (proc.) and raw (raw) winter rape combine harvester data in 2002 for the whole field and for single landform elements at the field site „Bei Lotte".

Several authors (Auernhammer et al., 1993, Grisso et al., 1999, Reyniers et al., 2001) have reported, that grain yields harvested by hand or small combine harvester deviate up to 10 % and even more from results obtained by combine harvest. For the field site "Bei Lotte" in 2000 the difference between hand and combine harvest was 0.3 tha^{-1}. The smallest difference between both sampling methods in the three years was found in the year 2000 at "Bei Lotte", however no linear relationship could be observed (n=192) between combine and hand harvest grain yield (Table 18). For the years 1999 and 2001 combine harvest yields were not processed due to various failures as described earlier on Page 25. For the field site "Sportkomplex" 1.3 tha^{-1} more grain yield was obtained by hand harvest in 2000 and 2 tha^{-1} in 2002, compared to the respective combine harvest data. The large differences in 2002 are mainly attributed to heavy rainstorms, which occurred between sampling dates for hand harvest (128 mm on August, 13. 2002) and combine harvest. For the field site "Sportkomplex", both years under investigation showed a coefficient of determination of 0.57 (2000) and 0.48 (2001) between combine and hand harvested grain yields (Table 18), which were similar to results found by Machet et al. (2001) for three years of yield data. Machet et al. (2001) sampled four microplots of 0.55 m^2 for one grid cell and concluded, that whole field yield results showed no differences between combine and manual harvesting methods. However, individual nodes by Machet et al. (2001) were overestimated by up to 11%.

Table 18. Differences between field mean grain yield values of hand harvest and combine harvest. In parentheses the CV in % is stated. NA - data not available.

Bei Lotte	**Year**	**Hand harvest**	**Combine harvest**	**Regression Equation**	R^2
spring barley (SB)	1999	6.64(13.6)	NA		
winter rye (WR)	2000	10.99(14.1)	10.66(20.1)	y = 0.0807x +9.7717	0.04
winter rye (WR)	2001	10.38(13.0)	NA		
Sportkomplex	**Year**	**Hand harvest**	**Combine harvest**	**Regression Equation**	R^2
winter wheat (WW)	2000	9.01(28.4)	7.71(16.3)	y = 0.3732x + 4.3482	0.57
winter wheat (WW)	2002	8.60(17.9)	6.58(7.1)	y = 0.0211x + 4.7689	0.48

These differences in mean grain yield can be attributed to different sources: I) occurrence of lodging areas II) small-scale heterogeneity in plant stands and III) uncertainty in the process of yield sampling and processing for hand and combine harvest, and (IV) differences between sampling dates (heavy rainfall), respectively. The first three sources will be discussed in the following paragraphs.

The general effect of the lodged areas is shown as an example in Figure 22. Differences between hand harvested and combine harvested grain yields in 2000 at the field site "Sportkomplex" were computed for every plot (n=64). The SD of these differences was used as a threshold against the computed difference to separate data points for over- and underestimation. Triangles represent areas, where combine harvest underestimates the grain yield in comparison to the hand harvest, whereas squares represent plots where an overestimation occurred. Apparently, the three plots, which overestimated grain yield for combine harvest, were in zero fertilization areas, which are characterized by low-density plant stands. The question remains open, why only three of the eight zero fertilization plots show this pattern, as the spike number at harvest shows no differences between the zero fertilization plots.

In contrast, the underestimated plots are mainly found in areas, where lodging of plants was observed (see Figure 22). Areas with lodging plants were manually recorded using a DGPS-

Field-GIS at the time of the hand harvest. In 19 of the 27 cases of underestimation, lodged areas were found inside the sampling cell. This could lead to the conclusion that lodged areas tend to be underestimated by combine harvesting, whereas low-density plant stands tend to be overestimated. It has to be noted that plots with severe lodging, were recorded, but no underestimation occurred (see Figure 22).

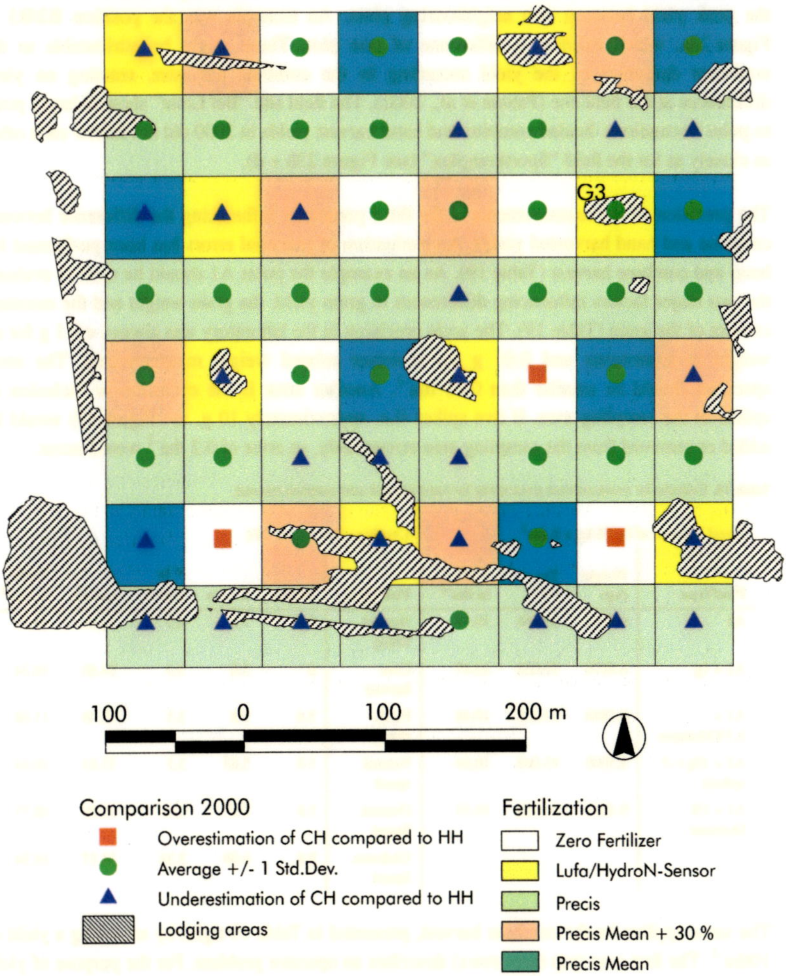

Figure 22. Differences between hand (HH) and combine harvested (CH) grain yield at the field site "Sportkomplex" in the year 2000. Lodging areas are shaded in gray. Additionally, five different fertilizer strategies are given.

The second reason, why the local small-scale variability cannot be assessed in depth is because no hand harvesting with distances smaller than 27 m x 27 m was performed. A combine harvester processes an area 12 times the width of hand harvesting at one time, therefore scale differences might occur. However, the scaled yields (see Wendroth et al., 2001 for method) of combine and hand harvests showed a close agreement in the direction of field operations (Figure 23a, east-west) and perpendicular to it (Figure 23c- south-north) for the field site "Sportkomplex" in 2000 (n=64). Differences between both harvesting methods were shifts of the peak yield between two neighbouring plots. An example was the position B2/B3 in Figure 23a, which exhibits a difference of one plot. These might be attributable to the nonlinear dynamics of the yield recording in the combine harvester, reacting on yield differences at the field site (Panten et al., 2002). The field site "Bei Lotte" showed larger point to point fluctuations. Scaled combine and hand harvest yields in 2000 did not follow each other as closely as for the field "Sportkomplex" (see Figure 23b + d).

The precision of the measurements is the third parameter influencing the difference between combine and hand harvested yields. An evaluation of potential errors has been performed for hand and combine harvest (Table 19). As an example the point A1 should be used to evaluate the two major factors influencing differences in grain yield: the grain weight and the moisture content of the grain (Table 19). The scale precision in the laboratory was always < 0.1 g for all weight measurements and 0.01 g for moisture related weight measurements. The error specified should be smaller than 0.01 tha^{-1}. Another error is the exclusion or inclusion of spikes in the sampling area. If five spikes (i.e. approximately 10 g, see Figure 49) would be added or removed from the sampling area erroneously, an error of 0.2 tha^{-1} would occur.

Table 19. Example for measurement uncertainty for hand harvest and combine harvest.

Hand Harvest of 0.486 kg x 0.5m^2				**Combine Harvest of 10 t**					
Plot/Type	**Weight (kg)**	**Trs. (%)**	**Yield in tha^{-1}**	**Plot/Type**	**Width**	**L in m**	**V in kmh^{-1}**	**qm(6s)**	**Yield in tha^{-1}**
A1	0.4860	93.068	10.45	Normal filling	5.8	5.8	3.5	33.64	10.70
A1 + 1g	0.4870	93.068	10.47	Core harvest	6	5.8	3.5	34.80	10.34
A1 + 0.1%Moisture	0.4860	93.168	10.46	Partly filling	5.6	5.8	3.5	32.48	11.08
A1 + 10g (~5 spikes)	0.5000	93.068	10.66	Normal speed	5.8	5.83	3.5	33.83	10.64
A1 + 1% Moisture	0.4860	94.068	10.55	Overest. Speed	5.8	5.67	3.47	33.50	10.75
				Underest. Speed	5.8	6.00	3.54	34.17	10.54

The same applies for the combine harvest, presented in Table 19 right, by assuming a yield of 10tha^{-1}. The first case (top three rows) describes an operator problem. For the purpose of yield recording, the cutting table width is set to 5.8 m, whereas the total cutting table width is 6.0 m. If the combine operator performs a core harvest, the cutting table will be filled completely, which results in a difference of 0.3 tha^{-1} to the usually obtained yield under "normal" harvesting conditions. Respective differences occur, if the operator only fills the table partly. Generally, it should be noticed that these differences are currently an operational/human error. Guidance systems based on laser technique might help to measure the width of the cutting table filled. Another problem is the recording of grain yield along tramlines. In this case the cutting table might be filled properly; however, each of the tramlines might be around 0.3 m width. No solution can be provided for the moment. The second example shown in Table 19 (right bottom three rows) is a true technical problem of measuring the velocity for every second in time. Jürschik (2003) stated that the velocity error measured with a well-calibrated distance sensor under good field conditions should be less than 1%. This would lead to a possible calculated error of 0.1 tha^{-1} (Table 19).

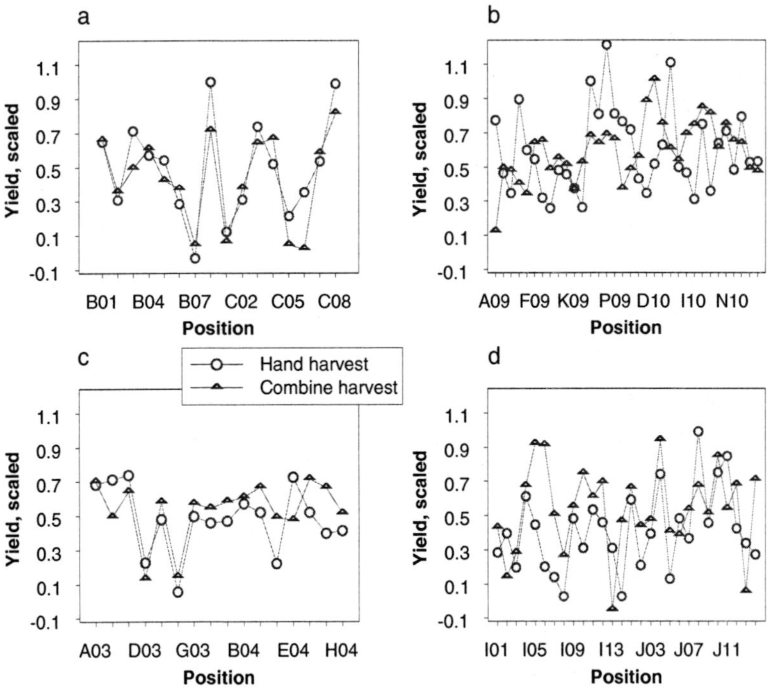

Figure 23. Differences between hand harvested and combine harvested grain yield in the year 2000 in direction of the harvest tracks (east-west) for the field site "Sportkomplex" (a) and "Bei Lotte" (b) and perpendicular to the direction of the harvest tracks (north-south) for the field site "Sportkomplex" (c) and "Bei Lotte" (d).

7.1.2. Grain yield semivariogram analysis

Semivariogram analysis of the hand-harvested yields at the field site "Bei Lotte" is shown in Figure 24. Spring Barley showed a range of approximately 82 m. In 2000 (WR), the range extended up to 120 m, while in contrast, in 2001 the range only reached a value of 60 m. The nugget variance for all three datasets was 0.56, 1.92, and 1.27 for each of the respective years using a spherical semivariogram model. The spatial dependency of the yields (NS-ratio) was low (0.64 in 1999, 0.8 in 2000, 0.64 in 2001). The smallest semivariogram range at the field sites "Bei Lotte" and "Sportkomplex" reached approximately 50 m. This minimum value corresponds to values reported in the literature, - Bakhsh et al. (2000) reported 39 m, Jaynes and Colvin (1997) 52 m, Miller et al. (1988) 80 m, Kerry and Oliver (2001) 37 m.

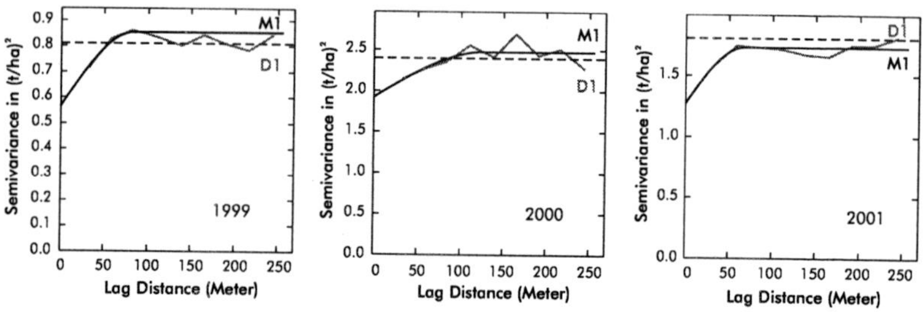

Figure 24. Semivariogram analysis for hand-harvested yields for the field site "Bei Lotte" for Spring Barley in 1999, Winter Rye in 2000 and Winter Rye in 2001.

As shown above, the semivariogram model ranges at the field site „Bei Lotte" exhibit no value shorter than 50 m. Possible explanations would be related to the range of relief parameters and/ or the range of soil texture (60 m). For example, profile curvature (second derivative of the elevation) showed a semivariogram range <50 m, whereas for slope (first derivative of the elevation) a semivariogram range of above 150 m can be observed (see Table 24). Topography and soil parameters represent intrinsic variables, which influence yield development. However, extrinsic variables such as management strategies, modify the spatial dependencies in addition to the impact of specific weather conditions in different years.

The observed difference in the semivariogram ranges of 60 meters for two different seasons leads to Figure 25 inspired by the work of Jaynes and Colvin (1997). The growing season precipitation was defined as the cumulative sum of precipitation between March and July in mm. The cumulative precipitation was plotted against the semivariogram model ranges of hand harvest yield. Datasets from the time period 1999-2001 of hand harvest (HH) for the fields "Bei Lotte" (BL) and "Sportkomplex" (SK) show a decrease in range with an increasing precipitation. A linear least square fit was performed yielding the regression line: Range (HH) = -0.5839 (Precipitation Sum of March to July) + 272.71 with an $r^2 = 0.6775$.

To validate the findings obtained with hand harvest, ranges were obtained from semivariogram fitting combine harvest (CH) grain yield for the time period 1998 - 2001 (Figure 26). Again a negative relationship could be observed with Range (CH) = -0.4278 (Precipitation Sum of March to July) + 205.32 with an $R^2 = 0.8703$. However, it has to be noted that two datasets were excluded, still shown in Figure 26. The CH at "Sportkomplex" in 1998 was identified as an outlier based on the semivariogram range using the Nalimov outlier test (Lozan and Kausch, 1998 with an $\alpha = 1\%$). The range of CH at "Bei Lotte" in 2000 was additionally excluded because such low range value (34 m) represents the lowest value ever observed in any CH dataset obtained for the Luettewitz region. Still, both values are shown in Figure 26, but were not included in the computation of the least square fit.

The negative relationship between yield range and precipitation sum was also observed by Bakhsh et al. (2000), who showed the range=-0.21 (Precipitation Sum of March to July in mm) +185.7 (R^2= 0.91, n=3). The yield semivariogram ranges were 44 m (1995), 39 m (1996) and 92 m (1997). However, our results disagree with Jaynes and Colvin (1997), who found a positive relationship (a=0.13 (Precipitation Sum of March to July in mm) -8.8; R^2= 0.79, n=6). The differences might be related to (I) the corn/soybean rotation grown there, or (II) the amount of precipitation for the March to July period started off with 400 mm of rain compared to 300 mm in Luettewitz.

Jaynes and Colvin (1997) could not explain the reason for a positive relationship between precipitation and semivariogram range. Their suggestion was that extended rainfall might mask the water holding capacity variations of the soil in the field, leading to smoother yield patterns. This hypothesis might be questioned for the Luettewitz case, as the water holding capacity of the soils found at both field sites did not differ (only two textural classes). Moreover, drillings to 2.5 m depth at the field site "Bei Lotte" (n=25) revealed uniform parent material conditions, which are known to influence the water holding capacity.

Rainfall as well as droughts pronounce the topographic effects of the water content distribution in soils and lead to a more differentiated yield pattern, which in turn influences the semivariance as well as the range of yield data (see 8.3.1 on page 136). The FS position might serve as a basic example at the field site "Bei Lotte". During a year with elevated precipitation, grain yield might be lower at the FS position due to air deficiency or plant lodging, whereas during a year with a severe drought plants will be able to sustain these conditions better than shoulder or backslope positions of the field. However, several other effects of plant growth are possible, which will not be discussed here.

As a result, for future yield sampling in that specific Loess region a sampling distance based on the precipitation sum of a period before harvest could be recommended. The preferable sampling distance is defined as distance where 67% of the semivariance between sill and nugget variance can be captured (Herbst et al., 2001). In wet years smaller distances with 17 m for sampling need to be chosen, whereas during dry seasons distances could be extended up to 33 m (2/3 of 100 meters)

The semivariogram range of grain yield differs from year to year due to the meteorological conditions (Figure 25 and Figure 26). Apart from differences in semivariogram ranges due to the type of harvest, different semivariogram ranges can be observed from two fields in one year, even if the meteorological conditions are assumed to be similar. The question arises, how

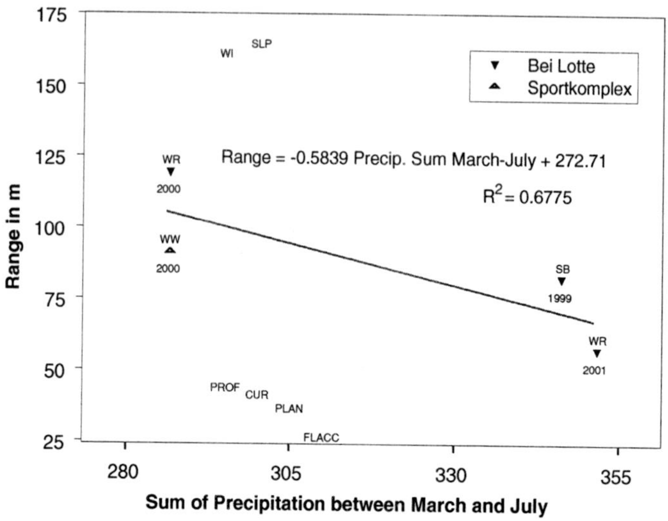

Figure 25. Relationship between Precipitation Sum for March to July (week 8 to 31) in mm and semivariogram model ranges of hand harvested yield at two field sites "Bei Lotte" and "Sportkomplex" for the years 1999, 2000 and 2001. The semivariogram model ranges of profile curvature (PROF), planform curvature (PLAN), curvature (CURV), flow accumulation (FLACC, slope (SLP) and the topographic wetness index (WI) are shown additionally.

the range parameter differs from field to field as seen in Figure 25 for the year 2000 with differences for hand harvest up to 30 m.

The differences in semivariogram range for single fields were shown using combine harvest data for the year 1998 in Table 20. First, different models were fitted to the available data. The five fields showed fluctuations of up to 80 m for that respective year. The semivariogram range of the same species at different fields varied between 50 m for spring barley and 10 m (first structure) for winter wheat. Semivariogram range at the field site „Am Teich“ showed a nested structure with an unbounded power model indicating a large-scale trend. The maximum differences between fitted semivariogram ranges observed extend the semivariogram range differences of 31 m found by Yang et al. (1998) for one given landscape and year. Nugget variances were found to be lowest and more similar for winter wheat and higher and more differentiated (up to 100 %) for spring barley. Nugget variances for winter wheat are in between values (0.05 - 0.2 $(tha^{-1})^2$, mean yield 5.9-8.3 tha^{-1}) observed by Yang et al. (1998). The nugget variances for spring barley exceed these values.

The different semivariances for the LF obvious from Figure 21 are presented in more detail in Table 21. Creating yield maps by kriging assumes stationary of the underlying data (e.g. the expected value and the variogram are constant within the neighbourhood used in the kriging process - see Armstrong, 1998). As already seen for the 2002 combine data, FS and BS contain higher semivariances than total field data. Semivariogram analysis was performed for different

LF-yield combinations using data from the hand harvest in the years 1999, 2000, and 2001. Only the first two lag classes were interpreted, due to the limited area extension of the LF elements. Semivariances for lag classes larger than 2 would result in comparison of two different clusters of data (i.e., SH positions left and right from the depression at "Bei Lotte").

Table 20. Variogram model parameters obtained from omnidirectional semivariogram fitting of grain yield of combine harvest data at the field sites Sportkomplex (SK), Bei Lotte (BL), Am Teich (AT), Kleinbahn (KB) and Bauhof (BH) for the year 1998. Lag distances were set to 12 m in all cases. (Sph = Spherical semivariogram model, Power - Power semivariogram model).

Field	Species	Nugget var. in $(tha^{-1})^2$	Range in m	Model	Sill in $(tha^{-1})^2$	Range in m	Model	Power	Sill in $(tha^{-1})^2$
AT	winter wheat	0.1426	35.38	Sph	0.3256	120.00m	Power	2.00	0.2415
BL	winter wheat	0.0833	44.22	Sph	0.0986				
BH	Spring barley	0.3869	14.47	Sph	0.1173	110.37m	Power	2.00	0.1612
SK	Spring barley	0.7464	90.32	Sph	0.5558				
KB	Triticale	0.1002	13.91	Sph	0.1091	66.23m	Sph		0.0644

Semivariance for Shoulder landforms is investigated more closely. SH-elements show a slightly below average (e.g. field) semivariance for 1999 (0.62 $(tha^{-1})^2$ to 0.72 $(tha^{-1})^2$ see Table 21). In contrast, in 2001 the semivariance at SH positions was above the field semivariance (2.1 $(tha^{-1})^2$ to 1.4 $(tha^{-1})^2$- notice: the highest yields were found at the LF in 2001). In the high yielding year 2000 no distinct difference of semivariance for SH positions at

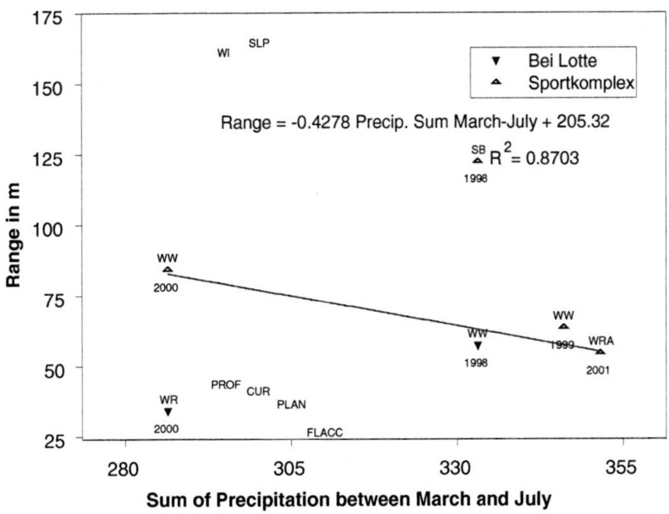

Figure 26. Relationship between Precipitation Sum for March to July (week 8 to 31) in mm and semivariogram model ranges of grain yield. The ranges were obtained for the years 1998 to 2001 for combine harvest (CH) data measured at two field sites "Bei Lotte" (BL) and "Sportkomplex" (SK). Specifically indicated data points in 1998 at "Sportkomplex" and in 2000 at "Bei Lotte" were excluded from the Least Square Fit. The semivariogram model ranges of profile curvature (PROF), planform curvature (PLAN), curvature (CURV), flow accumulation (FLACC, slope (SLP) and the topographic wetness index (WI) are shown additionally.

the first two lag distances can be found compared to the field average semivariance (2.1 $(tha^{-1})^2$ for field semivariance at the first lag to 3.5 $(tha^{-1})^2$ for SH-positions; 2.2 $(tha^{-1})^2$ for field semivariance to 1.8 $(tha^{-1})^2$ for SH-positions).

Footslope LF shows a different distribution in semivariance compared to the Shoulder-LF. In 1999 the highest semivariances of all LF are observed at the first lag distance (1.2 $(tha^{-1})^2$ for FS to 0.7 $(tha^{-1})^2$ for the field average), followed by low semivariances for the next lag distances (0.4 $(tha^{-1})^2$ at FS to 0.8 $(tha^{-1})^2$ for the field average). The semivariances in 2000 are higher than field level and in 2001 mixed compared to the field semivariances.

Table 21. Semivariance values in $(tha^{-1})^2$ of hand harvest grain yield at the field site „Bei Lotte" in 1999, 2000 and 2001 for the first four lags (lag distance was set to 27m), stratified for shoulder, backslope, footslope and level landforms. Additionally the values of the whole field is given. (spring barley (SB), winter rye (WR)).

SB 1999	SHOULDER	BACKSLOPE	FOOTSLOPE	LEVEL	FIELD
Lag 1	0.617	0.678	1.156	0.654	0.716
Lag 2	0.696	0.879	0.460	0.850	0.828
Lag 3	1.091	1.196	0.614	0.790	0.859
Lag 4	1.074	1.119	0.108	0.775	0.834
WR 2000	**SHOULDER**	**BACKSLOPE**	**FOOTSLOPE**	**LEVEL**	**FIELD**
Lag 1	3.562	1.360	2.255	1.797	2.117
Lag 2	1.842	1.868	2.301	1.923	2.254
Lag 3	3.184	1.932	1.250	2.129	2.342
Lag 4	8.644	3.535	3.927	2.139	2.563
WR 2001	**SHOULDER**	**BACKSLOPE**	**FOOTSLOPE**	**LEVEL**	**FIELD**
Lag 1	2.063	2.123	0.975	1.255	1.391
Lag 2	3.442	2.659	1.782	1.379	1.502
Lag 3	1.073	2.209	1.909	1.422	1.523
Lag 4	2.361	1.937	1.684	1.327	1.594

The hand harvest dataset contains only a limited number of points (n=192). In contrast, combine harvest delivers a larger number of yield values (n=4398 in 2002, n=4436 in 1998), which allows evaluation of the influence of LF elements more thoroughly. In Figure 27 the semivariograms for 13 LF units are shown together with the semivariogram for the total field. Notice that the area threshold in the classification process was increased for this analysis to 10.000 m^2 to have a sufficient number of combine harvest points for each lag distance. Whereas the nugget values were similar for all cases (approximately 0.1 $(tha^{-1})^2$), the shape and the sill differed between LF units and also between the LF units and the total field as seen in Figure 27. This conforms with observations by Yang et al. (1998), who found differences between 20 m and 110 m in range values for zones based on contour lines and sharply defined changes in topography. The ranges found for subunits were below the ranges found for the whole field in 10 of the defined 37 sub regions (Yang et al., 1998). Reasons for the smaller sub region spatial range compared to the field range are either (I) due to the arbitrarily selected zones, or (II) due to the lower numbers of pairs used for the calculation of the semivariogram.

Conclusions drawn from this are: (I) the grain yield semivariograms computed for entire fields differ from those for single landforms (Figure 27) and (II) the observed semivariances for single LF units were not stable in time (Table 21) as semivariogram shapes change from year to year due to different yield development based on weather conditions.

Different semivariogram shapes as shown in Figure 27 can be identified for certain LF elements. To be precise, one would need a semivariogram model for each landform element, and krige the yield for the respective element, and finally combine all elements to give an estimate for the whole field. However, this approach seems to be quite ambitious due to the amount of work needed (semivariogram fitting for each landform, what kind of landform elements to use).

One first step to overcome these differences would be the consideration of borderlines of large (in terms of area extensions) LF elements similar to the concept of geological faults (Bleines et al., 2000). Therefore, the borderlines of LF elements (Figure 28) would be used as faults to perform the semivariogram analysis and kriging. Faults can be used to limit the search radius in the kriging routines in ISATIS. An advantage is that only points for one landform are considered in the kriging process for the given point. Still, one disadvantage is the use of one common semivariogram model.

To test, if landform boundaries used as faults can help improve the results of kriging, two types of analysis were performed. Thereby, (I) hand harvest yields from 1999-2001 were evaluated using jackknifing with and without landform boundaries and (II) results of kriging itself again with and without landform boundaries. One large limitation exists with respect to the shapes of semivariograms at different LF as shown in Figure 28, which cannot be considered using a fitted semivariogram for each respective LF unit at the current time. Semivariograms for the whole field including landform boundaries as faults and semivariograms without faults yielded similar variances, as well as similar range parameters for hand harvested grain yield in 1999, 2000 and 2001 and for combine harvest grain yield in 1998.

More outliers were identified using cross-validation by including landform element boundaries as faults than without faults (Figure 28). A outlier was defined as exceeding a threshold of the 99% confidence interval of a normal distribution (value of 2.5 in ISATIS cross validation). For the year 1999 jackknifing identified six outliers instead of three without landform boundaries. In the year 2000 five outliers were identified by using landform boundaries as faults compared to zero outliers without any boundaries. Finally in 2001 five outliers could be identified by using landform boundaries instead of two without landform boundaries.

Results from the kriging process using LF element boundaries as faults and without faults are shown for hand harvest (2001) and for combine harvest (1998) in Table 22. A large increase in kriging SD is visible at the borders of LF boundaries (see Figure 28). This matches the statistical parameters (Table 22) for the kriging SD for four subsequent years, which show an

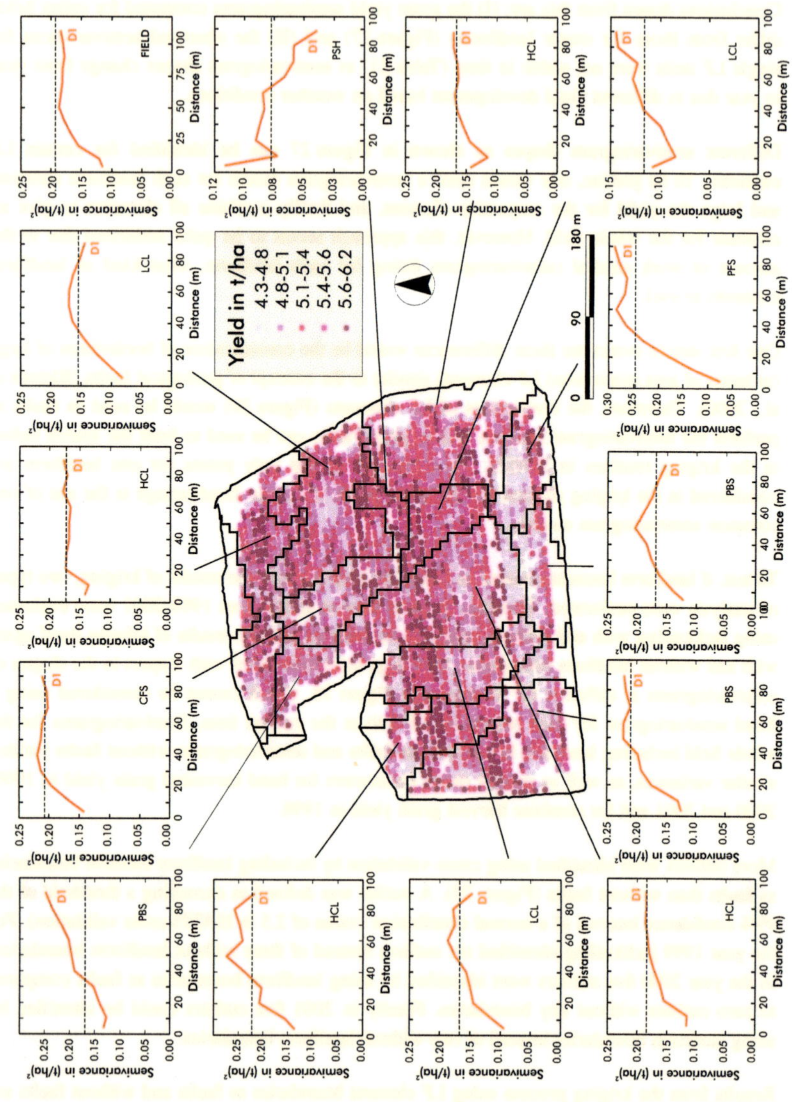

Figure 27. Semivariogram analysis for grain yield in tha^{-1} at the field site „Bei Lotte“ in 1998, seperated for different landform elements (Minimum area 10.000 m^2). The top rightmost variogram (FIELD) represents the average field conditions. Other variograms (PSH planar shoulder, PBS planar backslope, CFS convergent foot slope, LCL low catchment level, HCL high catchment level) show semivariances computed at different landform elements.

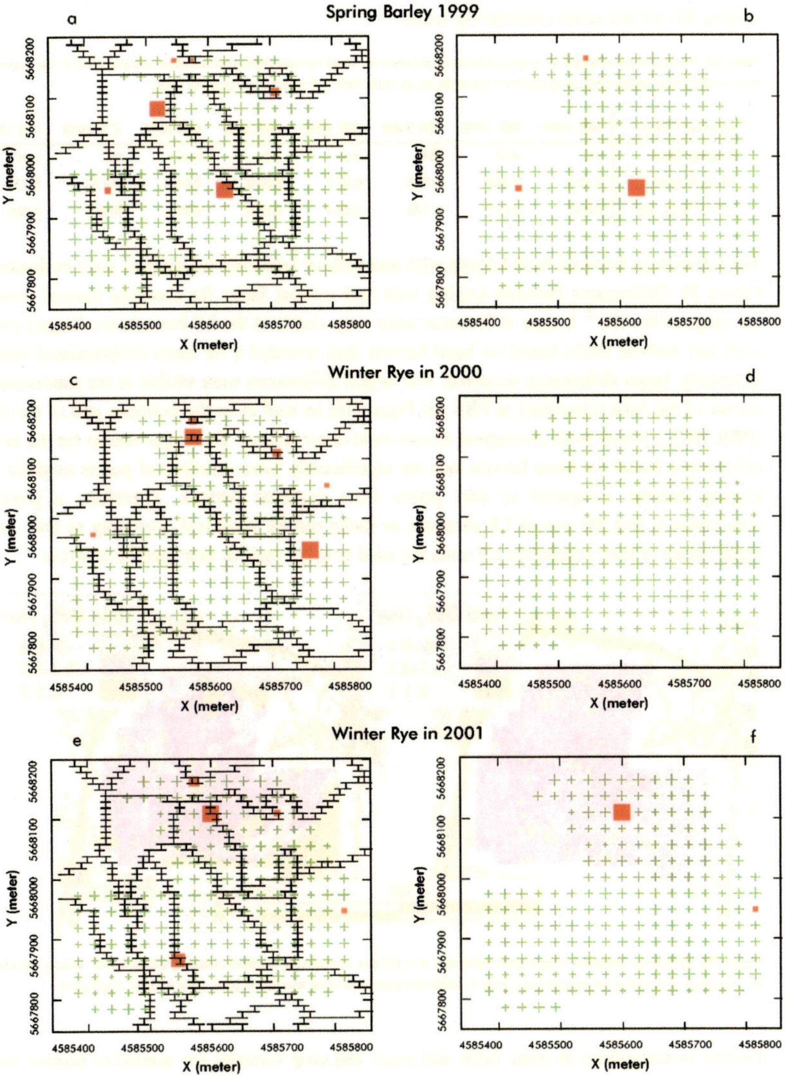

Figure 28. Outliers detected (marked with red squares) by cross validation with (left side, a, c, e) and without (right side, b, d, f) landform element boundaries as faults for the years 1999, 2000 and 2001 at the field site „Bei Lotte“ (n = 192). LF element boundaries are shown with black lines in the left column.

increase in kriging SD (see row mean in Table 22) as well as an increase in the SD of the kriging SD for the cases considering faults.

Table 22. Statistical parameters for kriging standard deviation for hand harvest (HH) and combine harvest (CH) in the respective year. Kriging was performed using landform boundaries as faults (wf) and without using faults (wof).

Harvest / Year	HH 1999	HH 1999	HH 2000	HH 2000	HH 2001	HH 2001	CH 1998	CH 1998
Kriging Type	wf	wof	Wf	wof	wf	wof	wf	wof
Mean Kriging SD	0.463	0.405	0.933	0.817	0.694	0.618	0.138	0.132
SD. of Krig. SD.	0.067	0.012	0.128	0.014	0.090	0.014	0.016	0.009

The differences between both kriging with and without landform boundaries were evaluated in Figure 29. Differences between kriging with and without faults for combine harvest were in the range of 0.1 tha^{-1}. Larger differences were only visible at the LF boundaries. Kriged yields with and without faults based on hand harvest data revealed a far more differentiated image. Generally, larger differences occurred. The largest differences were visible in the southwestern corner of the field (classified as PBS see Figure 16) as well as in the northern part of the field (PBS, HCL), which both correspond to own field observations. The major reason for the larger differences found for hand harvest was the significantly lower number of points used for the kriging process compared to data points from combine harvest. Therefore, a possible conclusion is that the use of LF elements as faults can be evaluated positively as only grain yield values of their respective LF units are used to krige yield at that specific location.

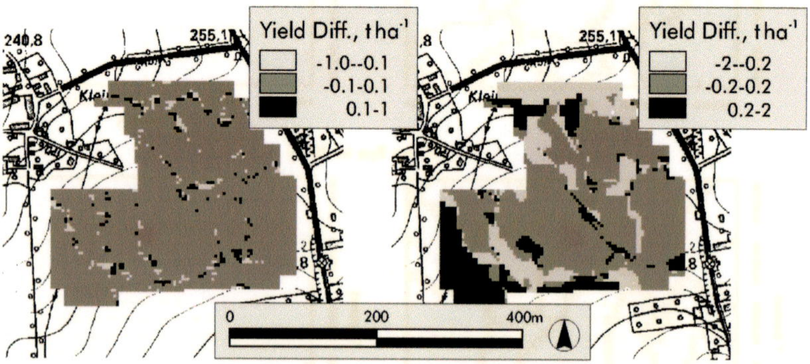

Figure 29. Difference between kriged grain yield with and without considering landform boundaries as faults. Values are shown for examples of combine harvested grain yield in 1998 (left part) and hand harvested grain yield in 2001 (right part).

Further investigations at other fields and years and crop varieties are needed to support these conclusion. Secondly, the application of different semivariogram models being representative for each respective landform for kriging, should be investigated more closely.

7.1.3. Correlations between grain yield components

The influence of weather condition on crop yield development and yield properties has been known for some time (see Pollmer, 1957, Anderl et al., 1981). Researchers tried to identify relationships between grain properties often in terms of correlation analysis for different plot experiments at several field sites. A general result is that thereby the variability of grain yield between different plots of the same application increases the difficulty of interpretation of the plot experiments. The use of such plot experiments was questioned already back in the 1960's (Collis-George and Davey, 1960). Knowledge about the general development of grain yield and their related yield components (Figure 30) was developed based on these plot experiments. Generally, the development of grain yield can be influenced at four time periods during vegetative growth (Figure 30). For an in-depth discussion see Roth et al. (1988).

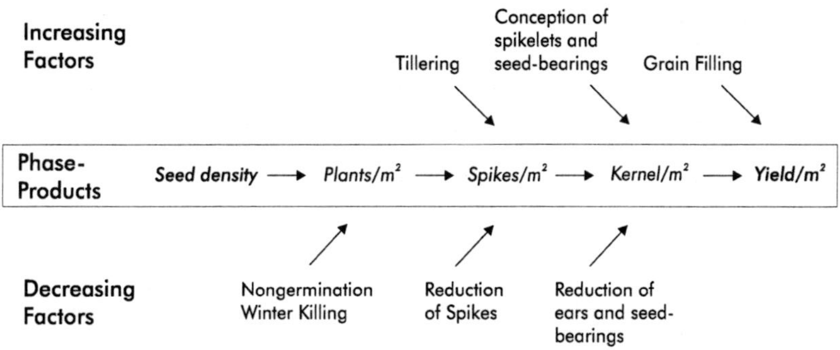

Figure 30. Ontogenesis of grain yield (adapted from Damisch, 1971).

Relationships between yield components were observed at plot experiments. The same can also be investigated at entire fields. However, little knowledge is available if and how such relationships changes from plot to field experiments. The following tables are shown as an example of field experiments. Please remember that during the three years a homogeneous soil and crop management was applied.

Spike number was correlated strongly to grain yield in 1999, 2000 and 2001 (0.78, 0.77, 0.59), as well as the kernel number with yield (0.9, 0.9, 0.78) (Table 23). This shows a much stronger relationship than described in Anderl et al. (1981), who showed values of 0.38 and 0.3, respectively (see Figure 31). Additionally, Anderl et al. (1981) showed a relationship between Thousand Kernel Mass (TKM) and grain yield of 0.44. However, the values found for in-field conditions at the field site „Bei Lotte“ were lower (0.18, -0.09, -0.03; see Table 23). Geisler (1983) stated that differences in TKM are mainly due to plant density, which showed a negative correlation to TKM (-0.23, -0.38, -0.22) in different years.

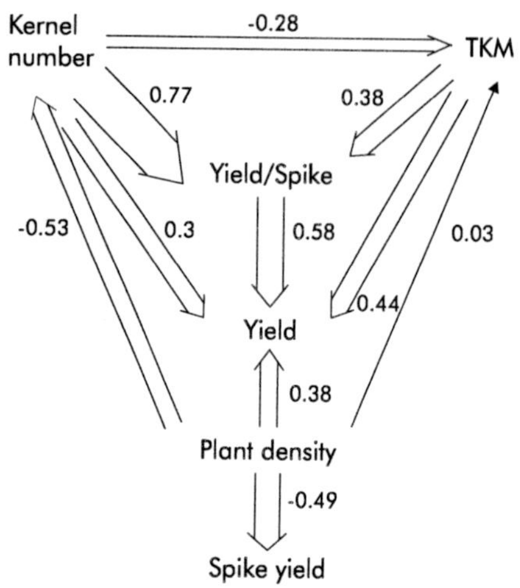

Figure 31. Correlation between yield components (adapted from Anderl et al., 1981).

Negative correlation coefficients of -0.59 in 1999, -0.2 in 2000 and -0.37 in 2001 were found between protein content and TKM (Table 23). Sufficient nitrogen supply is known for an extended vitality of plant organs. Under the assumption that mineral soil nitrogen content was never limited during time of development (see amount of mineralized nitrogen in spring in Table 40), the environmental conditions observed by the plants during the final kernel filling phase must have led to kernels receiving less assimilate, which led in turn to higher protein concentration (Geisler, 1983). This might explain the different strength in correlations between protein content and TKM. Positive correlation coefficients between protein and grain yield has been observed for the years 1999, 2000 and 2001, being 0.038, 0.206 and 0.134, respectively (Table 23). This is in contrast to observations by Jensen et al. (2001), who found negative relationships between grain protein content and grain yield. The differences observed might be attributed to the larger number of samples (n= 800) and the lower mean of the harvested yield (2.38 tha^{-1}) by Jensen et al. (2001).

Yield per Spike (YpS) showed a significant relation to TKM in 1999 (0.59) for spring barley, in 2000 (0.45) and 0.17 in 2001 for winter rye (Table 23). Anderl et al. (1981) observed a smaller correlation coefficient of 0.38. With the meteorological conditions affecting plant growth in 2001 (e.g. higher precipitation sum) no strong differentiation for YpS of winter wheat is found. However, as conditions are more limiting (e.g. the year 2000 contained drier conditions), the kernel mass differed more. Still, spring barley in 1999 showed the strongest relationship (Table 23), even with precipitation amounts similar to those observed in 2001.

The years 1999, 2000 and 2001 showed significant relationships for the entire field's dataset (n = 192) as described in the previous paragraphs. The conclusion can be drawn, that different years show similar negative or positive correlation coefficients for most of the cases. The strength of relationships differs from one year to the next due to changing environmental conditions. With the same species of crop grown in two consecutive years (i.e. winter rye in 2000 and 2001), the strength of the relationships changed quite strongly.

Table 23. Correlation analysis for different yield components for the year 1999, 2000 and 2001 at the field site „Bei Lotte" for 192 hand harvested plots (0.5 m^2). The following abbreviations are used: SB = grain yield of spring barley, TKM = thousand kernel mass; PROT = protein content; KernelN = number of kernels; SpikeN = number of spikes; KpS = kernel per spike; YpS = Yield per Spike; year 1999 = 99; year 2000 = 00; year 2001 = 01. ** 0.01 (2-tailed) significant * 0.05 (2-tailed) significant.

		SB99	**TKM99**	**PROT99**	**KernelN99**	**SpikeN99**	**KpS99**
TKM99	Corr.(Pearson)	.186	1.000	-.592	-.175	-.229	.146
	Sig. (2-tailed)	.010	.	.000	.015	.001	.043
PROT99	Corr.(Pearson)	.038	-.592	1.000	.299	.267	.015
	Sig. (2-tailed)	.605	.000	.	.000	.000	.831
KernelN99	Corr.(Pearson)	.910	-.175	.299	1.000	.881	.094
	Sig. (2-tailed)	.000	.015	.000	.	.000	.197
SpikeN99	Corr.(Pearson)	.781	-.229	.267	.881	1.000	-.375
	Sig. (2-tailed)	.000	.001	.000	.000	.	.000
KpS99	Corr.(Pearson)	.126	.146	.015	.094	-.375	1.000
	Sig. (2-tailed)	.083	.043	.831	.197	.000	.
YpS99	Corr.(Pearson)	.175	.595	-.345	-.086	-.462	.827
	Sig. (2-tailed)	.015	.000	.000	.236	.000	.000
		WR00	**TKM00**	**PROT00**	**KernelN00**	**SpikeN00**	**KpS00**
TKM00	Corr.(Pearson)	-.092	1.000	-.202	.280	-.375	.748
	Sig.(2-tailed)	.203	.	.005	.000	.000	.000
PROT00	Corr.(Pearson)	.206	-.202	1.000	.121	.356	-.273
	Sig.(2-tailed)	.004	.005	.	.096	.000	.000
KernelN00	Corr.(Pearson)	.929	.280	.121	1.000	.602	.345
	Sig.(2-tailed)	.000	.000	.096	.	.000	.000
SpikeN00	Corr.(Pearson)	.770	-.375	.356	.602	1.000	-.527
	Sig.(2-tailed)	.000	.000	.000	.000	.	.000
Kps00	Corr.(Pearson)	.071	.748	-.273	.345	-.527	1.000
	Sig.(2-tailed)	.325	.000	.000	.000	.000	.
YpS00	Corr.(Pearson)	.150	.456	-.264	.313	-.504	.930
	Sig.(2-tailed)	.037	.000	.000	.000	.000	.000
		WR01	**TKM01**	**PROT01**	**KernelN01**	**Spike01**	**KpS01**
TKM01	Corr.(Pearson)	-.035	1.000	-.370	-.310	-.216	-.097
	Sig.(2-tailed)	.630	.	.000	.000	.003	.179
PROT01	Corr.(Pearson)	.134	-.370	1.000	.355	.217	.106
	Sig.(2-tailed)	.063	.000	.	.000	.002	.144
KernelN01	Corr.(Pearson)	.788	-.310	.355	1.000	.634	.214
	Sig.(2-tailed)	.000	.000	.000	.	.000	.003
SpikeN01	Corr.(Pearson)	.591	-.216	.217	.634	1.000	-.591
	Sig.(2-tailed)	.000	.003	.002	.000	.	.000
KpS01	Corr.(Pearson)	.036	-.097	.106	.214	-.591	1.000
	Sig.(2-tailed)	.622	.179	.144	.003	.000	.
YpS01	Corr.(Pearson)	.153	.179	-.117	-.050	-.677	.817
	Sig.(2-tailed)	.034	.013	.105	.493	.000	.000

Up to now farmers have been using the knowledge about crop yield components and their interrelations to manage entire fields. However, knowledge is missing how these yield properties contribute to the yield development within subregions of entire fields. Researchers may need to develop and use models to understand yield development processes inside single fields, which are influenced by different environmental conditions.

For the purpose of modeling specific, possibly constant factors are applied in plot experiments to determine modelling parameters (see Primost, 1977, Roth et al., 1980, Damisch et al., 1984). As different areas (i.e. landform units) represent different growing conditions, relationships between grain yield properties remain unobserved, when the block variances overlap the landscape. Additionally, two assumptions are often violated for plot experiments: (I) the spatial independence of observations and (II) the normal distribution of observations (Nielsen and Wendroth, 2003).

To test, if using a raster based approach in a "real" landscape may provide similar results to plot experiments, the following example is provided in Figure 32 for winter rye in 2000 at "Bei Lotte". The number of spikes per m^2 was plotted against the grain yield in tha^{-1}, separated for below 33, above 36 and between 33 and 36 kernels per spike. Three regression equations with a linear least square fits yielded: $y = 0.018\ x + 0.6522$ with an $r^2 = 0.7834$ for KpS < 33; $y = 0.0169\ x + 0.2801$ with an $r^2 = 0.9431$ for KpS between 33 and 36 and $y = 0.0115\ x+2.855$ with an $r^2=0.642$ for KpS above 36. Autocorrelation was not significant for the parameters used in the graph, indicating spatial independence of observations. A similar graph was drawn

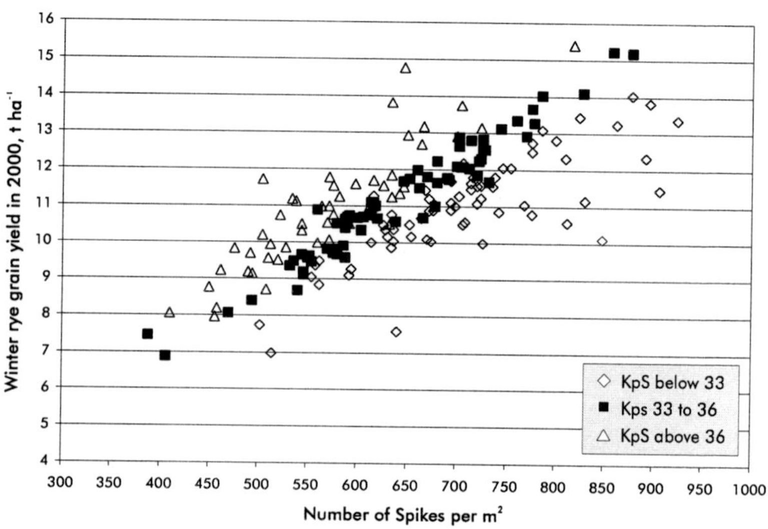

Figure 32. Relationship between Number of Spikes per m^2 and grain yield in tha^{-1} at the field site "Bei Lotte" for the year 2000, differentiated for Kernel per Spike (KpS): below 33 KpS, between 33 and 36 KpS and above 36 KpS (n=192).

using large plot experiments by Geisler (1983). This means, that by using hand harvest grain yields obtained at specific (low) sampling distances (grid) relationships can be observed which are similar to results drawn from small-scale plot experiments.

7.1.4. Temporal grain yield stability

Yield properties are correlated with each other with different strengths and in most cases with similar regression signs as shown in the section before. The reason for these differences is mainly due to the annual specific environmental conditions influencing plant development. In sight of these differences, the question exists, how stable the yield components are from year to year.

For this question long term field experiments are helpful (I) to assess the influence of management practices on crop yield and soil properties and (II) to evaluate the temporal stability of yield patterns. Eghball et al. (1995), for example, concluded that short-term variation was the major phenomenon (year to year variability). Moreover, they found that crop management may in some cases reduce the temporal variability of grain yield. Research performed by Pennock et al. (2001) supports this statement, showing that with increasing nitrogen fertilizer application a decrease of the CV of grain yield was observed.

Yield data were classified into five equally numbered classes for each year. Between years variability was evaluated in terms of differences between yield classes. If the class difference had been larger than one class, the position was assigned a value of one, otherwise a zero.

Figure 33 shows the temporal yield stability for the years 1999, 2000, and 2001 computed from hand harvest results for the field „Bei Lotte". Black cells show a different yield classification for each year, white cells a consistent yield classification in all three years. Areas with consistent yield classification appear clustered. Stable yield clusters reach a maximum distance of 81 m. Unstable areas extend up to a maximum distance of 189 m in NS direction and 135 m in WE direction. In general, unstable areas formed larger clusters, reacting differently each year due to different weather conditions than stable yield clusters. It has to be noted, that border areas around the edge of the field showed the highest number of yield classification differences from year to year (72% on a two year basis) compared to the rest of the field.

The percentage of cells with different classes was computed as total number of units as 100 % (n = 192) and the number of cells with a rank difference larger than one as x percent. For the analysis using landform units as separating elements, the total number of units was replaced by the actual number of LF elements (e.g. SH n = 17). Results from year to year variability were averaged for all three possible combinations (1999-2000; 1999-2001; 2000-2001) for aggregated landforms. Shoulder positions contained 5 % higher year-to-year variability than the field average (Figure 34 b) and Backslope positions a 2 % higher year-to-year variability. Level and Convergent Footslope positions showed below field average for year-to-year variability.

A further analysis of single LF's indicates that divergent positions showed a higher year-to-year variability than planar and level LF (Figure 34 a). Convergent positions exhibited different patterns: on back slopes the variability was 4 % higher than the field average whereas on footslopes the year-to-year variability was 3 % below average. Level landform units provided the most stable yield ranking over the period of three years. Less than fifty percent of the area contained an unstable yield. Shoulder and Backslope position showed the highest year-to-year variability with values up to 67%.

Another statistical measure to evaluate the temporal stability of grain yields are correlation coefficients. Large values indicate a similar spatial yield distribution. Correlation between spring barley in 1999 and winter rye in 2000 was 0.03, between spring barley in 1999 and winter rye in 2000 0.094, and between winter rye in 2000 and in 2001 0.148. The computed coefficients are in line with values observed by Grenzdörfer and Gebbers (2001) (0 - 0.48), Timlin et al. (1998) (0.061 - 0.287); Bakhsh et al. (2000) (- 0.69 - 0.66). Yield information at the field site "Bei Lotte" from one year to the next is highly unrelated to each other. Therefore, it might be questionable, if yield maps alone are a sound basis for future management decisions at the field site "Bei Lotte" and that region.

Figure 33. Yield classification for the field site „Bei Lotte". Black cells indicate no stable yield classification in all three years; white painted cells show a consistent yield classification during all three years.

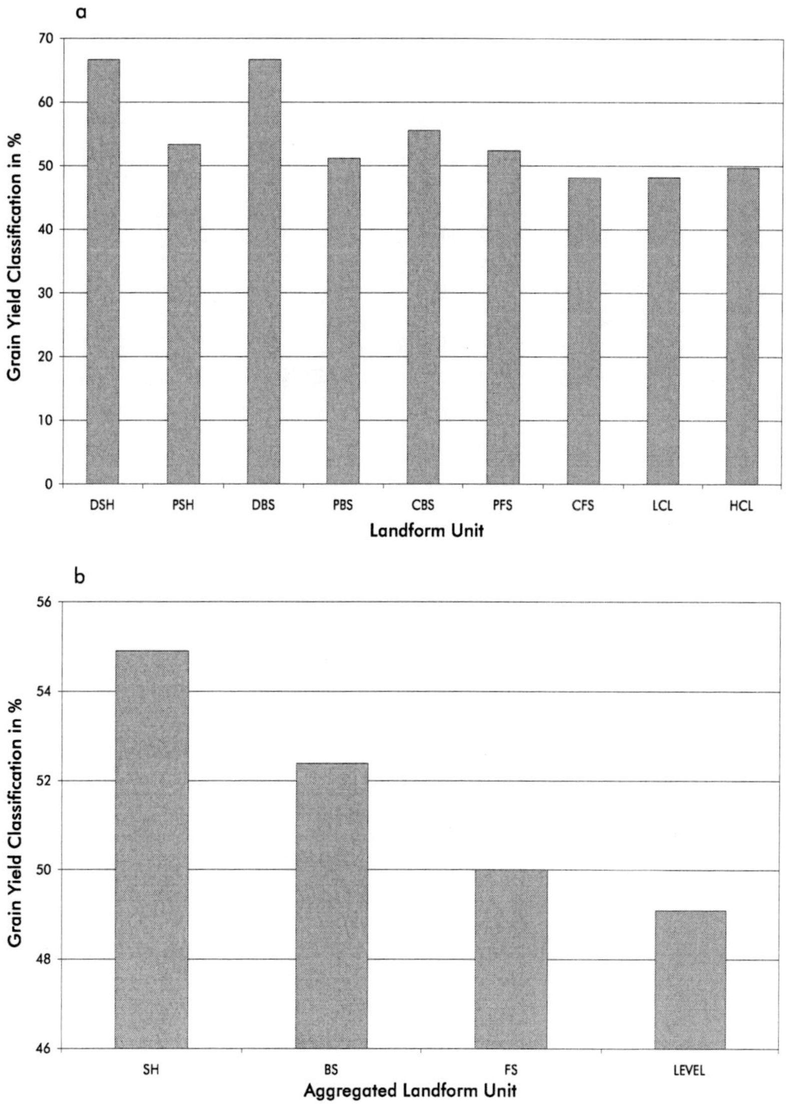

Figure 34. Average grain yield classification in percent for landform units (a) and aggregated landform units (b) (Shoulder SH, n = 17, Backslope BS, n = 49, Footslope FS, n = 16, Level LE, n = 110, left) at the field site "Bei Lotte" for the years 1999, 2000 and 2001.

7.1.5. Semivariogram analysis of grain yield components

The spatial variability of grain yield was discussed in detail in section 7.1.2 on page 65 ff.. However, little is known at the moment about the spatial variability of specific yield components, which influence the total amount of grain yield harvested (TKM, Yield per Spike, etc.). Meteorological, pedological and geomorphologic conditions are able to influence the development of yield components (see Figure 30), however little is known about the spatial semivariogram parameters (nugget variance, range and sill) of different yield components. The "Bei Lotte" dataset allowed to describe the spatial variability in grain yield development during a period of three years (see Table 24).

The spring barley in 1999 shows the strongest spatial dependencies for all three years. Spike per m^2 (111 m) and Kernel per m^2 (95 m) were the first properties, which develop early in the season. However, the secondary yield component Kernel per Spike showed no spatial dependency (see Table 24) at the field site „Bei Lotte". TKM developed later in the growing season and showed a range of 99 m. Lastly, the yield per spike is an integrative parameter, which showed a smaller range of up to 78 m.

In the year 2000 all yield components showed an increase in semivariance with lag distances larger than 200 m, except the kernel number (113 m).

The year 2001 showed a range of 151 m for the TKM, which is between the values observed in 1999 and 2000. The observed range of TKM agrees with values published by Thylen et al. (1999), who found a TKM range of 160 and 200 m.

The last yield property to be discussed in this spatial context is the protein content. A semivariogram range of 115 m in 1999 (Spring Barley), 75 m in 2000 and 110 m in 2001 could be fitted. The values observed at the field site "Bei Lotte" are in the range observed by Thylen et al. (1999), 110 m and 140 m.

The reason for the spatial variability of different yield properties lays in the different growing conditions experienced by the plants during the vegetation period. However, as information about the spatial extent of the development stage of the plants were not recorded, a further analysis in terms of climatic conditions during that respective developmental stage (for example the conception of spikelets and seed-bearings in Figure 30) similar to Damisch and Wiberg (1977) could not be performed. Manual recordings would be difficult to perform for such spatial extent, therefore observations by hyperspectral frequency analysers would be an asset for future observations (see Pattey et al., 2001, Zarco-Tejada et al., 2001, Gitelson et al., 2001, Goel et al., 2003).

The grain yield range (see Figure 24) showed different semivariogram ranges than yield components. In 1999, the yield range was similar to the observed YpS range; all other parameters extended up to twice the yield range (i.e. 100 m). In 2000, the range of the number of kernels agreed well with the yield range. Again, ranges of all other parameters extended to twice the yield range. The range of the number of kernels observed in the year 2001 showed a close pattern to the observed yield range followed by the TKM with again twice the observed yield range. Conclusions drawn from that are: (I) for winter rye in the years 2000 and 2001 the number of kernels showed a range similar to grain yield, regardless of the different amounts of precipitation (see Figure 25); (II) for spring barley the yield per spike range was the only yield property to be identified with a similar range to the observed grain yield range, (III) the range of other yield properties extended up to twice the range observed for yield, and (IV) distances for future yield property sampling (i.e. breeding) can oriented on the given values.

Table 24. Semivariogram parameters for different yield components in the years 1999-2001 at the field site "Bei Lotte". TKM = thousand kernel mass in g(1000 kernel)$^{-1}$; Protein = protein content on %; Kernel = number of kernels; spikeN = number of spikes; KpS = number of kernel per spike; YpS = Yield per Spike; NS-Ratio Nugget-Sill-Ratio.

1999	**Nugget**	**Model**	**Range**	**Sill**	**NS-Ratio**
KpS	1.618	Nugget	NA	NA	
Kernel	3570914.71	Spherical	95.24m	2518287.32	0.59
TKM	4.9329	Spherical	98.61m	1.4889	0.77
SpikeN	11903.9934	Spherical	111.22m	8046.0096	0.60
YpS	0.0038	Spherical	77.58m	0.0012	0.76
Protein	0.8522	Spherical	114.58m	0.3721	0.70
2000	**Nugget**	**Model**	**Range**	**Sill**	**NS-Ratio**
KpS	16.6342	Spherical	245.76m	5.3549	0.76
Kernel	9080026.6	Spherical	112.90m	1590880.03	0.85
TKM	3.7949	Spherical	217.17m	1.3527	0.74
SpikeN	8268.3126	Spherical	200.35m	2576.4818	0.76
YpS	0.0290	Nugget	NA	NA	
Protein	0.2505	Spherical	75.06m	0.2081	0.55
2001	**Nugget**	**Model**	**Range**	**Sill**	**NS-Ratio**
KpS	14.1188	Nugget	NA	NA	
Kernel	3130620	Spherical	69.17m	694767.448	0.82
TKM	4.767	Spherical	150.74m	1.5959	0.75
SpikeN	8141.6381	Nugget	NA	NA	
YpS	0.029	Nugget	NA	NA	
Protein	0.2361	Spherical	109.54m	0.1135	0.68

7.1.6. Semivariogram analysis of ear length and length of internodes

The spatial structure observed during grain yield development was discussed in 7.1.5 on page 82 for single yield components. However, information about the spatial patterns of the plant length growth itself might contain additional information on spatially varying growing conditions (Ciha, 1982, Vachaud and Chen, 2002). The semivariogram parameters obtained for ear length and internode length are presented in Table 25 for the years 1999, 2000 and 2001. For spring barley, a decrease in semivariogram range during the course of plant development can be observed for second internode to the fourth internode and even ear length, i.e. from 113 m to 110 m to 88 m to 73 m in the year 1999 (Table 25). Similar decreases, that are not so pronounced, however are visible for winter rye in 2000 and 2001. The differences in semivariogram range seem to be small. As the development of different internodes occurs very close in time, this was expected.

Geisler (1983) states that (I) the length growing of the first and second internode ends at the same time, (II) the third internode growth ends shortly after and III) top internode starts with the extension if the first and second internodes are finished. The second internode shows the largest range in the above average precipitation years (1999 and 2001 - see Table 25). In contrast, no range parameter could be determined for the year 2000.

The lowest, first internode showed range parameters of between 77 m and 100 m, well below the ranges observed for the second internode. If we assumed that semivariogram ranges are similar, this observation would be in contrast to the statement by Geisler (1983), that the first and second internodes finish their development at the same time. A possible explanation for the observed differences is that at the time of the development of the first internode, differences in plant growth due to limitations occur. However, as the second internode finished growth the major development occurred, differences in environmental properties did lead to larger spatial variation.

Irrgang et al. (2001) stated that plant height might serve as a good indicator for grain yield. For this reason, plant height measurements at a field scale at time of EC32 can use longer distances for sampling, whereas as crop development approaches EC60-69, distances between sampling points must be reduced to optimize sampling (Herbst, 2002).

Table 25. Semivariogram parameter of ear length and internode length for spring barley in 1999, winter rye in 2000 and 2001.

Type	Year	Nugget in cm^2	Range	Sill in cm^2	Model
Ear Length	1999	0.05	73.38m	0.05	Spherical
4.Internode	1999	0.65	99.45m	0.42	Spherical
3.Internode	1999	1.33	110.38m	0.53	Spherical
2.Internode	1999	0.72	113.74m	0.38	Spherical
1.Internode	1999	5.38	87.67m	3.83	Spherical
Ear Length	2000	0.33	87.67m	0.17	Spherical
4.Internode	2000	2.44	81.79m	1.59	Spherical
3.Internode	2000	3.37	89.36m	0.85	Spherical
2.Internode	2000	1.57			Nugget
1.Internode	2000	1.80	99.45m	0.53	Spherical
Ear Length	2001	0.45	NA	NA	Nugget
4.Internode	2001	2.13	66.65m	0.58	Spherical
3.Internode	2001	2.04	74.22m	0.86	Spherical
2.Internode	2001	0.74	133.08m	0.37	Spherical
1.Internode	2001	1.98	77.58m	0.52	Spherical

7.2. Crosscorrelation analysis between grain yield and grain yield components for different relief parameters, and versus soil texture, penetration resistance and NDVI values

Crosscorrelation coefficients (CC) were computed for the field sites „Bei Lotte“ and “Sportkomplex” for a variety of parameters as shown in Table 45 in Appendix Geostatistics. Only CC with separation distances larger than one class above the 95 % confidence interval were selected and will be discussed in the following paragraphs.

No significant crosscorrelation could be observed between hand harvest grain yields and yield components for the years 1999 - 2001 at the field site “Bei Lotte”. In contrast, two different groups of significant CC regarding the years could be observed for the field site “Sportkomplex“ (Figure 36). The years 1997, 1998 and 1999 were crosscorrelated with high covariances, lower crosscorrelations at lag 0 and significant crosscorrelations of up to 2 lag classes (54 m). The second group were the years 2000, 2001 and 2002 with generally lower covariances and higher CC at lag 0, both compared to the first group, but similar significant crosscorrelation distances (SCD) (54 m). Differences between both groups can be explained by the starting of a fertilizer field experiment at the field site “Sportkomplex” in 2000 (Schwarz et al., 2000). Additionally, a significant CC was determined between grain yields in 1999 and 2002, where winter rape was grown under a homogeneous management except for some zero fertilizer plots.

CC between relief parameters and grain yield and yield components showed that elevation was the relief parameter with the longest correlation length and the largest number of significant crosscorrelations (Figure 36 and Figure 37).

For the field site “Sportkomplex” the relief parameter elevation showed in 1998 and 2001 a SCD up to 6 lag classes, in 2000 up to 2 lag classes (Figure 37). The difference in SCD between 2000 with 2 lag classes were caused by the dry conditions, compared to the 6 lag classes observed in 1999 and 2001, where precipitation above average was observed. No significant CC for elevation could be observed for the years 1997 and 1999 for grain yield at “Sportkomplex”, as well as for results of hand harvest in the years 1999-2001 at the field site “Bei Lotte”. If a significant crosscorrelation was found at the field site “Sportkomplex”, crosscorrelation length was at least twice (108 m) compared to the 30 - 40 m by Li et al. (2001) for cotton yield.

The year 1998 at the field site “Sportkomplex” showed a large number of significant CC between relief parameters and grain yield as seen in Figure 37. However, CC above the 95 % confidence limit could not be observed in other years at “Sportkomplex” and at the field site “Bei Lotte”. Such results show, that under higher precipitation than average (as in 1998), relief parameters show a significant impact on yield development. Exceptions were the CC between grain yield in 1999 and slope at “Sportkomplex” (Figure 35). The CC in Figure 35 is shown together with the CC for the same year at the field site “Bei Lotte”. A CC with the same sign and a similar SCD, however well below the 95 % significance interval can be observed for the field "Bei Lotte".

Grain yields from hand harvest at "Bei Lotte" showed no significant crosscorrelation larger than one lag class, however single yield components did (Figure 38). The number of spikes in 1999 showed a positive CC up to 3 lag distances, whereas in 2000 a similar, but negative CC with a separation of one lag class is observed (Figure 38). CC between yield per spike in 1999 and 2000 at "Bei Lotte" show a similar pattern, however with less differentiation in SCD (Figure 38). The yield components kernel per spike, TKM, protein content and yield per spike in 2000 as shown in Figure 38 show different positive as well as negative crosscorrelations, however no significant CC could be observed between grain yield components and relief parameters in 2001.

Li et al. (2001) have shown a negative CC for lint yield and elevation for one year, findings indicate that the sign of the CC, the value of the CC at lag 0 and the separation distances may differ from year to year (see Yield per Spike and Number of Spikes versus elevation in 1999 and 2000 in Figure 38).

Applying cospectral analysis, Timlin et al. (1998) found no dominant component of variance of grain yield and curvature at any particular frequency, but a similar distribution of frequencies. Additionally, Timlin et al. (1998) mentioned, that grain yields were higher in normal to dry years at concave positions, whereas no correlations were seen for wet years. No significant crosscorrelation function could be identified between grain yield and grain yield components for the field sites "Sportkomplex" and "Bei Lotte" for years with different precipitation. One exception occurred for grain yield in the above average precipitation year 1998 at the field site "Sportkomplex, which did show significant crosscorrelation. This finding is therefore in contrast to the results of Timlin et al. (1998).

Crosscorrelations between grain yield and yield components in Figure 39 were computed for penetration resistance at "Sportkomplex" (n = 54) and "Bei Lotte" (n = 192). Penetration resistance in the upper horizons (0-10 cm, 10-20 cm) at "Bei Lotte" was found to be positively crosscorrelated up to a SCD of 3 lag classes with kernel per spike and yield per spike in 1999, and negatively crosscorrelated with the number of spikes (Figure 39).

Only these yield components of spring barley showed a significant CC, whereas no significant CC could be observed for grain yield and other yield components at "Bei Lotte" and "Sportkomplex" in all other years. Exceptions were the protein content in 2000 at "Bei Lotte" with penetration resistance in deeper soil layers (only results of 50-60 cm depth are shown in Figure 39). A SCD of 2 lag classes could be observed for winter rye. In contrast, the winter wheat in 2000 at the field site "Sportkomplex" did not show any significant crosscorrelation (Figure 39). However, the winter wheat in 2002 showed a CC between penetration resistance in deeper layers (60-70 cm) with a SCD of 54 m.

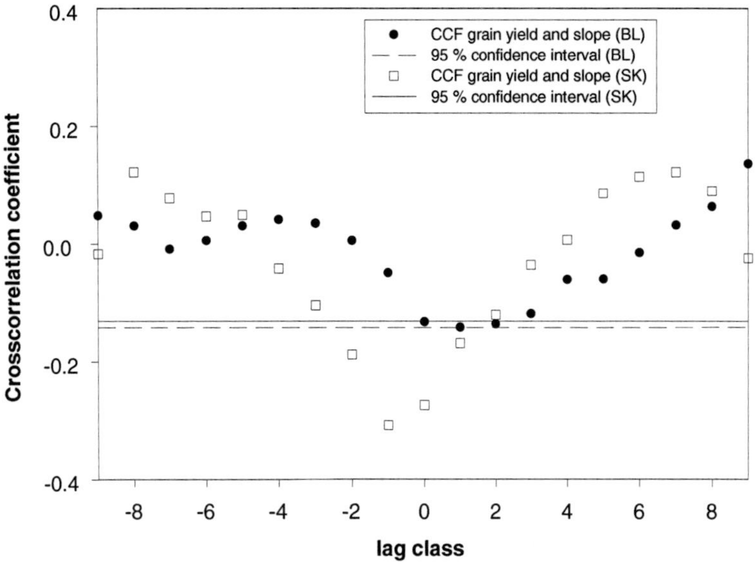

Figure 35. Crosscorrelation between grain yield in 1999 and the relief parameter slope at the field sites "Sportkomplex" (SK) and "Bei Lotte" (BL). The dashed line represents the 95% confidence interval. Lag Distance interval 27 m.

Significant CC between grain yield and silt content at the field site "Bei Lotte" and "Sportkomplex" are shown in Figure 40. Silt content showed SCD for grain yield in 1998 and 2000 for three depths (0-30 cm, 30-60 cm and 60-90 cm) of up top 2 lag classes, but not for the other years. This indicates that silt content is an important parameter for years with precipitation above average (e.g. 1998) as well as for years with precipitation below average (e.g. 2000), but not for years with average precipitation. Similar results could not be observed at the field site "Bei Lotte". As an example, the CC between silt content and grain yield in 2000 is shown in Figure 40 for the field site "Bei Lotte".

Sand content was found to be negatively crosscorrelated for the field site "Sportkomplex" in 1998 in 60-90cm depth, as well as in 2000 and 2001 in 0-30cm (Figure 41). A similar extent in crosscorrelation and SCD (up to 2 lag classes) could be observed for silt content (Figure 41).

Li et al. (2001) reported a negative crosscorrelation between clay content and lint yield. Similar results can be drawn for clay content in 1998 in 0-30 cm and in 30-60 cm at "Sportkomplex". However, no significant CC could be observed in any subsequent year there as well as at the field site "Bei Lotte". As the year 1998 provided the largest above average precipitation amount of all years investigated, clay content might be a yield influencing variable in "wet" years.

The significant crosscorrelations between NDVI and grain yield or yield components in different years are shown in Figure 42 and Figure 43. SCD varied between one to five lag classes for all significant CCs. Except for the protein content and the yield per spike, both in 1999 (Figure 42), no significant CC could be observed for the same time period. This means that for example a significant crosscorrelation was observed between protein content in 2000 and NDVI from May 2002, but not with an NDVI image in 2000 (Figure 42).

These findings should be highlighted. Specific years led to significant CCs between grain yield / yield components and same year NDVI images. However, NDVI - values of other years show also significant CC to this grain yield / yield components, but with a lower crosscorrelation at lag distance 0. This indicates that differences in plant development are visible from NDVI images in every year, however are not reflected in total grain yield with "enough" variability to show significant crosscorrelations.

These results are supported by results found for CC between grain yield and NDVI values at "Sportkomplex" (Figure 44). Grain yield in 1998 and 2000 shows the largest crosscorrelations at lag 0 between grain yield and NDVI images of the respective year. Significant CCs for NDVI values from different years (i.e. grain yield 2000 and NDVI 04/1999) show (I) smaller crosscorrelation at lag distance 0, (II) similar significant crosscorrelation distances and (III) no significant CCs to their respective year itself (i.e., NDVI from 2000 shows a significant CC to grain yield in 2001, but not to grain yield in 2000).

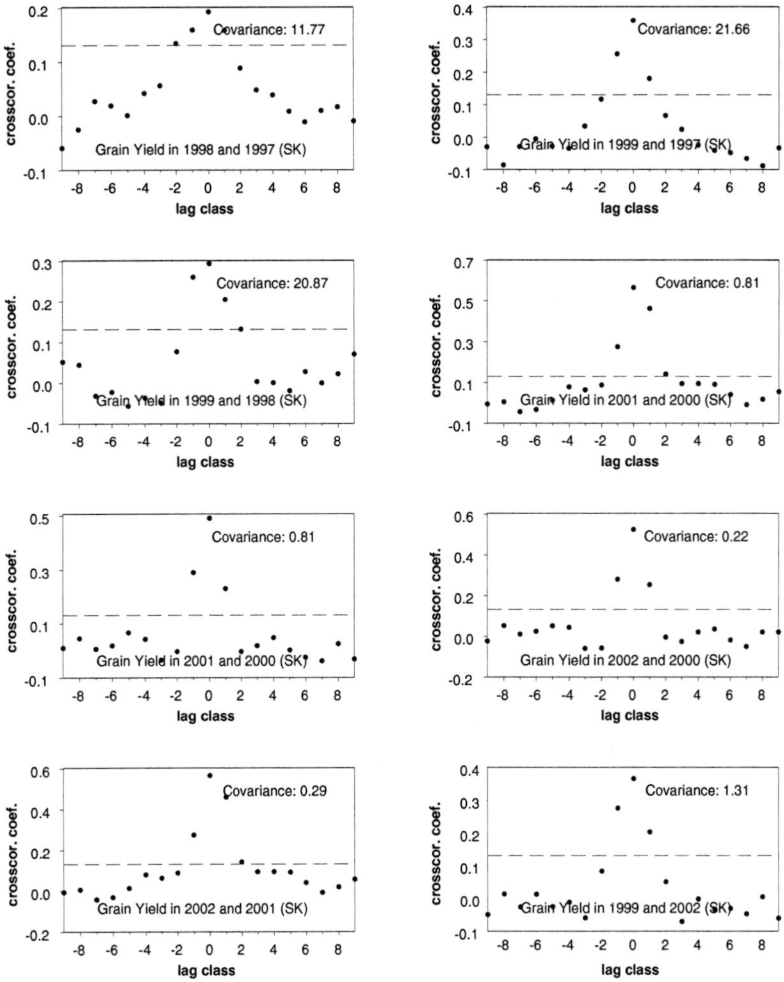

Figure 36. Selected significant crosscorrelations between grain yield at the field site "Sportkomplex" (SK) for the period 1997-2002. The dashed line represents the 95% confidence interval. Lag Distance interval 27 m.

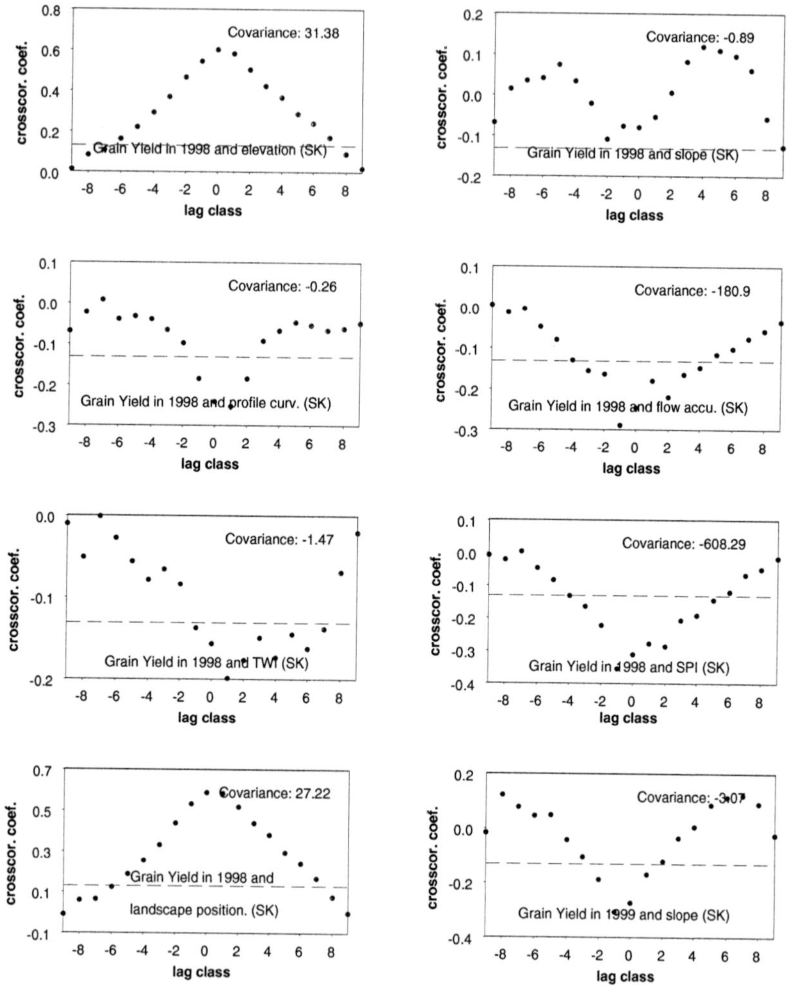

Figure 37. Selected significant crosscorrelations between grain yield at the field site "Sportkomplex" (SK) in 1998 and the relief parameters mean elevation, slope, profile curvature, flow accumulation, topographic wetness index, stream power index, and landscape position. The dashed line represents the 95% confidence interval. Lag Distance interval 27 m.

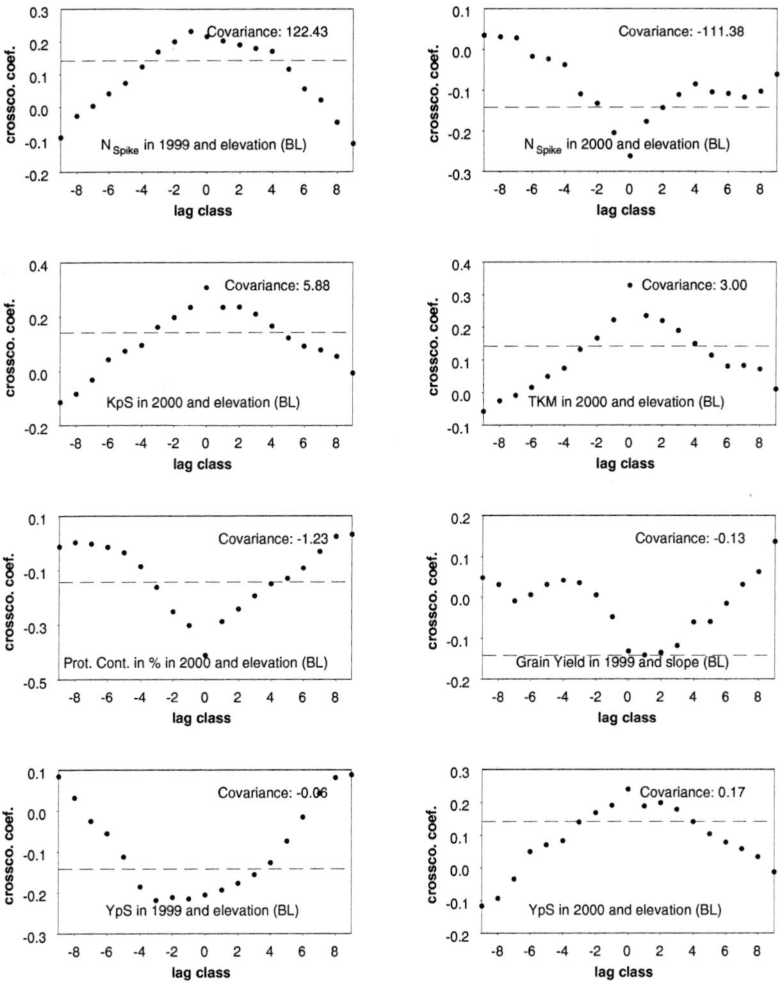

Figure 38. Selected significant crosscorrelations between number of spikes, kernel per spike, TKM, protein content and yield per spike and the relief parameters elevation and slope at the field site "Bei Lotte" (BL) in 1999 and 2000. The dashed line represents the 95% confidence interval. Lag Distance interval 27 m.

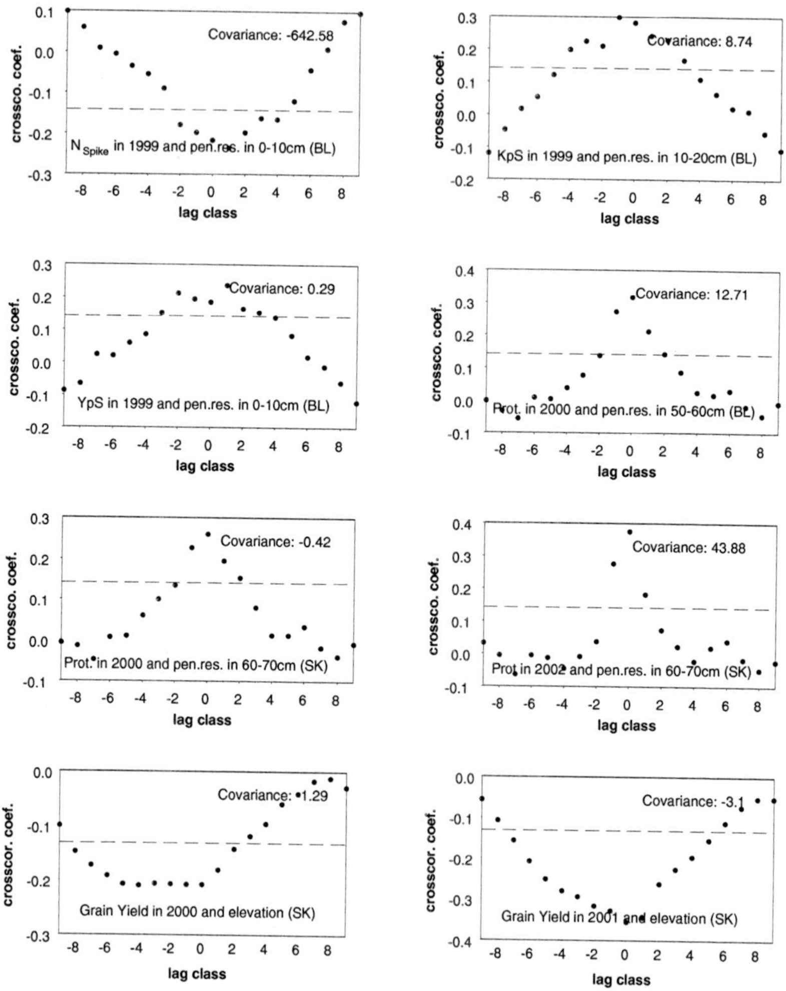

Figure 39. Selected significant crosscorrelations (CC) between number of spikes, kernel per spike, yield per spike and protein content and soil penetration resistance at different depths at the field site "Bei Lotte" (BL) in 1999 and 2000, Additionally CCs are drawn for the protein content at the field site "Sportkomplex" (SK) in 2000 and 2002 versus penetration resistance and secondly, between grain yield and elevation. The dashed line represents the 95% confidence interval. Lag Distance interval 27 m (54m for SK protein).

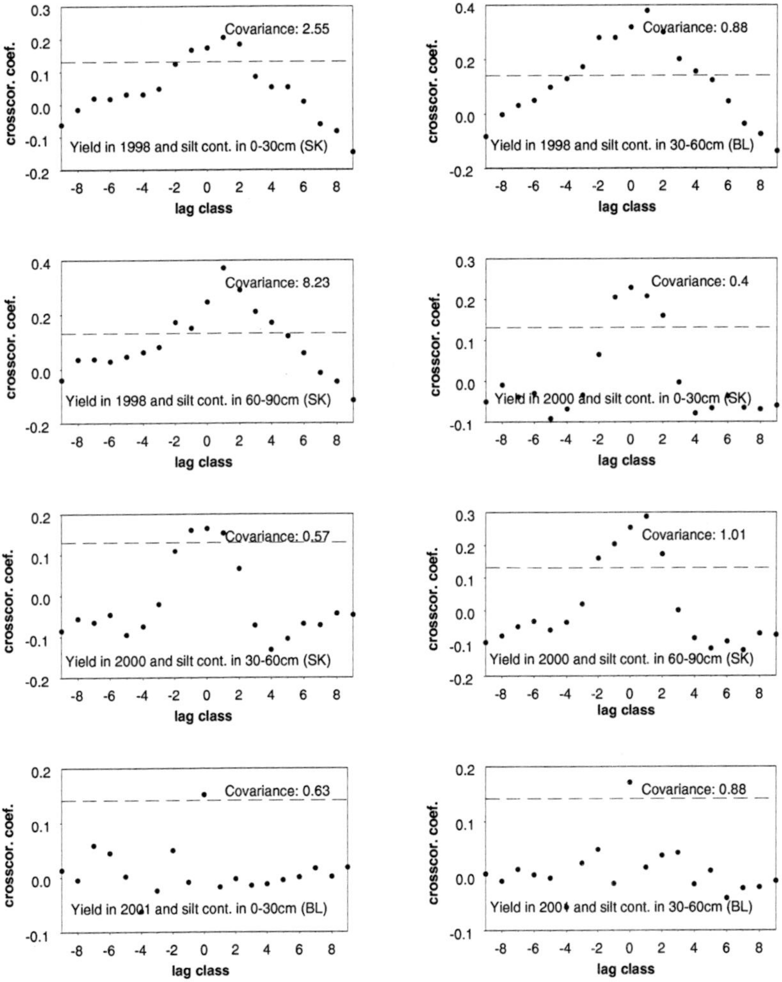

Figure 40. Selected significant crosscorrelations between grain yield and silt content at three different depths at the field site „Sportkomplex“ (SK) and "Bei Lotte" (BL). The dashed line represents the 95% confidence interval. Lag Distance interval 27 m.

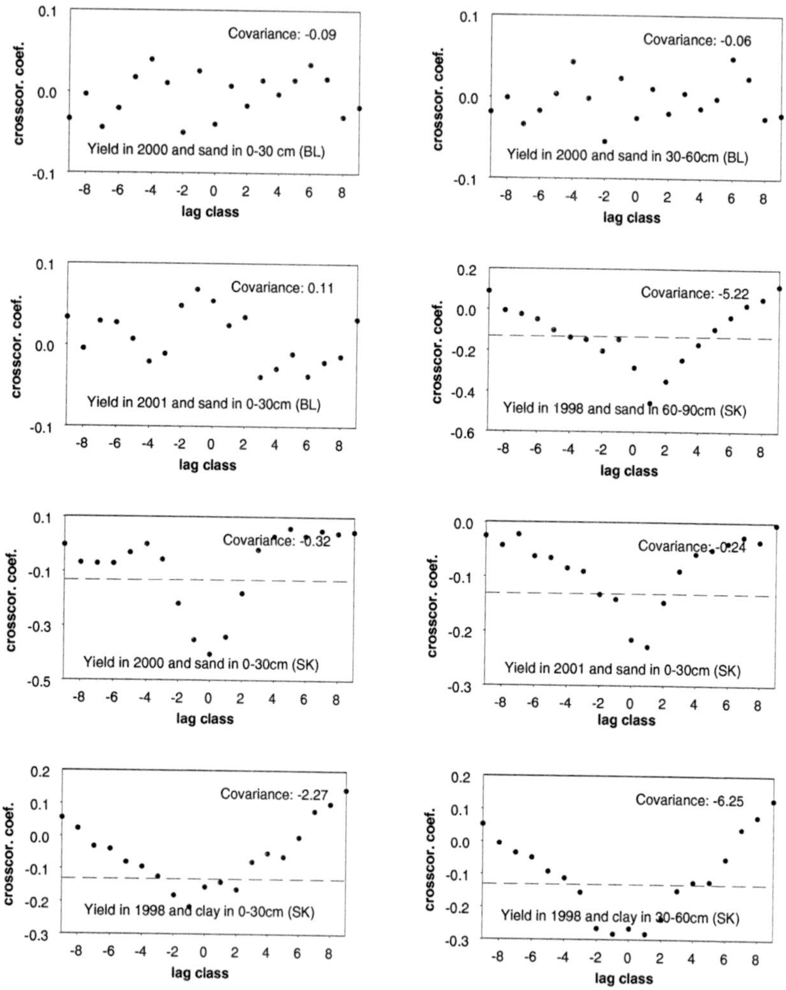

Figure 41. Selected significant and non significant crosscorrelations (CC) between grain yield and sand content at the field site „Sportkomplex" (SK) and "Bei Lotte" (BL) for the years 1998, 2000 and 2001. Additionally, the CCs between grain yield and clay content in 1998 at "Sportkomplex" are given. The dashed line represents the 95% confidence interval. Lag Distance interval 27 m.

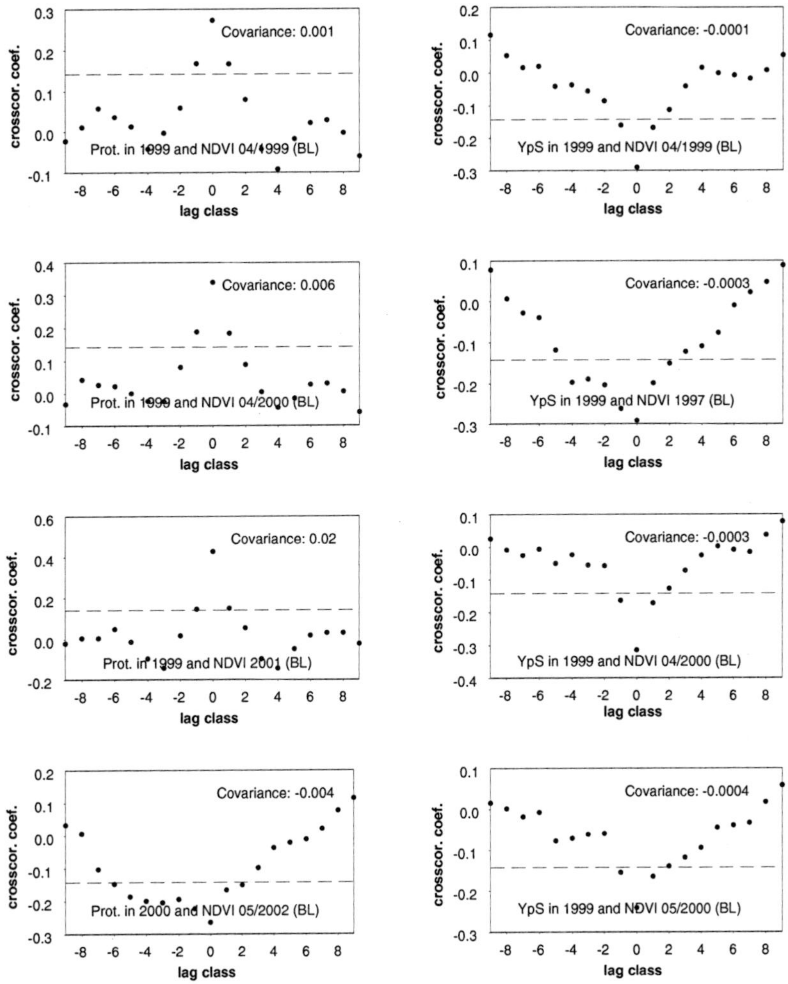

Figure 42. Selected significant crosscorrelations between grain yield and yield properties and NDVI at the field site "Bei Lotte" (BL). The dashed line represents the 95% confidence interval. Lag Distance interval 27 m.

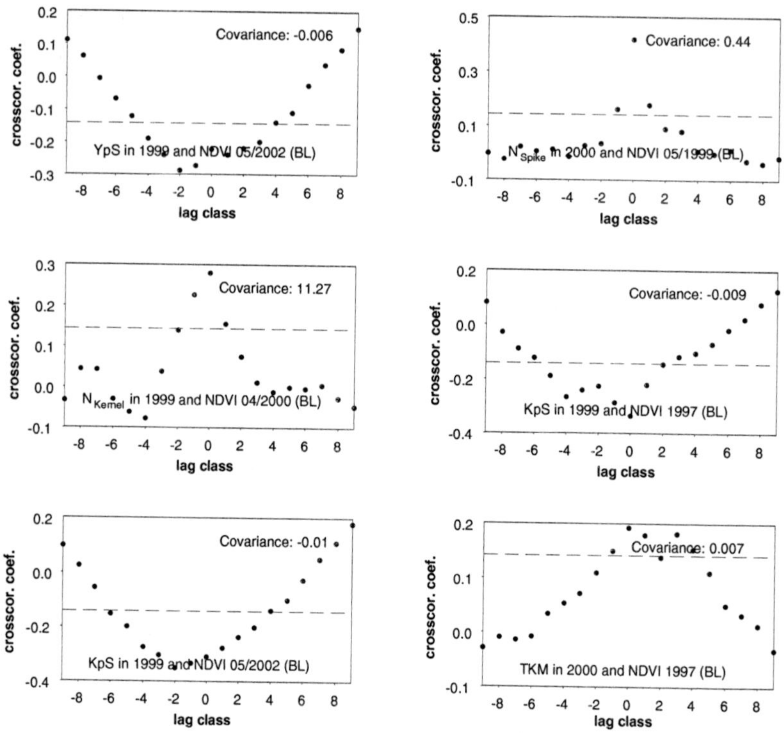

Figure 43. Selected significant crosscorrelations between grain yield and yield properties and NDVI at the field site "Bei Lotte" (BL). The dashed line represents the 95% confidence interval. Lag Distance interval 27 m.

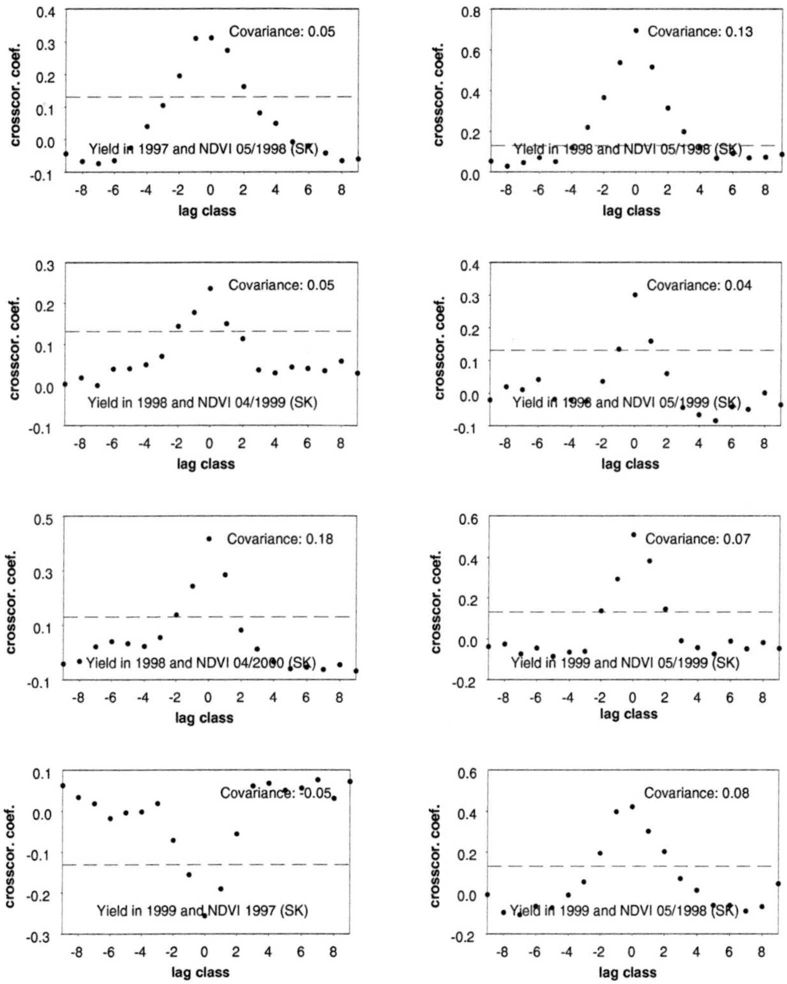

Figure 44. Selected significant crosscorrelations between grain yield in 1997, 1998 and 1999 and NDVI at the field site "Sportkomplex" (SK). The dashed line represents the 95% confidence interval. Lag Distance interval 27 m.

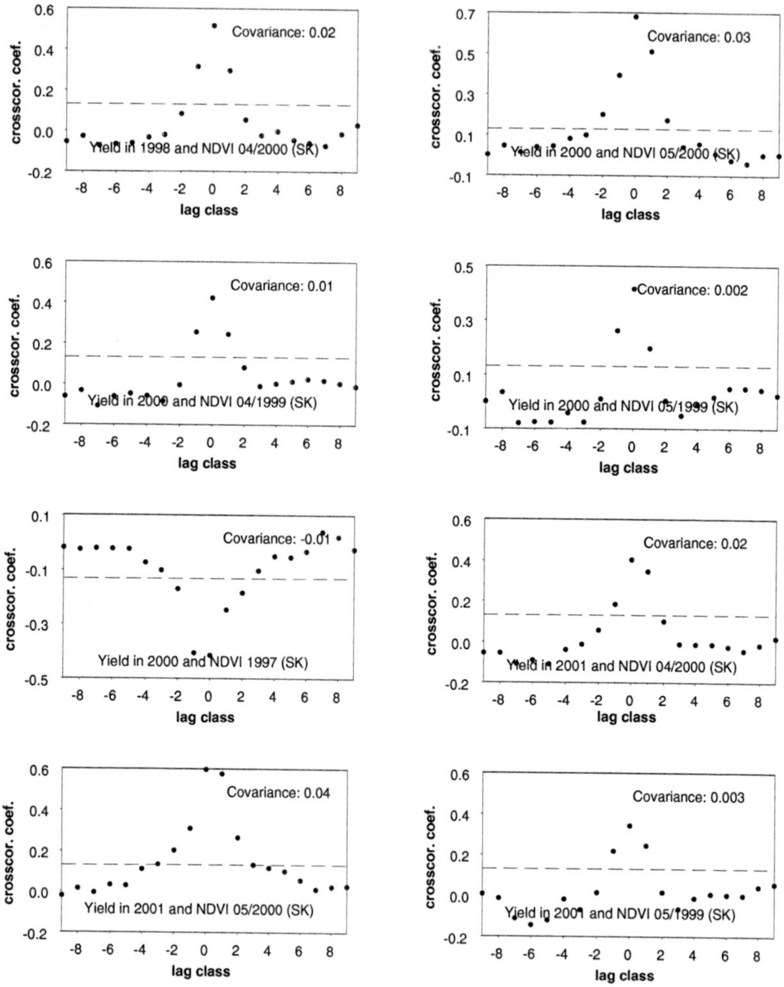

Figure 45. Selected significant crosscorrelations between grain yield in 2000 and 2001 and NDVI at the field site "Sportkomplex" (SK). The dashed line represents the 95% confidence interval. Lag Distance interval 27 m.

7.3. Grain yield components and plant properties for selected relief parameters

In the following section crop yield and yield properties are examined using the relief parameter profile curvature and landform elements.

The hypothesis is put forward that characteristic grain yield magnitude and grain yield components are affected by different landforms (LF), and each landform unit will reflect a characteristic yield development, both based on soil and meteorological conditions, as well as management history. Therefore, the question was to investigate, whether the observed grain yield patterns are related to specific LFs and whether LFs might be helpful to explain yield variability at the field scale. The same applies for the relief parameter profile curvature.

7.3.1. Grain yield development versus landform

The number of counted plants per m^2 together with the number of Spikes per m^2 is shown in Figure 46 for spring barley in 1999 and winter rye in 2001 (Figure 47) for each different LF. Plant numbers show a nearly homogeneous distribution (field average 254) for spring barley in 1999. The maximum differences of 11 plants per m^2 were observed between BS and FS landform elements. However, shoulder positions show the least number of spikes in 1999 and 2000, with a slightly increasing trend for the landforms backslope-footslope-level. No relationship can be observed between the number of plants per m^2 counted upon field emergence and the number of spikes per m^2.

The number of winter rye plants in autumn of 2000 showed shoulder positions with 12 plants less than the field average. Plant numbers at the other landforms were very similar. In contrast, the harvest in the year 2001 showed SH positions with the highest number of spikes. The least numbers of spikes were found at BS positions. Still, a slight increase in spike number with the landforms backslope - footslope - level was observed.

Figure 49 shows the results of the yield and yield properties for the field „Bei Lotte“, with each of the six drawings showing the results for the respective landform and the field average for the given year.

The number of yielding spikes per m^2 (N_{spike}) is the first yield component to develop in the growing season (Figure 49 B). SH positions contain approx. 60 spikes per m^2 in 1999 and 50 spikes per m^2 in 2000 less than the field average. In contrast, number of spikes was found to be highest at SH positions in 2001. Additionally, a slightly increasing trend in N_{Spike} can be found for the LF positions BS-FS-Level for all three subsequent years (Figure 49 B). Differences in environmental conditions (e.g. precipitation) may be the reason for the year-to-year grain yield variability within a respective LF. Aufhammer (1973) stated that a reduction in N_{Spike} could be attributed to (I) differences in growth response and (II) soil nutrients. However, soil nutrients are found to be sufficiently high across the field site "Bei Lotte"(see Table 43 and Table 44).

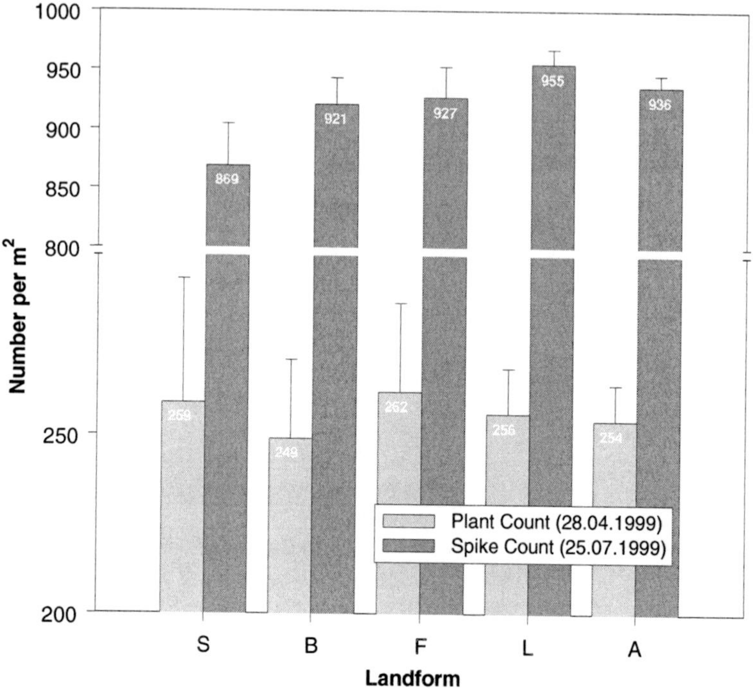

Figure 46. Number of spring barley plants per m^2 at 28.04.1999 and spike numbers per m^2 at harvest (25.07.1999) for the landforms Shoulder (S, n = 17), Backslope (B, n = 49), Footslope (F, n = 16), Level (L, n = 110) and the field (A, n = 192) at the field site "Bei Lotte". Bars show the standard error.

Therefore, the observed differences in growth responses would supposedly be due to different environmental conditions at different LFs. Depression areas like FSs might provide air deficiency stress in spring to plant growth during a moist growing season as in 2001, whereas SH positions might enhance a drought stress to plant growth during a dry growing season such as 2000. Roth et al. (1988) state that dry conditions at the time of heading leads to an unusually high reduction of spikes, which could possibly occur at SH positions. A similar pattern with equal number of plants in spring and less spikes at harvest would be expected for the year 2000, but cannot be shown due to missing plant count numbers at plant emergence.

The number of kernels (N_{Kernel}) is the second yield property to develop in time (see Figure 30). It shows decreased values at SH positions in 1999 and 2000, whereas in 2001 the largest N_{Kernel} was observed (Figure 48 D). In addition, Figure 48 E gives an impression of the grain yield development based on the Number of Kernel and N_{Spike}. In 1999 and 2000 SH positions showed the largest number of kernels per spike (KpS), and the lowest number is found for FS positions in 2000. In contrast, during the relatively wet year 2001 the highest KpS

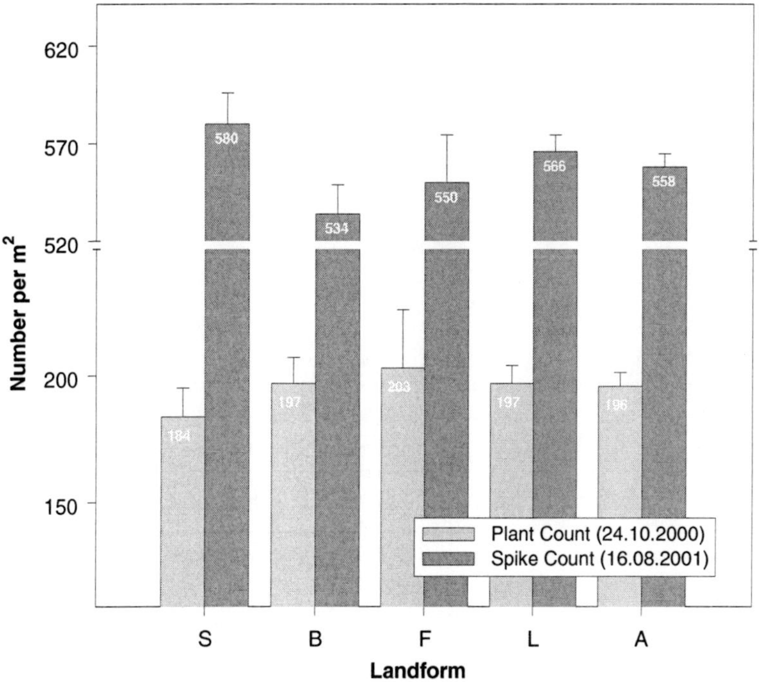

Figure 47. Number of winter rye plants per m^2 at the 24.10.2000 and spike numbers per m^2 at harvest (16.08.2001) for the landforms Shoulder (S, n = 17), Backslope (B, n = 49), Footslope (F, n = 16), Level (L, n = 110) and the field (A, n = 192) at the field site "Bei Lotte". Bars show the standard error.

was found at the FS position and the lowest at the SH positions. If soil water content increases, get saturated in the worst case scenario, extended respiration occurs in the roots. Therefore a high demand for O_2 exists in the soil for plants as well as for microorganisms (Marschner, 1995). Michael (1990) showed that with a decrease in soil O_2, an increase in CO_2 and ethylene occurred. Ethylene is known for the depression of root elongation and the enhancement of root development. Additionally, Smit et al. (1989) reported a suppression of leaf elongation under an decrease of soil O_2.

The third yield component - the TKM, provides an image of the environmental conditions at the time of the kernel-filling (Figure 48 F). Spring Barley at SH positions in 1999 yielded approximately 1 g higher TKM than in other LFs. In the year 2000, only small differences in TKM are found across LF's, in contrast to the decrease of 1.5 g found at FS compared to all other LF's in 2001. A decrease in TKM is reported due to increasing N_{Spike} by Darwinkel (1980), as well as due to a prolonged time of dryness during kernel-filling by Roth et al. (1988). Such results are found in 1999, with an increase in TKM together with a decrease in

N_{Spike} at SH positions (see Figure 48 F +B). However, results for SH positions in 2000 show no increase, even if N_{Spike} was decreased again. Additionally, TKM is reduced at FS in the wet year 2001, indicating stress during the grain - filling period (Entz and Fowler, 1998). As soil water content after harvest was almost evenly distributed across all LF (see Figure 72), differences are probably due to the observed lodging or different states of ripeness of grain yield.

Yield per Spike is another yield component (Figure 48 C). A more homogeneous distribution across all LFs can be seen, indicating a high potential of plant compensation. Generally, SH positions in 1999 and 2000 show the largest values for that parameter, whereas FS positions in 2001 showed the lowest yields per spike.

Finally, all yield components discussed so far are combined in the yield itself (Figure 48 A). In the moist year 2001 SH positions exceeded all other LF's by approx. 0.75 tha^{-1} probably due to more favourable growing conditions. In contrast in the dry year 2000, plants at SH positions suffered during the growing season, yielding approx. 0.75 tha^{-1} less than the field average. The year 1999, shows an interesting phenomenon for the observed yield at the SH position. Although KpS and N_{Spike} show their lowest values during the development of the spring barley, the plants compensate this deficiency with an increase in TKM during the grain filling phase so that any differences in grain yield developments were minor (see Figure 48 D, B, F). Hence a similar grain yield level was reached as observed at FS positions. As the plants are able to compensate to some extent due to differences in observed environmental conditions due to intra-plant yield components, one important parameter that influences yield is N_{Spike}.

Pollmer (1957) stated that the fixation of plant density occurs during the time of shooting. At the start of that time period, the maximum number of spikes is developed - depending on environmental conditions- followed later by spike reductions. Plant density is therefore a fixed factor - sometimes limiting - in this landscape.

Generally, if weather conditions were known beforehand (see Stone et al., 1996) several management activities could be applied to influence grain yield development. Examples are: I) seeding a specific grain variety adapted to the expected weather conditions (Ciha, 1982), II) applying additional irrigation to influence crop development (Roth et al., 1980), III) distributing nitrogen fertilizer spatially accordingly to push or limit the yield development (Pennock et al., 2001) and the related root growth, and IV) distributing growth retardants spatially accordingly to enhance or limit the yield development. Examples are given without any reference to specific LFs, because such task has yet to be performed in another research project. Further research could test how these management options could be implemented.

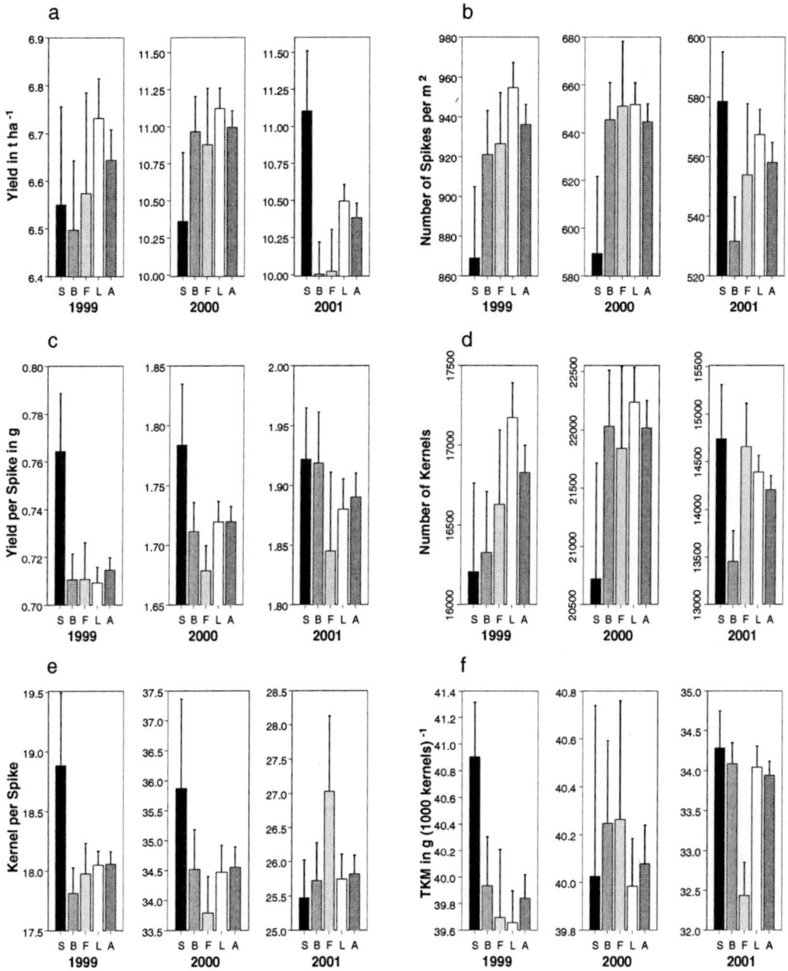

Figure 48. Grain yield in tha^{-1} (a) and the yield properties number of spikes m^{-2} (b), yield per spike in g (c), number of kernels per m^2 (d), number of kernels per spike (e) and TKM in g(1000 kernels)$^{-1}$ (f) for the field site "Bei Lotte" for the landforms Shoulder (S, n = 17), Backslope (B, n = 49), Footslope (F, n = 16), Level (L, n = 110) and the field (A, n = 192). Bars show the standard error.

The analysis of LF units versus Protein content for three different years is shown in Figure 49. Whereas single yield components react at different LF, during three years and two different plant species, protein content remains fairly constant given the environmental conditions observed.

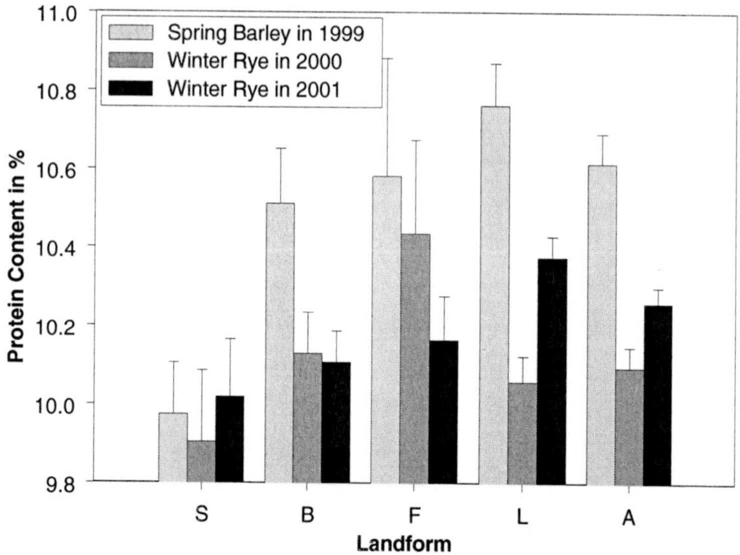

Figure 49. Protein Content in 1999, 2000 and 2001 at the field site "Bei Lotte" for the landforms Shoulder (S, n = 17), Backslope (B, n = 49), Footslope (F, n = 16), Level (L, n = 110) and the field (A, n = 192). Bars show the standard error.

The protein content as a quality parameter was analysed for the four major LF elements. For winter rye protein contents should be high for a high baking quality. For spring barley it should be low if grown for malting purposes. In general, shoulder positions contained the least protein content during all years, followed by BS, FS and Level positions in increasing order, even regardless of weather conditions, or the amount of grain yield harvested at that location.

An exception occurred in the year 2000 with the protein content at footslope and level landforms in reverse order. Fiez et al. (1994) stated that late-season stress increased grain protein independent of plant N-responses. Own observations showed that at FS positions, straw was still green, whereas the rest of the field was considered fully ripe. In contrast, combine grain moisture measurements showed a slight decrease at footslope positions.

Results for protein content for different landforms are in contrasts to results in the literature. Fiez et al. (1994) observed the highest grain protein at shoulder positions, whereas the lowest was observed at north exposed backslope positions. Manning et al. (2001a) observed the lowest protein content at FS positions and attributed that to the greater biomass production at that landform together with a dilution effect. One possible explanation of such difference between results at the field site "Bei Lotte" and the literature might be attributed either (I) to differences in the availability of soil nitrogen to the plants (Jowkin and Schoenau, 1998), or (II) attributed to late season moisture stress (Fiez et al., 1994).

Additionally to the conclusions drawn above for the yield development, special consideration should be taken to influence the protein content. A possible working hypothesis would be that different amounts of nitrogen fertilizer for late-season fertilizer application increase the protein content of grain yield at footslope landforms. However, such recommendations must depend on the expected weather conditions. Secondly, special attention must be paid, that the farmer receives a reliable estimate of nitrogen protein from the grain yield harvested, because of the amount of money paid for the special qualities. Thylen and Algerbo (2001) have already shown that measurements of grain yield nitrogen can be performed during harvest, and therefore it will be an opportunity to sort the grain into different protein qualities.

7.3.2. Grain yield development versus profile curvature

The relief parameter profile curvature is known to have influence on grain yield and soil properties (Timlin et al., 1998, Sinai et al., 1981).

Convex (-0.2 to -0.1 1/100 m) and concave (0.2 to 0.1 1/100 m) landscape positions showed slightly lower yields in 1999 and 2000 than planar sections of the field (-0.1 to 0.1 1/ 100 m)(Figure 50 A). The highly concave area (0.3 to 0.4 1/100 m) showed a low yield of 5 tha^{-1} in 1999, compared to the field average of 6.6 tha^{-1}. In contrast during the dry season of 2000, the highest yielding areas were found at the highly concave position with a value of 11.2 tha^{-1}, however with a large standard deviation. Differences in yield between the five profile curvature classes are larger than described by Timlin et al. (1998), who showed differences of up to 0.2 tha^{-1} for two curvature classes for corn grain yields for a mean of 4.69 tha^{-1} in 1984 and 4.4 tha^{-1} in 1985.

The crop yield at the field site „Bei Lotte“ obtained in the year 2001 indicates a completely different pattern to the years 1999 and 2000. In that growing season, yield on convex (-0.2 to -0.1 1/100 m) and concave areas (0.1 to - 0.2 1/100 m) reaches up to 10.8 tha^{-1}. Lower yields (10.4 tha^{-1}) are found at planar areas (-0.1 to 0.1 1/100 m) as well at the two points representing strong concave sections (signed 5 in Figure 50). The question occurs, why the concave sections yield similar results to convex parts of the field. Profile curvature is expected to characterize the water redistribution due to topography in a landscape - diverging water away from convex positions and collecting of water at concave positions. However under dry conditions, the distribution of soil water depends on soil textural differences. The snapshot of soil water content after harvest (Figure 71) as well as during spring sampling (Figure 70) showed similar results for convex (-0.2 to -0.1 1/100 m) and concave profile curvatures (0.1 to 0.2 1/100 m) in question.

Again, profile curvature is assumed to characterize water distribution. Therefore, different grain yield development across the observed profile curvature would be expected. As TKM, kernel number and spike number developed during different sections of plant growth, different reactions would be expected at specific levels of development (see Figure 30).

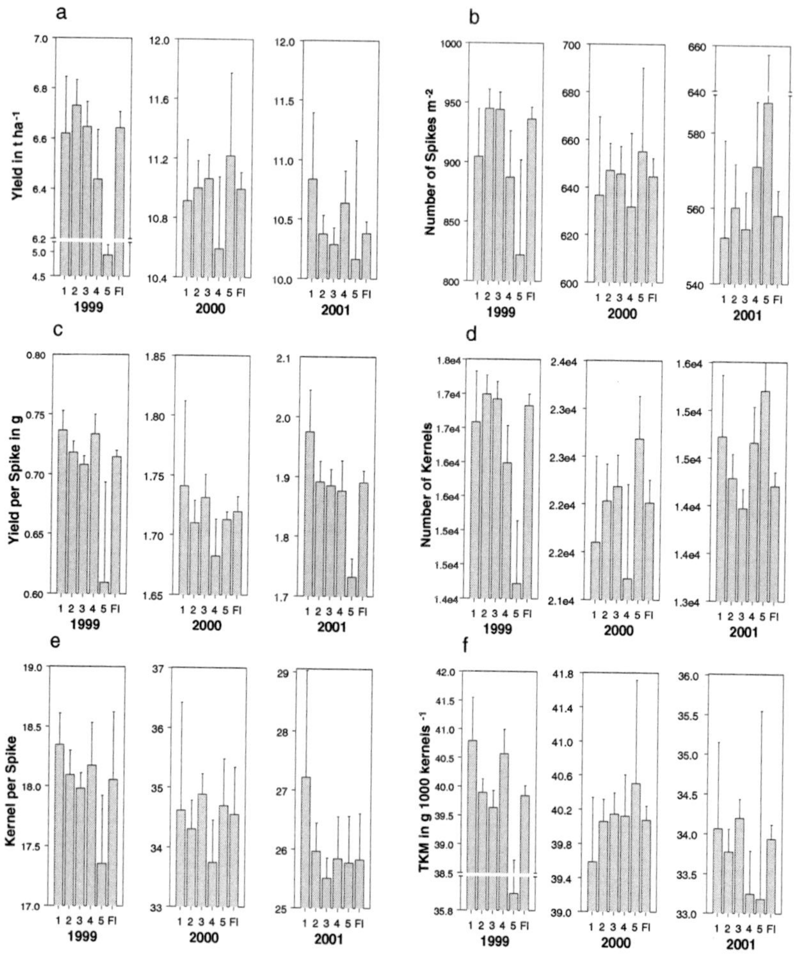

Figure 50. Grain yield in tha^{-1} (a) and the yield properties Number of Spikes m^{-2} (b), Yield per Spike in g (c), Number of Kernels per m^{2} (d), Number of Kernels per spike (e) and TKM in g(1000 kernel)$^{-1}$ (f) for the field site "Bei Lotte" for the profile curvature classes 1 (-0.2--0.1, n=12), 2 (-0.1-0.0, n=74), 3 (0-0.1, n=89), 4 (0.1-0.2, n=15), 5(0.3-0.4, n=2) and the field (Fl, n = 192). Bars show the standard error.

The number of spikes per m^2 shown in Figure 50 B shows that at planar landforms (-0.1-0.1) around 40 spikes per m^2 more in 1999 and 20 spikes per m^2 more in 2000 are observed compared to more pronounced profile curvatures (class 1, 4, 5 in Figure 50). A higher number of spikes per m^2 could not be observed for the year 2001. However, it has to be noted that the differences observed in 2001 were far less than in the year 1999 and 2000. The pattern observed in 2001 is similar to results by Fiez et al. (1994), who observed differences in spike

m^{-2} of around 100 for convex to mid to concave positions in 1990 and 1991. Still, the question remains open, why these differences between profile curvature classes occur.

The differentiation between planar (-0.1-0 and 0-0.1, see Figure 50 D) and convex and concave positions is also visible for the number of kernels per m^2. For the years 1999 and 2000 around 500 kernels more than at convex and concave positions are observed, whereas in 2001 the opposite is the case (Figure 50 B).

An interpretation of kernel per spike and yield per spike to the relief parameter profile curvature seems to be difficult, as no clear pattern can be seen (Figure 50 C+E).

One of the latest yield parameter in the growing season is the TKM (Figure 50 F). Spring barley in 1999 shows a similar distribution as observed for kernel and spike number. The year 2000 shows class 1 with a TKM 0.5 g lower than the rest of the field - TKM increased with decreasing profile curvature. This would indicate that during a dry year at concave positions the kernel mass increases compared to the rest of the field. Results are similar to Ciha (1982), who found an increase from convex to mid to concave positions for TKM (0.2 g difference). Even larger differences for TKM were found by Simmons et al. (1989) with up to 1.2 g difference in 1984 and 1.8 g in 1985. In contrast, the year 2001 shows a completely different pattern. At profile curvature class 4 (0.1-0.2) around 1.5 g less TKM was observed compared to the rest of the field for the above average precipitation year.

7.3.3. Analysis of length of internodes and length of ears for landforms

Landform units are known to exhibit different environmental characteristics (see also 8.3.1 on page 136). Based on that assumption, a different yield development is expected at a given field site. Three different growth components, the ear length, the internode length and the total plant length were analysed. The relationships between classified LF units versus these properties for the years 1999, 2000 and 2001 at the field site “Bei Lotte” are shown in Figure 51.

In 1999 the 1. Internode (Figure 51 E) shows the smallest growth at SH positions with 9.5 cm in length compared to a field mean of 10 cm. These differences are still visible at latter development stages, however with less distinction. In general, spring barley showed distinct increase of internode length from SH, BS; FS up to the largest length observed at Level positions, except for the differences observed at the 3. Internode (Figure 51 D, C, B). This would indicate the application of growth regulators at this development stage.

The least growth of the 1. Internode observed at Shoulder positions in 1999 is also visible in the 2000 and 2001. In the year 2000 the largest internode distances were measured at BS position. In contrast, the FS positions yielded the second lowest growth (after SH). As precipitation was below average that year, a larger internode growth would have been expected at footslope positions compared to other landform units.

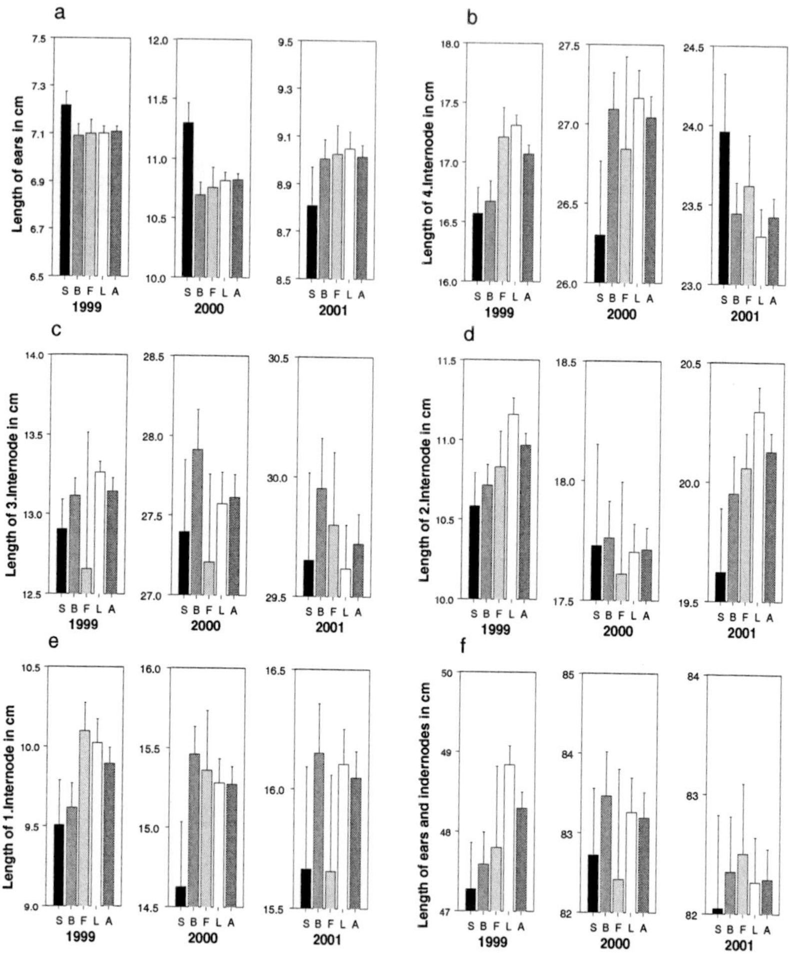

Figure 51. Ear Length (a), Length of the 4.Internode (b), Length of the 3.Internode (c), Length of the 2.Internode (d), Length of the 1.Internode (e) Sum of ear length and the length of the top three internodes (f) for the field site "Bei Lotte" for the landforms Shoulder (S, n = 17), Backslope (B, n = 49), Footslope (F, n = 16), Level (L, n = 110) and the field (A, n = 192). All values given in cm. Bars show the standard error.

The first internode in 2001 showed the smallest development (0.3 cm) for SH and FS positions below the field average. However, during the vegetation period plants at FS positions increased their growth and showed the largest plant length across all landforms (Figure 51 F).

Ear length was larger than the field average in the year 1999 and 2000 for shoulder positions. In contrast, in 2001 a significant decrease in ear length was visible at shoulder positions. The length of ears showed similar values at all other LF positions. Still, a slight trend with BS-FS-LE in all three years could be observed. The development of the ear length occurs mainly during the time of shooting (Figure 30). The question is now, why ear length at shoulder positions showed larger values than the rest of the field. Figure 51 B showed a lower number of spikes for SH, however, it can be postulated that due to less competition in later periods of plant development, all conceptual spikelets develop. At other landform units, competition in dense plant stands led to a reduction in number of spikelets and therefore to a reduction in spike length (Figure 30). Several authors reported that the tallest plants revealed the greatest grain yields (Ciha, 1982, Vachaud and Chen, 2002). Such relationship could not be observed for different landform elements for plant length and ear length as seen from Figure 48 A and Figure 51 A+F.

7.3.4. Analysis of length of internodes and length of ears for profile curvature

The relationship between the relief parameter profile curvature and the plant length components are shown in Figure 52. The largest development of plant growth was observed at planar positions with up to 1 cm more growth compared to convex and concave positions for fourth, second and first internode (Figure 52 B, D, E). An exception occurs at the 3. Internode (Figure 52 C). On the other hand, the growth at the extremely concave areas (0.3 - 0.4 1/100 m) is strongly reduced at the first internode, whereas later on (4. Internode) the highest internode lengths are found at that position.

In 2000 growth is reduced at all divergent positions (signed 1 in Figure 69). This seems to be reasonable due to the dry weather conditions. Concave positions (0.1 - 0.2 1/100 m, signed 4 in Figure 69) shown at the second internode (Figure 52 D) and the third internode (Figure 52 C) an increase in development (0.5 cm), compared to the other positions. However, as the fourth internode finished growth, planar positions showed up to 2 cm larger internode growth than convex and concave positions.

Spatial differences in internode growth were not so pronounced in 2001. At concave positions (0.1-0.2 1/100m- signed 4) the largest development can be observed with up to 1.5 cm larger plant growth than the field average, except at the second internode (Figure 52 D). The convex position showed the smallest extension with a growth of 14 cm at the first internode, whereas planar and concave sections reached up to 16 cm. The reason for the variation observed between positive planar (15 cm length) and negative planar sections (16.5 cm) remains unknown. A possible explanation might be shallow groundwater present during certain times and parts of the field and therefore might influence yield development.

As this field was managed homogeneously, the question of the differences of yield development and yield components inside a field can be assessed. These differences are attributed to different development stages and related to the influence of environmental conditions for the plants. As shown in Figure 52, differences in the observed growth may lead to the question, if the application time of fertilizer and growth retardants is critical. A possible, but at the moment impractical solution, would be to apply growth retardants only partly in time

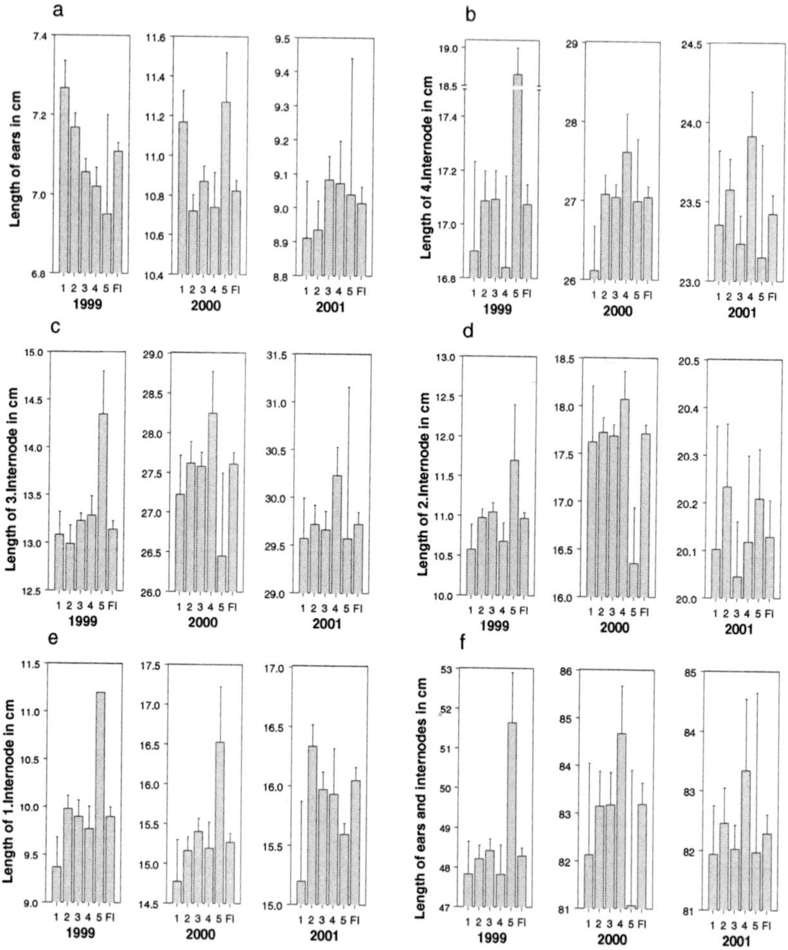

Figure 52. Ear length (a), Length of the 4.Internode (b), Length of the 3.Internode (c), Length of the 2.Internode (d), Length of the 1.Internode (e) Sum of ear length and the length of the top three internodes (f) for the field site "Bei Lotte" for the profile curvature classes 1 (-0.2-0.1 n=12), 2 (-0.1-0 n=74), 3 (0-0.1 n=89), 4 (0.1-0.2 n=15), 5 (0.3-0.4 n=2) and the field (Fl, n = 192). All values given in cm. Error bars show the standard error.

or with different amounts at specific landforms. However, both approaches would require a sensor to define application rates (Dammer et al., 2001). Again, it was tested if the tallest plants and additional the longest spikes revealed the greatest grain yields (Ciha, 1982, Vachaud and Chen, 2002). Such relationship could not be observed for profile curvature classes (Figure 50 A and Figure 52 A+F).

CHAPTER 8

Results and Discussion for soil properties

8.1. Statistical and geostatistical analysis of soil properties

Understanding the spatial, temporal and functional dynamics of soil is one of the most fascinating challenges, in soil science. The identification of soils and determination of their characteristics is only possible at a discrete number of points in space and time. Consequently only limited knowledge exists about soil properties across the entire domain of consideration. In this context, the statement by Beckett and Webster (1971) that up to half the variance within any field may already be present within any square metre of the field, gains special importance. The questions remain, (I) how to predict and describe the soil properties and (II) how to predict and describe soil variability both qualitatively and quantitatively. Heuvelink and Webster (2001) described different models that can be used to predict soil properties. Before utilizing these models, a preliminary investigation of soil properties needs to be performed to describe the spatial and temporal variabilities of the soil.

The descriptive statistics of soil properties at the field site "Bei Lotte" are shown in the tables in Appendix Statistics. Soil surface water content (SSM) in 1999 and 2000 varied between 11 - 26 g(100g soil)$^{-1}$, with a coefficient of variance (CV) between 3.8 and 11.9% (Table 41). The values of these CVs are larger than the values observed by Röver and Kaiser (1999), probably due to the larger sampling area considered in this study (20 ha instead of 0.68 ha). Soil water content (SM) obtained in the fall of 1998 - 2001 shows CVs between 4.7 and 31%. Generally, the CV of soil water content decreased with depth for all years. Differences in the CV for the same soil layer were smaller than 3% in the years 1998, 1999 and 2001. An extreme outlier was the CV in 2000 which had a value of 31%. Heavy rain occurred during the sampling campaign in the period Aug. 17 and Aug. 23, 2000 (see Figure 60 on page 123 middle row). Mean soil water contents and CVs for the first sampling period (starting Aug. 17) are given in parentheses in Table 41 and were similar to CVs determined during other years of the study.

Soil textural analysis of the field site „Bei Lotte" (see additionally Figure 53) showed high silt contents in the 0-30 cm, 30-60 cm and 60-90 cm depth (Table 42). The major textural component was silt (80%), followed by clay contents which reached a maximum value of 19.5%. Classification using the KA4 classification system (AG Boden, 1994) leads to the classification into the major textural categories UT3 and UT4 (see Figure 53). The average soil organic carbon (SOC) content of the 0-30 cm depth layer was 1.1%. CVs varied between sand,

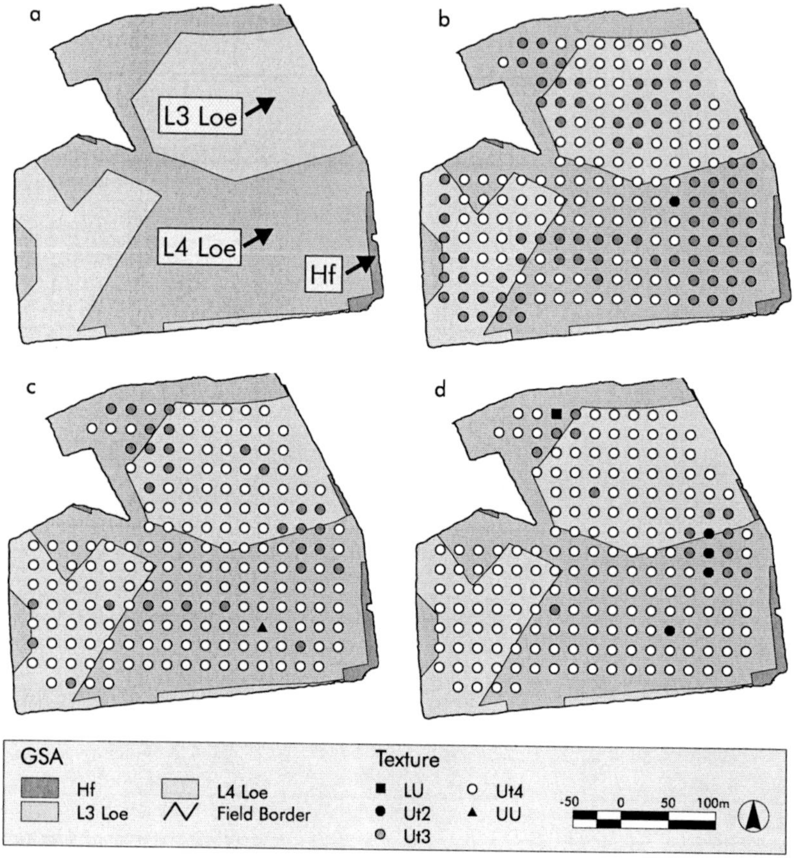

Figure 53. Soil information for the field site „Bei Lotte" according to the German Soil Appraisal (a), for the depth 0-30 cm according to KA4 (AG Boden, 1994)(b), for the depth 30-60 cm (c) and for the depth 60-90 cm (d). Black lines characterize the field borders.

silt and clay fractions. The major textural component silt had CV values between 3 and 6%. Larger values were computed for clay (15%) and sand (as high as 70%). Generally, the CV increased with depth for texture and SOC (Table 42). This is consistent with the observations of Bakhsh et al. (2000), who found that the CV decreased within three depth layers: sand varied with depth from 25% to 24.8% to 20.8% and clay from 12.4% to 15% to 18.1%.

The change in CV of SOC with depth is consistent with the results obtained by Pennock and Corre (2001), who found that the CV of SOC increased with depth from 20% (0-15 cm) to approximately 57% for deeper soil layers (30-60 cm) for a wheat field and from 30% to 74% for a canola field. Results for SOC and texture were also supported by the findings of Schmidt

(1984). In a north-east Germany glacial till landscape, consisting of three field sites, Schmidt (1984) observed that CVs for the fine particles (clay and fine silt) for the A_h horizon ranged between 24 and 27% (32 and 51%) and SOM from 16 to 21%. The data for the subsoil are given in parentheses.

Average soil P and K contents and variability decreased with depth (Table 43). The CV values for P and K exceeded the CV values observed by Tsegaye and Hill (1998) for P (14.7%) and K (16.6%), however they agree well with the values observed by Schmidt (1984) for P, who found a CV of 63-75% for surface soil conditions and 67-91% for subsoil conditions. The results for the soil nutrients P and K obtained by EUF-Analysis showed similar values for all years for CV (Table 43). Average P and K EUF-values for the first and second fraction illustrate the complex pattern of soil nutrient dynamics in time. No P and K fertilizer was applied during the time period of the investigation. The first fraction of EUF values represents the more available nutrient pool, whereas the second fraction represents the long term nutrient potential. An example are the EUF-K_1 fractions which started from a value of 5.15 mg (100g soil)$^{-1}$ in 1999, decreased to 4.5 mg(100g soil)$^{-1}$ in the dry year 2000 and then rose up to 5.0 g(100g soil)$^{-1}$ in 2001 again. The opposite pattern was visible for the EUF-K_2 fractions, which increased from 2.5 mg(100g soil)$^{-1}$ in 1999 to 2.57 mg(100g soil)$^{-1}$ in 2000, and decreased to a value of 2.44 mg(100g soil)$^{-1}$ in 2001.

Soil pH-value decreased from 1998 to 2000 by 0.25 units (Table 42). The CV of pH (2.8 - 4.8%) were below values observed by Schmidt (1984) in a glacial till landscape in north-eastern Germany (10 - 15 % for topsoil and 9 - 11 % for deeper soil layers).

The average residual (post - harvest) N_{min}-content showed higher values (70 kgha^{-1} for 0-90 cm) in 1998 and 1999 (74kgha^{-1}) compared to the years 2000 (38 kgha^{-1}) and 2001 (45 kgha^{-1}). Interestingly, the CV observed in the top 0 - 30 cm provides similar values for the years 1998-2000 at the field site „Bei Lotte“, but not for the year 2001 (Table 44). Deeper soil layers showed differences in the CV between years as high as 40 %. The different CVs are most likely an indication of (I) differences in nitrogen translocation due to precipitation, (II) differences in nitrogen uptake by plants due to environmental conditions, and (III) due to differences in mineralization rates.

8.1.1. Spatial analysis for N_{min} versus distance to tram lines

A homogeneous fertilizer distribution by the fertilizer spreader is assumed for all farm research. According to Weltzien (2002), the maximum allowed coefficient of variation for fertilizer spreading is set to 15% to ensure a homogeneous treatment. Residual N_{min} after harvest showed CVs between 38 % and 78 % (Table 44) and spring N_{min} for the 0 - 30 cm depth ranged between 12 % and 38 % (30 - 60 cm 11 % - 75 % and 60 - 90 cm 12 % - 90 %).

To test for the existence of a systematic error, introduced by the fertilizer application process, tramlines were recorded during fertilizer spreading in 2001 using the ACT-board computer. The distance between the fertilizer spreading plate (i.e. the tractor position or the tramlines) and the sampling points (n=192 for residual N_{min}, n=12 for spring N_{min}, n=120 for residual N_{min} obtained in five nests) were determined and aggregated to six classes (2-2.5m each) and average N_{min} contents computed (Figure 54). Due to the lower number of sampling points in the spring sampling period no standard errors were given.

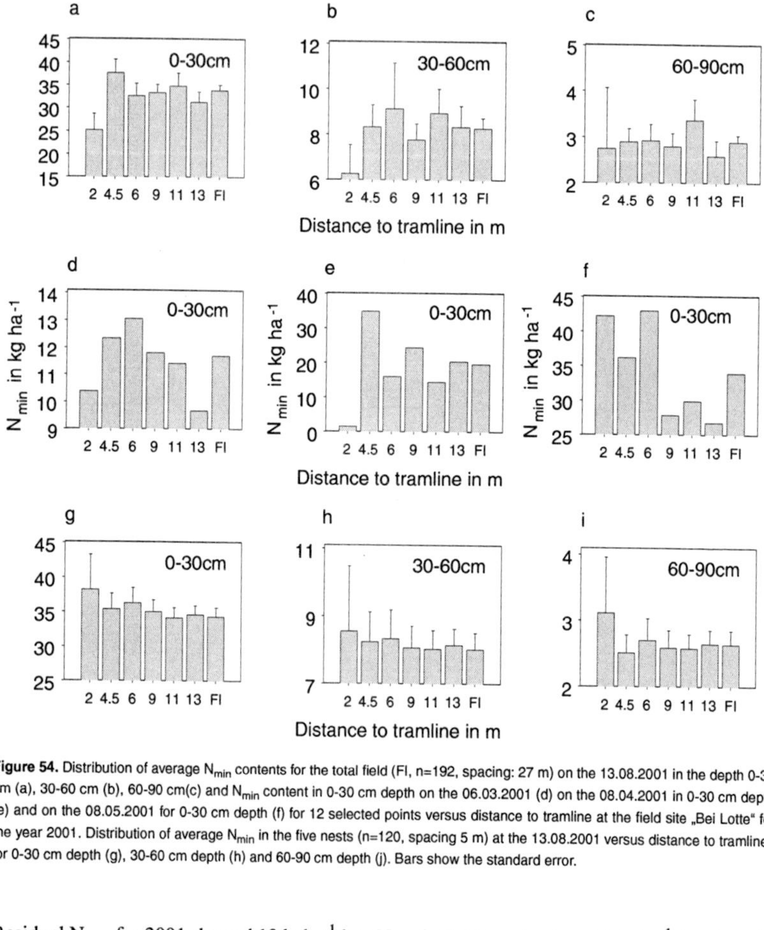

Figure 54. Distribution of average N_{min} contents for the total field (Fl, n=192, spacing: 27 m) on the 13.08.2001 in the depth 0-30 cm (a), 30-60 cm (b), 60-90 cm(c) and N_{min} content in 0-30 cm depth on the 06.03.2001 (d) on the 08.04.2001 in 0-30 cm depth (e) and on the 08.05.2001 for 0-30 cm depth (f) for 12 selected points versus distance to tramline at the field site „Bei Lotte" for the year 2001. Distribution of average N_{min} in the five nests (n=120, spacing 5 m) at the 13.08.2001 versus distance to tramlines for 0-30 cm depth (g), 30-60 cm depth (h) and 60-90 cm depth (j). Bars show the standard error.

Residual N_{min} for 2001 showed 10 kgha^{-1} less N_{min} in 0-30 cm depth and 2 kgha^{-1} less N_{min} in 30-60 cm depth for all sampling points up to 2 m apart from the tramline compared to the field average. The rest of the distance classes showed fairly similar N_{min} distribution (Figure 54 A, B, C). Sampling before the first nitrogen fertilizer application (Figure 54 D) revealed again the lower N_{min} at 0-2 m as well as from 11 - 13 m from the tramline with low differences of up to 2 kg N_{min}ha^{-1}. Sampling of N_{min} before the second nitrogen fertilizer application revealed 18 kgha^{-1} less N_{min} at 0-2 m and 15 kgha^{-1} N_{min} more at 2-4.5 m than the field average, whereas, the rest of the field ranges around +/-5 $kgha^{-1}$ of the field average. The decrease in N_{min} for distances between 11-13 m diminished (Figure 54 E). The last sampling before the third application revealed that the distance classes 0-2 m and 4.5-6 m showed a 10 kg N_{min} increase in N_{min}-content, whereas all distances less than 6 m show approximately 6 kg N_{min} less than the field average (Figure 54 F). In contrast to the N_{min} distribution of the entire field, the amount of residual N_{min} in the five nests (see Figure 6 for the locations of the nests) showed the 0-2 m and the 4.5-6 m class with larger N_{min}-values (2 kg) at distances further away from the tramlines (Figure 54 G, H, I).

Results in Figure 54 showed differences with respect to tramline distance of up to 18 $kgha^{-1}$ N_{min}. Results shown here provide some periodic fluctuations, similar to observations by Böttcher et al. (1988). However distance classes are four times larger, therefore possibly masking more small scale occurrences.

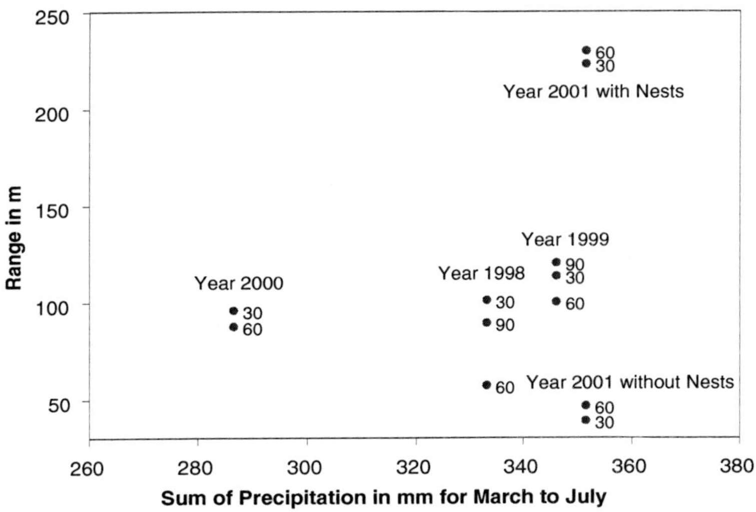

Figure 55. Relationship between the total precipitation in the period from March to July and the semivariogram model range of N_{min} for three depth layers (0-30 cm (30), 30-60 cm (60) and 60-90 cm depth (90)) for the field „Bei Lotte". For data point 1998 in the depth 0-30 cm outliers were removed for semivariogram processing using cross validation (99% of normal distribution). For data points in 2001 the two largest N_{min} samples (175 kg and 78 kg Nha^{-1}) were removed for semivariogram processing. Additional, the range of N_{min} in 2001 is shown for a dataset combining the standard raster (n=192) and five small scale nests (n=120).

8.1.2. Semivariogram analysis for soil mineralized nitrogen

The variogram model parameters manifesting the spatial structure of mineralized nitrogen (N_{min}) obtained during the post-harvest sampling campaigns of 1998-2001 at the "Bei Lotte" field site is presented in Table 26.

Between 1998 and 2001, temporally varying spatial ranges were observed for single layers. Ranges for N_{min} for a given year differed by up to 50 m for the three depth layers. In general, N_{min} semivariogram ranges decreased with depth from 0 - 30 cm to 30 - 60 cm in three of the four years. This fits within the results obtained by Cambardella and Karlen (1999). Fitted semivariogram ranges are larger than those by Tsegaye and Hill (1998), who calculated semivariograms for NO_3 with a range of 9.9 m, however, values computed here agree with the work conducted by Wendroth et al. (2001).

Table 26. Variogram model parameters for N_{min} in the time period 1998-2001 at the field site "Bei Lotte". Number in parentheses in 1998 show range and sill for an outlier removed dataset by cross validation.

Year	Depth in cm	Nugget in $(kgha^{-1})^2$	Range in m	Sill in $(kgha^{-1})^2$	MODEL	NS-Ratio
1998	0-30	188.94 (124.8)	223.89(101.12)	121.91 (49.59)	SPHERICAL	0.72 (0.71)
1998	30-60	104.42	57.40	119.36	SPHERICAL	0.47
1998	60-90	36.37	89.36	15.16	SPHERICAL	0.71
1999	0-30	255.53	113.74	189.53	SPHERICAL	0.57
1999	30-60	32.68	100.29	15.66	SPHERICAL	0.68
1999	60-90	7.38	120.47	3.62	SPHERICAL	0.67
2000	0-30	96.54	96.08	33.47	SPHERICAL	0.74
2000	30-60	9.06	87.67	1.17	SPHERICAL	0.89
2000	60-90	8.13	NA	NA	NUGGET	
2001	0-30	131.19 (104.20)	218.85(39.81)	56.1(48.31)	LINEAR (SPHERICAL)	
2001	30-60	30.64(16.12)	229.78(46.65)	13.6933(5.15)	LINEAR (SPHERICAL)	
2001	60-90	4.74(4.00)	NA	NA	NUGGET	

Ranges in grain yield showed a dependency on the total precipitation between March and July. As plant growth requires nitrogen, years with a larger variation in crop yields may lead to shorter semivariogram ranges due to the removal of different amounts of nitrogen from the soil. To test this hypothesis, the range of N_{min} was plotted against the total precipitation received from March to July in Figure 55. Points for the 60-90 cm depth in 2000 and 2001 are missing due to pure nugget effects found in the semivariogram analysis. Similar semivariogram ranges of N_{min} for different amounts of precipitation in the years 1998, 1999 and 2000 can be seen. The range values observed for the standard raster in 2001 are based on a linear semivariogram function (Table 26), which is unbound in the field borders of the field „Bei Lotte". By including 120 additional points obtained in five small scale nests, a spherical semivariogram with a small range could be fitted. The second structure, as observed by the standard raster, was still present.

No relationship between fitted semivariogram ranges and the total precipitation received between March and July was observed, as shown, for grain yield (Figure 25). One reason may be the amounts of nitrogen applied as fertilizer (1999 - 111 kg Nha^{-1}, 2000 - 140 kg Nha^{-1}, 2001 - 173 kg Nha^{-1}) at the field site „Bei Lotte“, which may mask the spatial dependency of post-harvest nitrogen.

8.1.3. Spatial analysis for texture

Spatial variability analysis was performed for soil texture for three different depths at the field site "Bei Lotte". Before interpreting the results of Table 27, the impact of data density on the resulting semivariogram should be examined.

Semivariance was calculated for samples taken 54 m apart from each other (Figure 56a). A pure nugget effect with a semivariance of approximately 8.3 $(g(100g\ soil)^{-1})^2$ can be observed in Figure 56c. Adding 15 samples in between the 54 m sampling grid (Figure 56b), allowed the determination of a semivariance value for a lag distance of 27 m for some samples (Figure 56d). After the recalculation of the semivariogram, a nugget of 1.65 $(g(100g\ soil)^{-1})^2$ (Figure 56 top left) and a semivariogram with a range of 67 m (Table 27) was observed.

A nested sampling design can be used to identify different spatial structures and can allow to scale measured or simulated properties in space and time, along with several other applications. An advantage of multi-scale sampling is the identification of short and long range variations (Saldana et al., 1998). The transect approach shown in Figure 56 is a special type of a nested sampling design, defined by a subdivision of two different classes with distinct stages occurred (Webster et al., 1990). However, it is not a regular design as seen in Saldana (1998), but allows the identification of spatial variability across short lag distances (half of the regular sampling distance).

Table 27. Semivariogram model parameters for soil organic carbon (SOC) and texture (Sand, Silt, Clay) in three different depths at the field site "Bei Lotte".

Pro-perty	Depth in cm	Nugget in $(g(100g)^{-1})^2$	1.Range in m	1.Sill in $(g(100g)^{-1})^2$	1.Model	2.Range in m	2.Sill in $(g(100g)^{-1})^2$	2.Model	NS-ratio
SOC	0-30	0.0036	80.03	0.0115	Spherical				0.24
SOC	30-60	0.0197	76.68	0.0056	Spherical				0.78
SOC	60-90	0.0007	69.97	0.0109	Spherical				0.06
Sand	0-30		65.78	2.2977	Spherical	165.52	0.6738	Order-1	
Sand	30-60		60.76	3.9049	Spherical	211.61	0.9632	Order-1	
Sand	60-90	0.8728	50.70	7.3424	Spherical	209.10	2.9909	Order-1	
Silt	0-30	1.6574	67.46	7.2412	Spherical				0.19
Silt	30-60	8.4766	57.40	3.1559	Spherical				0.31
Silt	60-90	9.802	NA	NA	Nugget				
Clay	0-30	3.2971	57.40	3.7285	Spherical				0.47
Clay	30-60	9.9614	NA	NA	Nugget				
Clay	60-90	4.1628	57.40	5.4163	Spherical				0.43

In general a decreasing range can be observed with increasing depth for SOC and sand content (Table 27). The silt content showed strong spatial dependencies for the two top soil layers with a range of 60 m. Clay content showed ranges of 57 m for 0 - 30 cm and 60 - 90 cm depth, however no spatial dependency could be observed for the 30 - 60 cm depth. Semivariogram model ranges for flow direction and classified landforms reached similar semivariogram range values (Table 14) compared to texture and SOC.

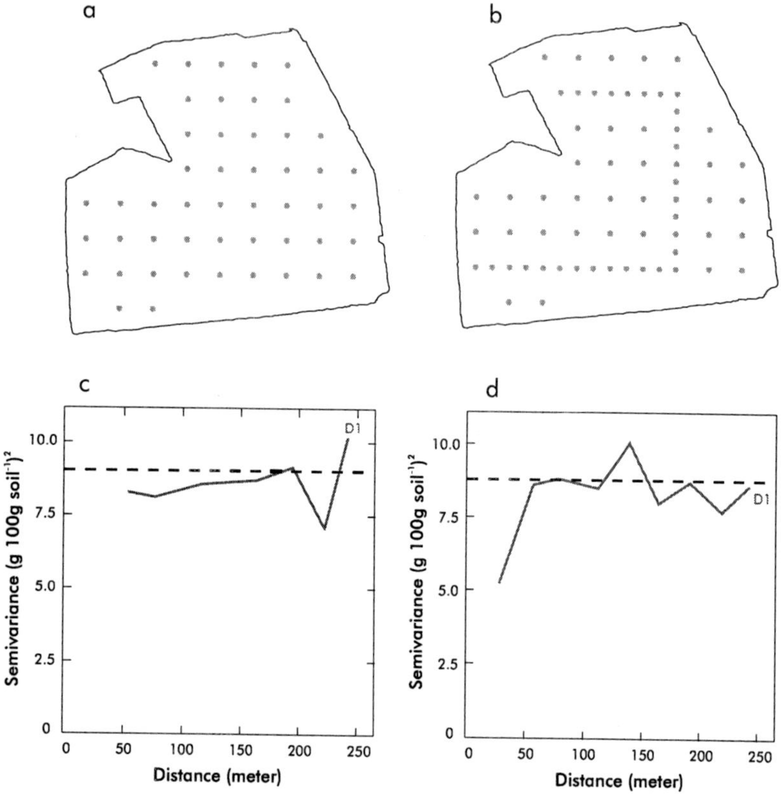

Figure 56. Sampling points for silt content at 0-30 cm depth sampled in a 54 m raster (a). The same raster with additional samples spaced at 27 m is shown in (b). The semivariogram for the dataset (a) is shown in (c), for the extended dataset (b) in (d).

8.1.4. Semivariogram analysis for soil Phosphorus and Potassium

Results for the spatial variability of soil phosphorus and potassium content are shown in Table 28 and Table 29. In general a decrease in the total variance of phosphorus in 1999 was observed with depth for $P_{(DL)}$ in 0 - 30 and 30 - 60 cm depths. The range of 96 m for $P_{(DL)}$ in 0 - 30 cm doubled with depth for the field site "Bei Lotte".

The ranges reached for the readily available EUF-P_1 fraction was approximately 185 m; in the more strongly bound EUF-P_2 fraction (the "potential") a range of up to 228 m was observed in 1999. For the soil layer 0 - 30 cm, P-EUF preserves twice the range compared to $P_{(DL)}$ in 0-30 cm. The EUF-P_2 fraction (0-30 cm) and $P_{(DL)}$ taken in the 30 - 60 cm depth showed similar spatial ranges. However, it has to be noted that these semivariograms are unbound.

The ranges of the EUF-P_1 fraction over three years were stable with 185 m for the years of 1999 and 2000. An increase in the semivariogram range of 20 m was observed in the year 2001. Generally, a decrease in total variance was observed for the EUF-P_1 fraction. In contrast, the EUF-P_2 fraction showed no decrease in total variance, however the fitted semivariogram showed different ranges. Samples taken in 1999 led to ranges of up to 230 meters, followed by the smallest ranges in 2000 (190 m) and ranges increased to 210 m in 2001 (Table 28). The question arises, why the ranges differ during three subsequent years for the EUF-P_1 and the EUF-P_2 fractions. Generally, phosphorus is needed for photosynthesis (Marschner, 1995) and has not been limiting according to classification of the VDLUFA (Kerschberger et al., 1997). Years with a higher spatially different biomass production would show pronounced differences in the amount of P - therefore probably leading to a decrease in semivariogram range. The year 2000 for example, showed a high coefficient of variation as well as a large grain yield (see Table 17) and showed the smallest semivariogram range.

Table 28. Semivariogram model parameters for $P_{(DL)}$ and EUF-P_1 and EUF-P_2 fractions in the time period 1999-2001 at the field site "Bei Lotte".

Type	Year	Nugget Variance in $(mg(100g)^{-1})^2$	Range in m	Sill in $(mg(100g)^{-1})^2$	Model	NS-ratio
$P_{(DL)}$ in 0-30 cm	1999	0.8805	96.1	19.4942	SPHERICAL	0.04
$P_{(DL)}$ in 30-60 cm	1999	2.3474	213.79	6.6959	LINEAR	
EUF-P_1	1999	0.1532	185.21	0.1575	SPHERICAL	0.49
EUF-P_1	2000	0.1375	186	0.0703	LINEAR	
EUF-P_1	2001	0.1124	204.55	0.0786	LINEAR	
EUF-P_2	1999	0.0285	228.1	0.0387	SPHERICAL	0.42
EUF-P_2	2000	0.0544	189.42	0.0375	LINEAR	
EUF-P_2	2001	0.0410	208.76	0.0245	LINEAR	

Scheffer and Schachtschabel (1992) reported that phosphorus availability is increased under anaerobic conditions, e.g., after flooding. Therefore, the question presents itself , if areas with higher levels of phosphorus correspond to elevated levels of soil water content. Such a question is especially important due to sampling considerations. Figure 57 provides such a relationship for EUF-P_1-content in 1999, a 3D View of the distribution is shown in Figure 58. Similar results can be drawn for $P_{(DL)}$ and EUF-P_1 and EUF-P_2 in different years at the field "Bei Lotte". Different reasons lead to elevated P-contents: (I) locations related to the field border (A, D in Figure 58), (II) locations with increased colluvium at FS positions (C in Figure 58) or (III) at positions where drainage problems occur (B in Figure 58).

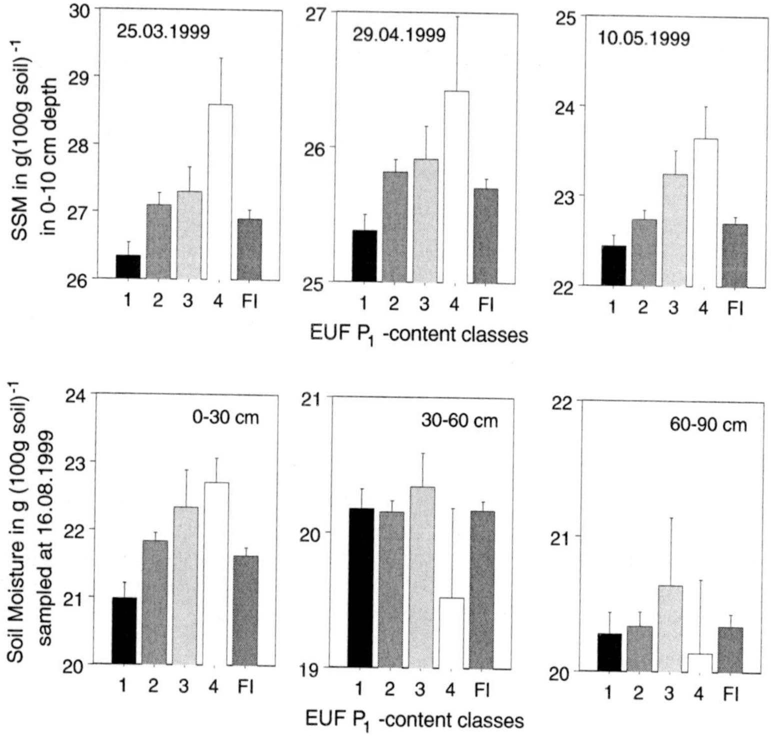

Figure 57. EUF-P_1-content classes (1 = 0-1 mg(100g soil)$^{-1}$, 2 = 1-2, 3 = 2-3, 4 = 3-4, FI = field average) versus spring soil surface moisture sampling (top row, three different dates) and soil water content from fall sampling in 1999 (bottom row, three different depths). Error bars show the standard error.

Potassium contents and variogram model parameters are shown in Table 29. Potassium ($K_{(DL)}$) showed a decrease in total variance with depth in 1999. Simultaneously, a slight reduction in the semivariogram range of 20 m was observed. Since clay particles are able to fix potassium (Scheffer and Schachtschabel, 1992), it mayght be expected that both clay content and potassium contents would show similar semivariogram ranges. However, ranges observed for clay are approximately one fourth of the observed ranges for potassium (see Table 29).

Table 29. Semivariogram model parameters for $K_{(DL)}$ and EUF-K_1 and EUF-K_2 fractions for 1999-2001 at the field site "Bei Lotte".

Type	Year	Nugget in (mg(100g soil)$^{-1}$)^2	Range	Sill in (mg(100g soil)$^{-1}$)^2	MODEL
$K_{(DL)}$ in 0-30 cm	1999	1.4898	178.02m	7.7242	LINEAR
$K_{(DL)}$ in 60-90 cm	1999	2.3328	158.06m	3.2668	LINEAR
EUF-K_1	1999	2.7767	207.08m	4.4395	LINEAR
EUF-K_1	2000	2.2506	176.81m	1.5804	LINEAR
EUF-K_1	2001	2.3350	199.51m	2.0026	LINEAR
EUF-K_2	1999	0.5716	191.10m	0.4616	LINEAR
EUF-K_2	2000	0.5079	134.76m	0.3276	LINEAR
EUF-K_2	2001	0.4015	201.19m	0.2029	LINEAR

EUF-K_1 and EUF-K_2 fractions in the year 1999 provide ranges 20 m larger compared to the semivariogram ranges from $K_{(DL)}$. The ranges of the EUF-K fractions were similar to those for the years 1999 and 2001, with a value of approximately 200 m. In the year 2000, observed ranges decreased around 20 m for the EUF-K_1 fraction and by approximately 60 m for the EUF-K_2 fraction. This decrease in range may be related to increased plant uptake of potassium during a dry vegetation period (see Marschner, 1995). Interestingly, the semivariogram ranges in the year 2001 again reached the values found in 1999, when no potassium fertilizer was added. It could be postulated that a potassium - equilibrium between soil available and soil potential material was reached again during that period. This is supported by the change in average potassium content (Table 43). This would indicate that sampling for potassium should not be performed in years with below average precipitation, e.g., „dry" years. Two limitations

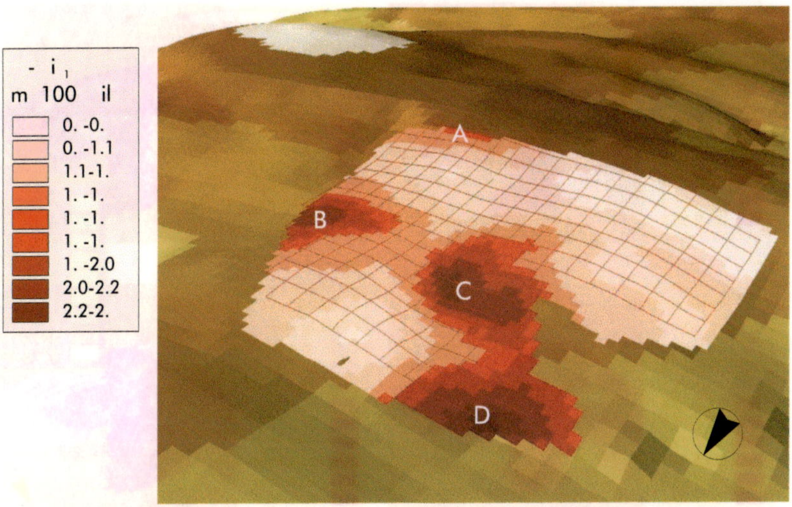

Figure 58. Distribution of the EUF-P_1 fraction in 1999 for the field site "Bei Lotte". A) closely related to a small farm yard, B) field entrance C) footslope positions with increased colluvium, D) represents both field entrance as well as increased colluvium.

have to be mentioned: first, this is only valid for EUF-K fractions, as no time series was obtained for $K_{(DL)}$ sampling and secondly, differences in semivariogram ranges were observed for linear semivariogram models, which where unbound for the field site "Bei Lotte". Further investigations at other field sites need to be performed to verify these findings.

8.1.5. Semivariogram analysis for soil water content

The spatial distributions of surface soil water content (0 - 10 cm) in the field "Bei Lotte" on four different dates are shown in Figure 59, viewed from northeast to southwest.

Areas with higher soil water content are visible at different zones throughout the field site. Soil surface water content (Figure 59) and soil moisture at sampling dates occurring after harvest (SM)(Figure 60) were relatively high in the low elevation area (represented by 1 in Figure 59 C). A zone with similar soil water content is located on the left side (west) of the field at a field entrance (2 in Figure 59c), where drainage problems exist. For the later sampling dates in 1999, the area with elevated soil water content extends further up the hillslope (to the south), probably indicating some subsurface flow (Figure 59b+c). This assumption is further supported by the sampling in the dry year of 2000 (Figure 59d), which showed elevated soil water contents at this location and on post harvest sampling dates in all three years (Figure 60).

Another pattern can be seen at the eastern part (3 in Figure 59c) of the field. As shown in Figure 16 this is classified as a BS position. For the soil water content measurements taken on

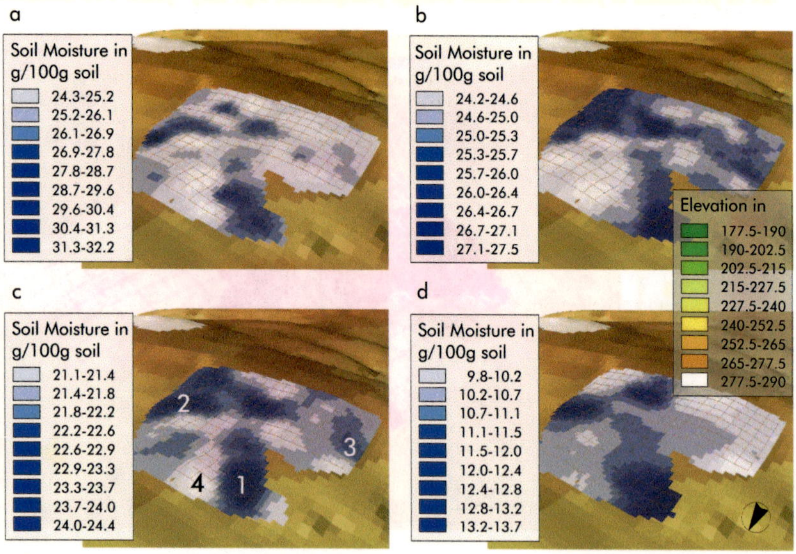

Figure 59. Spatial distribution of soil water content in 0-10 cm depth for the dates 25.03.1999 (a), 29.04.1999 (b), 10.05.1999 (c) and 12.05.2000 (d) for the field site "Bei Lotte".

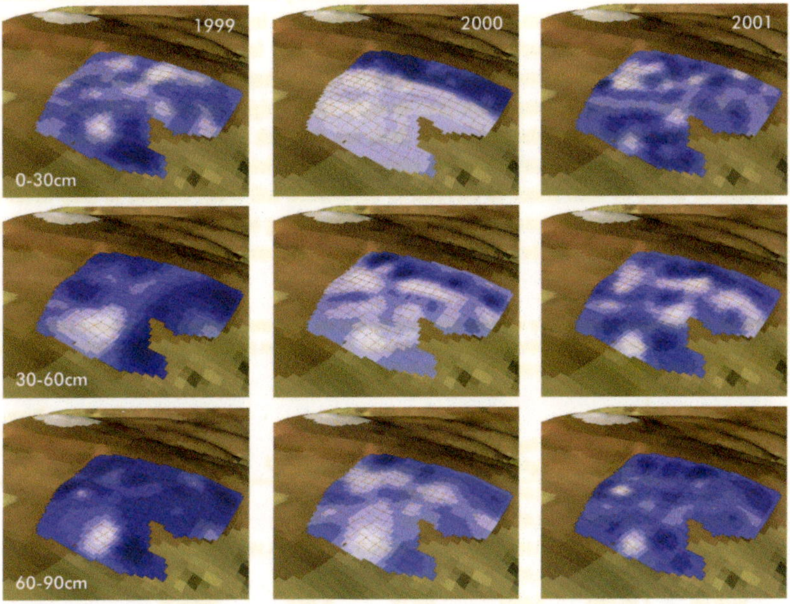

Figure 60. Soil water content distributions for post-harvest sampling for 0-30 cm, 30-60 cm and 60-90 cm depth for the years 1999, 2000 and 2001 at the field site "Bei Lotte". Range of values can be seen from Table 41.

25.03.1999 only slightly elevated soil water contents were observed at this landform. However for the dates of 29.04.1999 and 10.05.1999 soil water contents at the upper level of the range of the observed values were measured. A further support for subsurface flow is the slightly higher soil water contents measured during the fall sampling campaign (Figure 60), which is located south from the occurrence in Figure 59.

Additional proof of the special conditions at the BS positions described here are the differences in grain yield development for the three subsequent years (see Figure 48 on page 103). The same soil texture, meteorological and morphological conditions were available, and similar management practices were applied. However these areas showed different yields in 1999 and 2000 compared to other landforms.

Additionally, the area west of the depression (represented by 4 in Figure 58c) showed the lowest soil water content in 1999 for spring soil water content measurements, whereas in 2000, the areas at the top of the field (southern border) showed the lowest water contents. Considering that the fall sampling results for all three years investigated in this study resulted in low soil water contents, the area west of the depression (4 in Figure 59c) may be prone to having the lowest soil water content during harvest time.

A time series of soil water measurements can help identify areas of elevated soil water content and drought areas in field sites, which may cause differences in yield development. A relatively low cost method of soil surface water content sampling in the 0 - 10 cm depth can be performed. Results may be used from two aspects. On one hand, grain yield development can be influenced by crop management (i.e. drought areas might receive a different nitrogen fertilizer rate compared to the rest of the field, see Pennock et al., 2001). Second aspect is related directly to the practical testing for grain ripeness to determine time of harvest. Farmers should be aware of such zones of elevated or diminished soil water content in their fields to avoid wrong decisions, for example, in respect to the timing of the harvest. These differences in grain ripeness could also be identified by hyperspectral remote sensing as shown by Goel et al. (2003).

Table 30. Semivariogram parameter for soil water samples at the field site "Bei Lotte". Additional semivariogram parameters for four soil surface sampling dates at "Sportkomplex" are presented. Semivariogram parameter for the year 2000 are computed without data from the second sampling period and shown in parentheses.

Depth	Date	Nugget in $(g100g^{-1})^2$	Range in m	Sill in $(g100g^{-1})^2$	Model	Precp. Sum in mm for 3 days	Precp. Sum in mm 14 days
0-30 cm	08.1998	2.6418	57.40	1.5121	Linear		
0-30 cm	08.1999	1.0517	82.63	1.2775	Spherical		
0-30 cm	08.2000	3.6933(1.7)	186.06(84.3)	16.4373(1.19)	Spherical		
0-30 cm	08.2001	1.0716	66.65	1.5642	Spherical		
30-60 cm	08.1998	3.4243	122.15	0.7883	Spherical		
30-60 cm	08.1999	0.5887	147.38	0.3301	Spherical		
30-60 cm	08.2000	3.6636(1.62)	77.58(74.2)	2.4873(1.79)	Spherical		
30-60 cm	08.2001	1.7896	85.15	1.1688	Spherical		
60-90 cm	08.1998	2.2183	64.97	1.3888	Spherical		
60-90 cm	08.1999	0.7411	78.42	0.599	Spherical		
60-90 cm	08.2000	2.2553(1.95)	119.63(78.4)	1.5036(1.39)	Spherical		
60-90 cm	2001	1.2574	69.17	1.1386	Spherical		
0-10 cm (BL)	25.03.1999	0.8953	73.38	2.4016	Spherical	4	6.9
0-10 cm (BL)	29.04.1999	0.3204	63.68	0.298	Spherical	1.9	26.5
0-10 cm (BL)	10.05.1999	0.407	90.20	0.6534	Spherical	15.5	18.7
0-10 cm (BL	12.05.2000	1.1205	91.88	0.4679	Spherical	0	0.2
0-10 cm (SK)	02.04.1998	2.5558	132.05	2.4407	Spherical	6	6.7
0-10 cm (SK)	04.05.1998	1.9709	87.91	1.5918	Spherical	5.5	9.3
0-10 cm (SK)	25.03.1999	2.0005	71.14	1.8106	Spherical	4	6.9
0-10 cm (SK)	27.04.1999	2.027	79.7	0.3838	Spherical	8	27.5

Semivariogram analysis was performed for the years 1999-2001 to obtain the soil water content distribution maps. The semivariogram ranges varied between 57 - 85 m for the 0 - 30 cm layer, 74 - 147 m for the 30 - 60 cm and 64 - 78 m for the 60 - 90 cm depth layers. Differences between ranges for soil water content obtained for the same year decreased with depth (Table 30).

Soil textural classes as seen in Figure 53 were shown to be homogeneous. Therefore, the topsoil water content is influenced by meteorological conditions, plant water uptake and lateral subsurface flow. The last two properties are usually difficult to measure, whereas meteorological conditions can be observed more readily.

Therefore the existence of a relationship between semivariogram ranges and total precipitation for 3 days and 14 days periods prior to sampling were tested. No relationship was observed (Table 30). For the month of May, differences of 15 mm for the sum of precipitation in the previous 14 days are shown with similar ranges (see Table 30). Still, a relationship between the average soil water content (Table 41) and fitted semivariogram ranges (Table 30) was observed.

Range and range differences between fields and years were minimal in March with up to 10 m. Larger differences were observed in April, at which time differences up to 25 m were observed. Sampling performed on 2.4.1998 was excluded as 6.7 mm of rainfall fell during sampling.

For sampling purposes the following conclusions were drawn. No rainfall should occur during the sampling period. Secondly, according to Herbst et al. (2001), the preferable sampling distance is defined to be that which captures 67 % of the structured semivariance. If this rule of thumb (Lamp et al., 2001) is used, preferred sampling distances would vary between 20 m in March and 30 m in May, similar to the sampling distances used in this study.

Western et al. (1999) investigated the spatial variability of soil water content for one year (13 dates) in a 10.5 ha catchment. CV ranged from 8% up to 13%. For one of the dates a semivariogram is given which indicates a range of 120 m. Values in the Lüttewitz dataset do not exceed this value, except in one case. Western et al. (1999) considered four parameters to be important for soil water content: (I) at a scale <10m micro topographic variations due to disturbance of cattle and small scale soil heterogeneity, as well as vegetation patches (1-10m), (II) variation due to topographically routed lateral flow, (III) hillslope (e.g. different aspects) and (IV) the land use. The first two factors can explain up to 50% of the variance.

8.2. Crosscorrelation analysis between soil properties, different relief parameters and sensor values

8.2.1. Significant crosscorrelations of soil water content at different times

Crosscorrelation coefficients (CC) were computed for the field site „Bei Lotte“ for a variety of values as shown in Table 45 in Appendix Geostatistics. Only CC with separation distances larger than one class above the 95% confidence interval were selected and will be discussed in the following analyses. All crosscorrelations discussed are computed for the field site “Bei Lotte”, if not stated otherwise.

Soil surface water content measurements obtained in 1999 (Figure 61) showed a positive crosscorrelation coefficient (CC), with a low significant crosscorrelation distance (SCD) (1 lag distance). Soil surface measurements showed SCDs for the samples obtained between the 25.03.1999 and the 24.09.1999; between the 24.09.1999 and the 10.05.1999, and between the 25.03.1999 and the 10.05.1999. Additional samples obtained on 12.05.2000, however, showed a significant CC only with respect to the samples obtained on 25.03.1999 and not to the other two dates. These findings are supported by those of Tomer and Anderson (1995), who showed that the shape of the autocorrelation function between different dates did not change and the differences in soil water storage between several sampling dates showed no spatial structure. A significant crosscorrelation coefficient (CC) for post harvest soil water content between soil layers could only be identified between the 30-60cm depth and the 60-90cm in 1999.

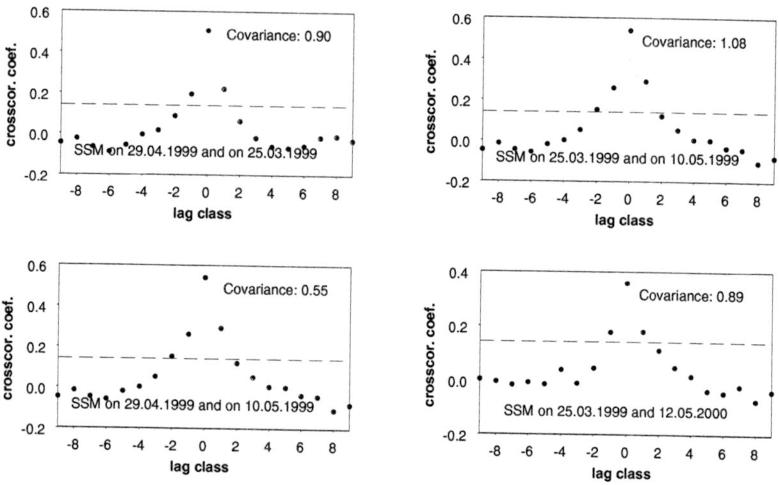

Figure 61. Significant crosscorrelations between soil surface water content measurements in 1999 and 2000 at the field site "Bei Lotte". The dashed line represents the 95% confidence interval.

8.2.2. Significant crosscorrelations of soil surface water content and different parameters

Soil surface water content (SSM) in the 0 - 10 cm depth was crosscorrelated to several relief parameters based on elevation (i.e. range, deviation, pctg, mean - see Table 4 for abbreviations). However, the topographic wetness index (TWI) was the only relief parameter which showed a significant CC for three of the four sampling dates (Figure 62). Such a pattern was expected, as it proves the validity of using the Topographic Wetness Index to characterize the soil surface water content distribution. SSM did not show significant CC to planform curvature, profile curvature or curvature as observed by Boyer et al. (1990).

The HydroN-Sensor is utilized to obtain the recommended amount of nitrogen fertilizer, based on the N-status of the vegetation. Aggregated HydroN-recommendations in the years 1998-

2000 were not found to be crosscorrelated except for one case (between 04.May 1999 and 11.May 2000). Another parameter closely related to crop productivity is the governing process of plant soil water uptake. Therefore, the values obtained by the HydroN-Sensor (a sensor) and the soil water content (a soil property) would be expected to be crosscorrelated. Generally, SSM showed significant CC versus HydroN-Sensor values for three cases: SSM on 10.05.1999 and HydroN on 04.05.2000, SSM on 29.04.1999 and HydroN on 04.05.2000 and SSM on 29.04.1999 and HydroN on 11.05.1999 (Figure 63). However, given the five different dates of HydroN-Sensor collections, this result seems to be disappointing as only two dates of HydroN - measurements showed a significant CC, and more significant CC were not observed. SSM and HydroN measurements were not obtained at the same time, which somehow limit the results drawn from that analysis. The conclusion would be that sensor observed nitrogen plant nutrition demands at the field site "Bei Lotte" appears to be independent of soil surface water content as determined by crosscorrelation analysis. Only when differences were more pronounced (e.g., drought), SCD can be observed.

Color InfraRed (CIR) images are taken by aerial photography. NDVI- values are calculated from these images and provides a additional variable. The analysis showed significant CC between NDVI calculated in April and May of 1999 versus SSM. The NDVI from different years (1997-2001) did not show significant CC to SSM in 1999 and neither did results from the analysis between grain yield and NDVI. Furthermore, SSM in 2000 showed no significant CC to NDVI (1997-2001).

Different amounts of biomass are produced depending on the limiting environmental conditions for plant development. One of the limiting factors for plant growth is water. Therefore, crosscorrelations for SSM might serve as an indicator for the use of different amounts of nutrients or different soil chemical properties. A significant CC was observed between EUF-P_1 in 2000 and SSM in March of 1999. No significant CC was observed for EUF-P_1 in 1999 and 2001, as well as for phosphorus content determined using the DL-method. The "potential" EUF-P_2 fraction shows in the years 1999, 2000 and 2001 a significant CC to soil surface water content in 2000. It is of note that SSM values in May 2000 were very low, compared to the SSM-values in 1999. Significant CCs were also observed between SSM sampling in March 1999 and EUF-P_2 for 2000 and 2001.

Potassium content using the DL-method showed significant CC to SSM on three of the four sampling dates for the 0 - 30 cm and 30 - 60 cm depth. The significant CC between SSM in May 2000 and $K_{(DL)}$ in the 30 - 60 cm depth are shown as an illustration in Figure 65. In contrast to the many significant CC observed between SSM and EUF-phosphorus, the EUF-potassium showed only significant CCs between SSM at 12.05.2000 and the EUF-K_1 and EUF-K_2 fraction in 1999.

Soil pH values in 0 - 30 cm and in 30 - 60 cm were positively cross correlated to SSM contents obtained on the 29.04.1999 (Figure 65), however not for the other three sampling dates. No significant CC could be observed for SOC, clay, silt, and sand. EM38 did not show significant CC for any of the four dates. ECa values showed a significant relationship for one out of four sampling dates; for ECa in the 0 - 30 cm to SSM on the 10.05.1999 and ECa in 30-60 cm depths to SSM at the 25.03.1999.

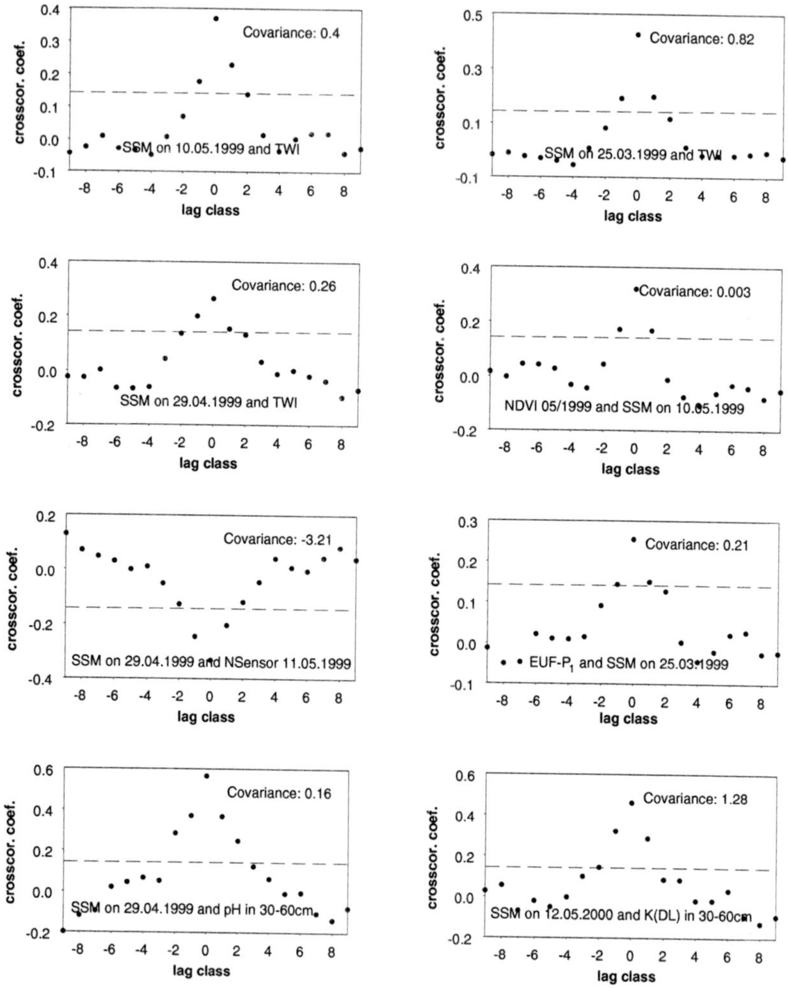

Figure 62. Significant crosscorrelations between soil surface water content (SSM) at different dates versus Topographic Wetness Index (TWI), HydroN-Sensor at the 11.Mai 1999, EUF-P_1 in August of 2000, pH in 30-60 cm in 1998, and $K_{(DL)}$ in 30-60 cm in 1998. The dashed line represents the 95% confidence interval.

8.2.3. Significant crosscorrelations of soil water content for different years and different parameters

Soil water contents after harvest in the 0 - 30 cm depth were negatively crosscorrelated in 1999, 2000 and 2001 with elevation (Figure 63). In 1998 the elevation deviation (a relief parameter directly obtained from elevation) was found to be negatively crosscorrelated (Figure 63). Deeper soil layers did not show significant crosscorrelation. The southern six rows of soil moisture data were not used during this analysis, as rainfall occurred during the sampling campaign in 1998 and 2000. If all values were used, several strong CC were observed, which are artifacts due to the rainfall occurred during sampling.

Additionally, two significant CC were observed for soil water content in the 0 - 30 cm in 1999 versus clay content in the 30 - 60 cm depth and versus soil pH in 0 - 30 cm depth (Figure 63).

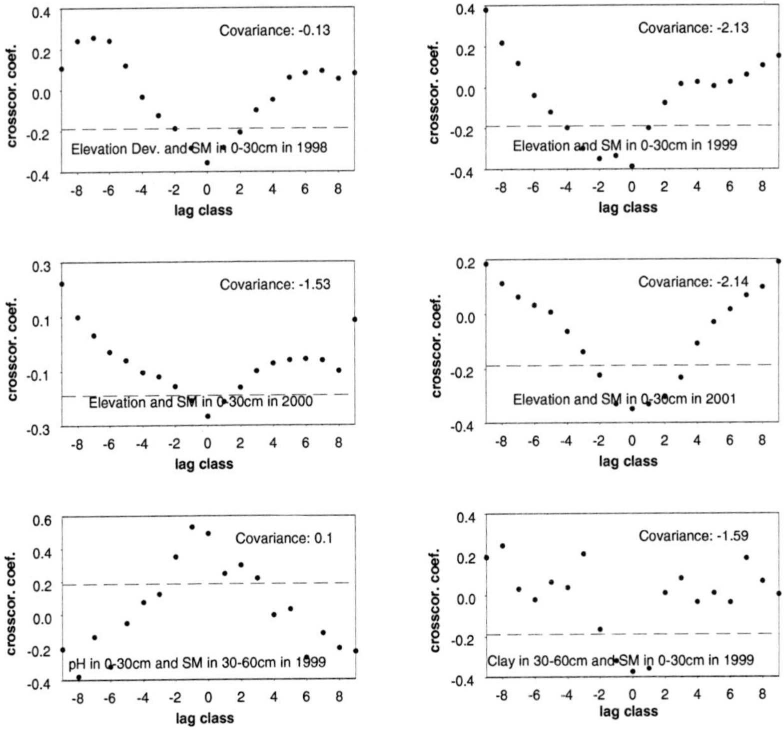

Figure 63. Crosscorrelations between soil water content after harvest (SM) for 0-30 cm depth and relief parameters or HydroN-Sensor. The dashed line represents the 95% confidence interval.

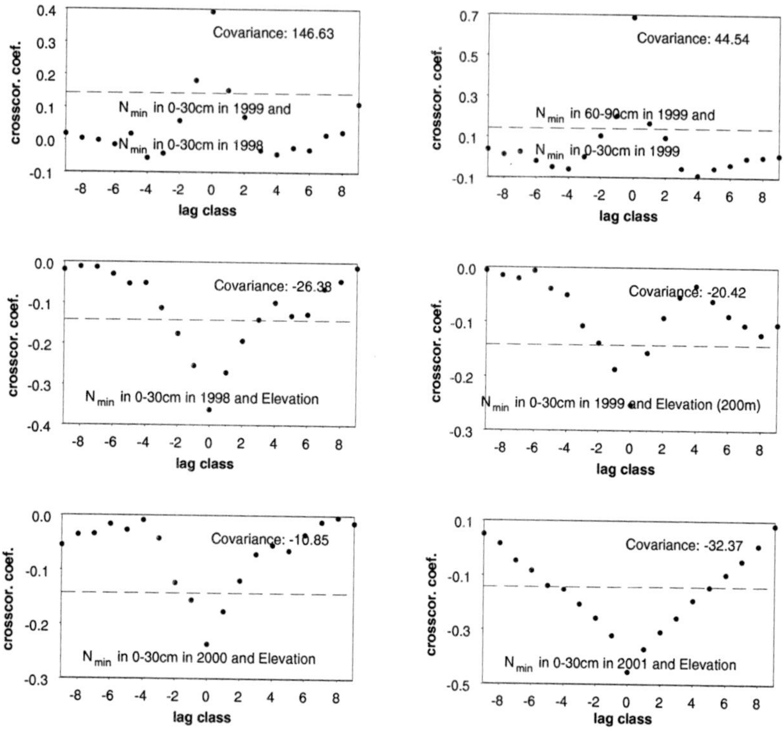

Figure 64. Crosscorrelations between residual N_{min} after harvest itself and between N_{min} after harvest versus elevation. The dashed line represents the 95% confidence interval.

8.2.4. Significant crosscorrelations of N_{min} after harvest for different years and with different parameters

Residual N_{min} after harvest showed only two significant CC with each other. The first SCD was found between N_{min} in the 0 - 30 cm depth in 1999 and in 1998; the second between N_{min} in 0 - 30 cm depth and in 60 - 90 cm depth in 1999.

In each year at least one negative SCD could be identified between the relief parameter elevation and the N_{min} content in the 0 - 30 cm depth (Figure 64). Significant negative CCs existed between 1 and 4 lag distances. The N_{min} content in deeper soil layers (30-60cm and 60-90cm) were crosscorrelated with elevation only in 2000 and 2001. This proves and extends the results of Li et al. (2001), who showed a negative CC between NO_3-N and elevation for only one year. Significant CCs were found between N_{min} and NIR, and between HydroN-Sensor and NDVI as shown in Appendix Geostatistics. These will not be further discussed.

8.2.5. Significant crosscorrelations between soil textural properties and with different parameters

Sand, Clay and silt were crosscorrelated with each other, however the only significant CC in respect to relief parameters were found for sand in the 0 - 30 cm depth and 30 - 60 cm depth and elevation. Negative CC for up to 2 lag distances were observed.

SOC, clay and silt content were not significantly crosscorrelated with different relief parameters. No significant CC - results were observed between soil properties and ECa and EM38-values. For example, the crosscorrelation between silt content in the 30 - 60 cm depth and elevation, silt content in the 30 - 60 cm depth and EC in the 0 - 30 cm are presented in Figure 65.

Soil pH values in the 0 - 30 cm depth showed a significant positive crosscorrelation of up to 4 lag distances for the elevation range, but no significant CC between soil pH values and elevation was observed. A significant negative CC with a NIR image was observed with the pH-values obtained in 30 - 60 cm depth.

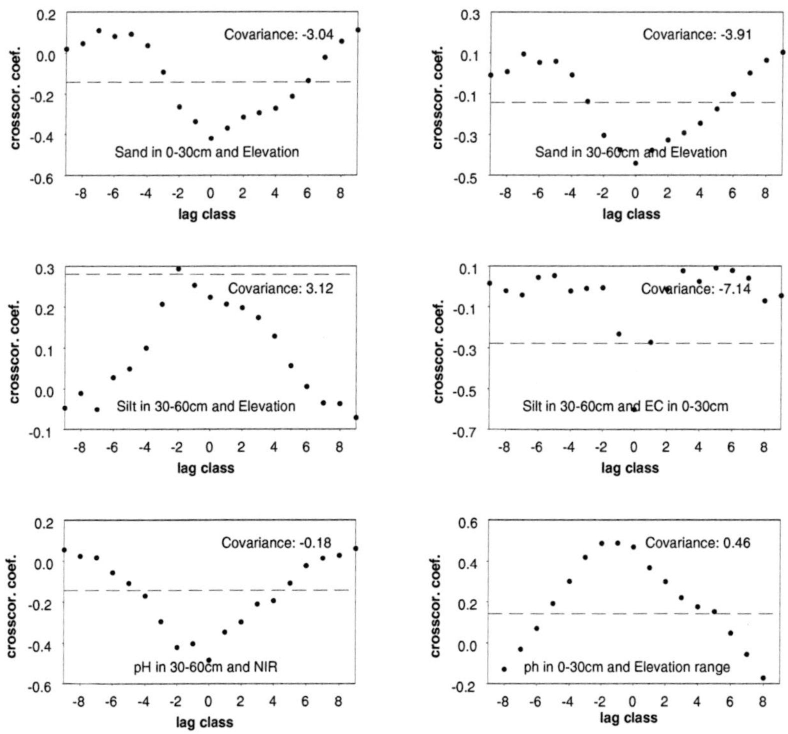

Figure 65. Significant and nonsignificant crosscorrelations between texture and chemical soil properties (sand, silt and pH) and elevation and NIR. The dashed line represents the 95% confidence interval.

8.2.6. Significant crosscorrelations of soil penetration resistance with depth

Penetration resistance in different depths were all found to be significant positively crosscorrelated at the field site “Bei Lotte”, except for one layer pair. Significant lag distances were four lag distances in the topsoil layers (0 - 10 and 10 - 20 cm) and one lag distance in the 10 - 20 cm and 20 - 30 cm depths. This depth coincided with the depth of the chisel plow in the no-tillage system. No significant crosscorrelation could be observed between penetration resistance in 20 - 30 cm depths and 30 - 40 cm depths. This depth (i.e. 45 cm) coincided with the depth of the subsoiler. Soil layers below the 30 - 40 cm depths again showed a significant positive crosscorrelation with each other.

8.2.7. Significant crosscorrelations between soil penetration resistance and relief / sensor values

Penetration resistance in different depth layers was tested for significant cross correlations with different relief parameters. Negative crosscorrelations were observed for the 10-20 cm depth with the elevation range. A second SCD was found with elevation for the 40-50 cm depth. Note that this was also the working depth of the subsoiler. In contrast, positive CCs of up to 2 lag distances were observed for flow accumulation and stream power index, indicating a close relationship between processes related to water and sediment transport.

Penetration resistance showed significant CC to different yield properties (Table 45). It may be asked, if such significant CC may also be observed during the growing period by sensor observation (i.e. NDVI, HydroN-sensor). To test this, penetration resistance in 2001 was compared against NDVI measurements (Table 31). In the dry year of 2000, the penetration resistance crosscorrelated with NDVI for the 0-20 cm depth and 50-70 cm depth (Figure 66), however were not crosscorrelated for the 30-40 cm depth layer. That same layer, 30-40 cm however (Figure 66), showed some significant crosscorrelations in the above average precipitation years 1999 and 2001 to NDVI. In all years the depth layer 40-50 cm did not show any significant crosscorrelation to NDVI. The same is true for the NDVI values in May 1999 and April 2000. If a crosscorrelation between penetration resistance and NDVI in May 1999 and April 2000 can be observed, even if they are dissimilar, reactions in the same direction can be observed for April 1999 or May 2000 (Figure 66). Based on these observations, the conclusion can be drawn that at least one soil layer showed a significant CC to NDVI values depending on the meteorological conditions observed. In years with below average precipitation, deeper soil layers seem to gain importance for crop vitality (NDVI), whereas for years with above average precipitation the depth of the tillage tools appears to lead to significant CC and therefore processes.

Crosscorrelations calculated between penetration resistance for different depth layers and HydroN-sensor values showed significant negative crosscorrelations for samples obtained on May 11, 1999 for the 30-40 cm layer, whereas for samples obtained on May 4, 2000 a positive SCD was observed. The opposing signs of the CC versus NDVI (negative versus positive crosscorrelation coefficients at lag distance 0) is an interesting phenomenon and needs further investigations.

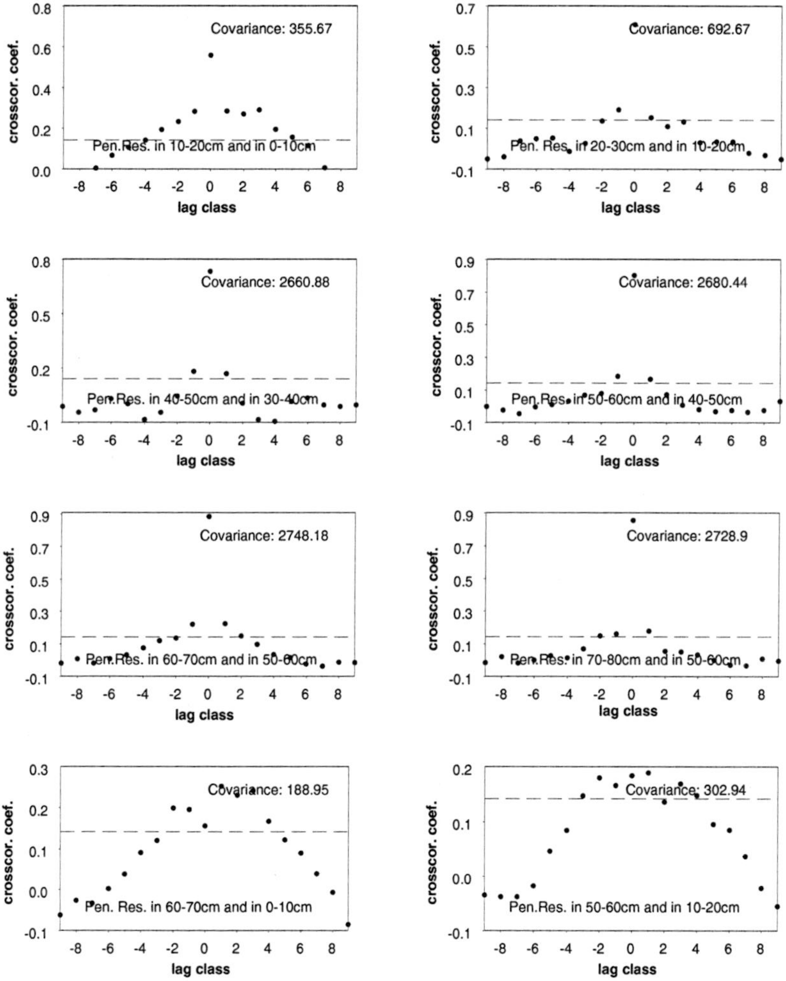

Figure 66. Significant Crosscorrelations for penetration resistance in different depths. The dashed line represents the 95% confidence interval.

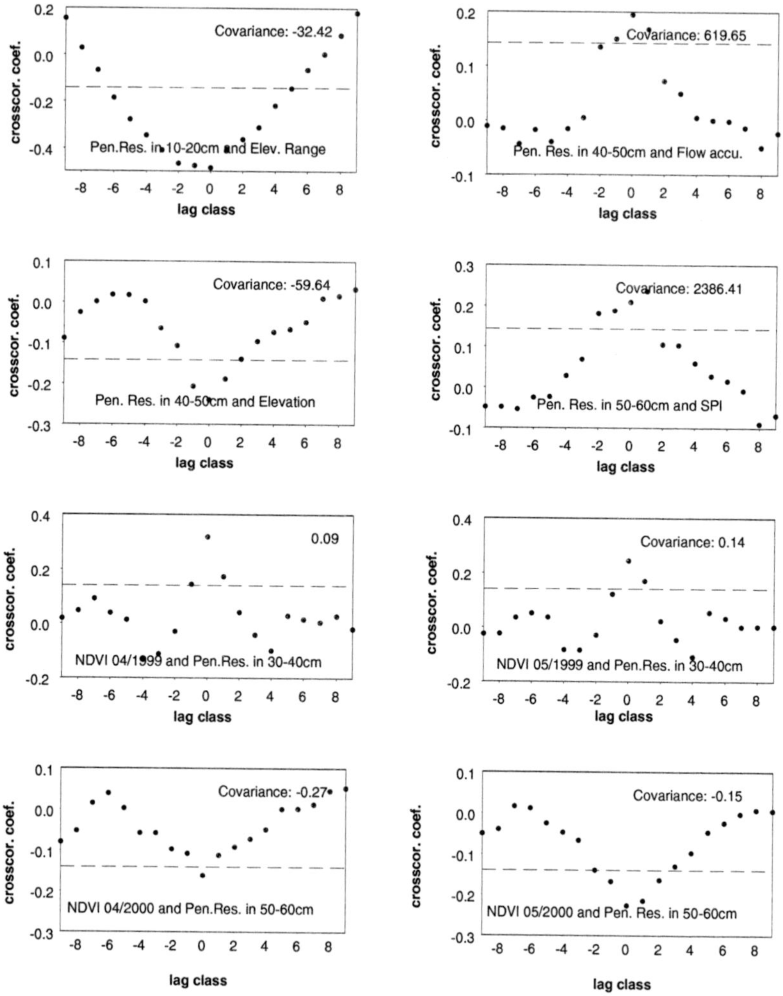

Figure 67. Significant crosscorrelations between penetration resistance in different depths versus relief parameter or versus NDVI. The dashed line represents the 95% confidence interval.

Table 31. Significant crosscorrelations between penetration resistance in spring of 2001 in different depth layers and NDVI measurements for the time period 1999 to 2001. A '+' indicates a positive CC at lag distance 0, a '-' a negative CC. 'NA' means no value can be identified as values are below a value of 0.1. 'Sign.' means significant cross correlations above the 95% confidence interval for at least one lag distance around lag 0.

Depth of Penetration Resistance	NDVI					HydroN-Sensor			
	04/1999	05/1999	04/2000	05/2000	05/2001	11.05. 1999	10.06. 1999	18.04. 2000	04.05. 2000
0 - 10 cm	-Sign.	NA	NA	-Sign.	+	+	NA	-	+Sign.
10 - 20 cm	-	+	NA	-Sign.	+	-	NA	-	+Sign.
20 - 30 cm	NA	+	NA	-	+	-	+	-	+
30 - 40 cm	+Sign.	+	NA	NA	+Sign.	-Sign.	+	-	-
40 - 50 cm	+	+	NA	-	+	-	+	-	NA
50 - 60 cm	-	NA	-	-Sign.	+	NA	NA	NA	+
60 - 70 cm	-	NA	-	-Sign.	NA	NA	NA	NA	+
70 - 80 cm	NA	NA	NA	NA	NA	NA	NA	NA	Na

8.2.8. Significant crosscorrelations for soil nutrients at different times and with different parameters

The nutrients potassium and phosphorus were not found to be crosscorrelated with depth with data from the DL-method. Additionally, no significant CCs could be observed with respect to potassium and phosphorus obtained using the EUF-method.

The parameters EUF-K_1, EUF-K_2, EUF-P_1 and EUF-P_2 were found to be crosscorrelated in 1999, 2000 and 2001. Significant crosscorrelations were also determined between years and within each year for EUF-K_1, EUF-K_2, EUF-P_1 and EUF-P_2.

Crosscorrelations were computed between the nutrients potassium and phosphorus against texture, however no significant CCs could be identified. Furthermore, no significant crosscorrelations were found for $K_{(DL)}$ and $P_{(DL)}$ with respect to different relief parameters.

The EUF-K fractions and EUF-P fractions exhibited significant crosscorrelations with respect to different relief parameters. A negative relationship up to 4 lag classes was observed in the years 1999, 2000 and 2001. For example, the crosscorrelations between EUF-K_1 and EUF-K_2 with respect to elevation are shown in Figure 68. Another significant crosscorrelation could be observed for flow accumulation, which showed a positive CC in all three years (Figure 68). Other primary relief parameters based on elevation as minimum, maximum, range, deviation showed many significant cross correlations. No significant CC could be observed for planform and profile curvature, or slope. The TWI, as a secondary relief parameter based on slope and flow accumulation showed some significant CC versus EUF-P in 1999 and 2001.

Significant crosscorrelations were observed for other variables like HydroN-sensor (Crop-N status), NIR, NDVI, soil chemical properties and penetration resistance were identified as significant, but will not be discussed here due to time constraints. Computed Crosscorrelation data are shown in Appendix Geostatistics.

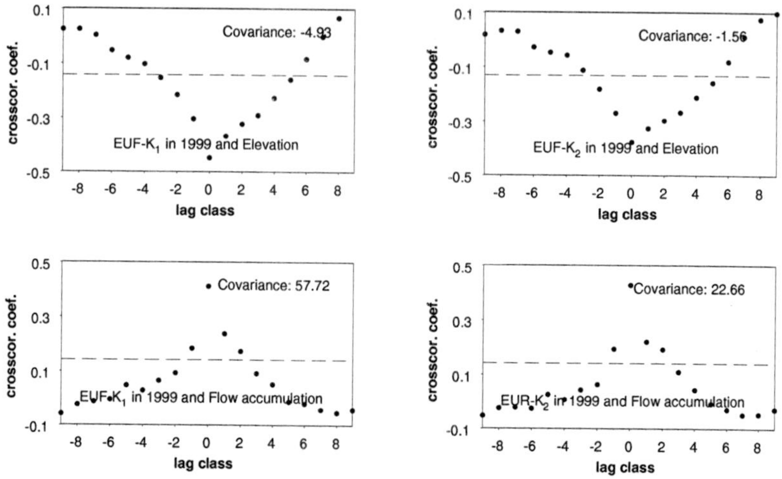

Figure 68. Significant Crosscorrelations between EUF-K_1 and EUF-K_2 fractions and the relief parameter elevation and flow accumulation. The dashed line represents the 95% confidence interval.

8.3. Soil properties for selected relief parameters

8.3.1. Soil water content distribution versus profile curvature and landform elements

The distribution of soil water content relative to landform and profile curvature classes for spring soil surface (0-10 cm) water content conditions (SSM) will be discussed in the following section. Figure 69 shows the distribution of profile curvature aggregated from a 10 m by 10 m DEM using a majority function for each sampling cell. The profile curvature values are classified as follows: values in the range -0.2 - -0.1 1/100 m were identified as class 1, -0.1 - 0 as class 2, 0 - 0.1 as 3, 0.1 - 0.2 as 4 and 0.3 - 0.4 as 5. No values existed for profile curvature between 0.2 and 0.3 for the 27 m aggregation. Spring soil surface water content (SSM) for 1999 and 2000 versus profile curvature are shown in Figure 70. Results for soil water content in 0 - 30 cm and 30 - 60 cm depth after harvest (SM) in the years 1999, 2000 and 2001 are presented in Figure 71. Additionally, Appendix Landforms contains the values for the 60 - 90 cm layer (Table 52, Table 54).

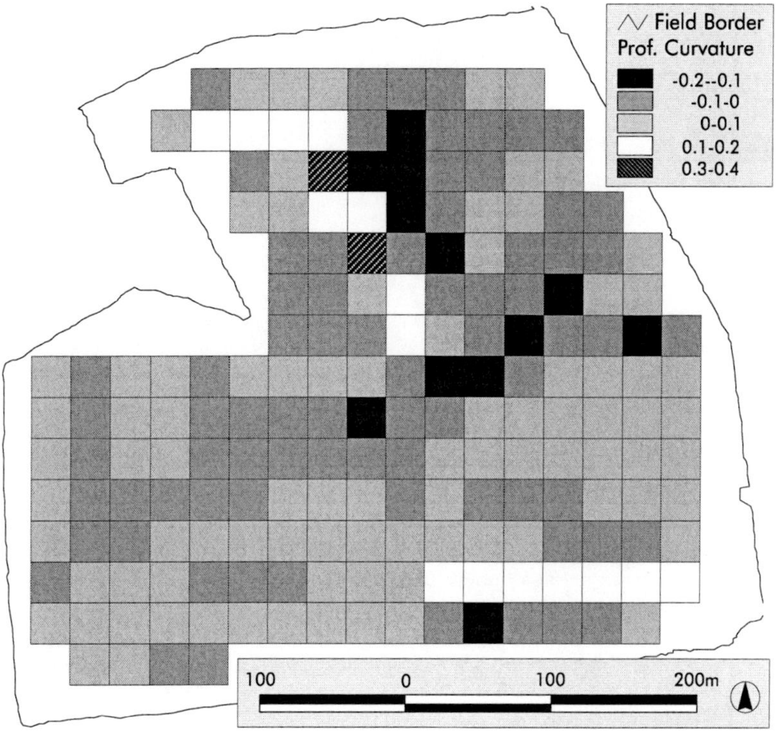

Figure 69. Average profile curvature for the 27 m by 27 m raster at the field site "Bei Lotte", aggregated from a 10 m by 10 m DEM.

The spring soil surface water content differences between convex and concave positions ranged between 5 g(100g soil)$^{-1}$ in March to approximately 1.5 g(100g soil)$^{-1}$ in April and May. These differences diminished in the drier year of 2000, when soil water content in the top 10 cm showed a maximum difference of 0.6 g(100g soil)$^{-1}$ (Figure 70).

The mean SM was found to increase with profile curvature in every year for the 0 - 30 cm depth. In 1999 convex areas (class 1 in Figure 71) in the field showed the lowest soil water content values, whereas concave parts (4) showed the highest values in all three years. These findings agree with results observed from soil surface water content measurements in spring 1999.

The 30 - 60 cm depth layer (Figure 72 right hand side), in contrast, showed that planar areas (2, 3) contain up to 1g(100g soil)$^{-1}$ less water in the years of 2000 and 2001 than the strong convex (1) and strong concave areas (4, 5). This may be related to the specific location of divergent positions, which appear within the planar sections of the field, followed directly by convergent areas. Therefore, moisture may be translocated via subsurface flow from planar areas over divergent areas to convergent areas. Such pattern was not found in 1999 in the 30 - 60 cm layer. The higher soil water content generally observed in the profiles in 2000 and 2001 as compared to that of 1999, might lead to lateral flow due to higher hydraulic conductivity.

Generally, the increasing soil water content with increases in profile curvature are seen in the topsoil layer (0-30 cm) are also present in the deeper soil profile (60-90 cm - see Appendix Landform), however with less distinction.

Convergent positions (represented by class 5 in Figure 70 and Figure 71) exhibited the highest soil water content throughout all observed years and all depths.

These results confirm the findings of Sinai et al. (1981), that an increasing curvature increases the soil water content. Sinai et al. (1981) found a strong linear relationship between surface curvature and soil water content. The curvature in their paper was in the range of -0.12 to 0.12. According to Ruhe (1960) this type of relief would be classified as a planar site. The strong relationship found in the Negev desert described by the equation: c = 8.68 + 50.4 (surface curvature) with an R^2 of 0.9 would certainly not be expected under central European conditions. Correlation coefficients between profile curvature and soil surface water content measured at four dates at the field site „Sportkomplex“ were calculated and yielded results of 0.32, 0.29, 0.32 and 0.07 for the respective date. This indicates a less pronounced dependency of surface soil water content on profile curvature in this study site than under the desert conditions.

Boyer et al. (1990) analysed curvature at various scale lengths (1.5 m to 30.5 m) and related it to the volumetric water content. The authors state that curvature had larger effects on soil moisture with increasing scale. As the resolution in the presented case with the 27 m by 27 m aggregation data approaches the upper resolution limit of the Boyer et al. (1990) investigations, scale differences in relation to profile curvature were not investigated further (see additionally Figure 85.

The distribution of soil water content in the spring of 1999 and 2000 for the major landform elements is shown in Figure 72. FS positions contained for all four dates the highest surface soil water contents. The opposite occurs at the SH positions which would be expected to represent dry conditions, as is the case for the year 1999 (above average precipitation) with values between 22g(100g soil)$^{-1}$ and 29g(100g soil)$^{-1}$. However, in the dry year of 2000, SH positions seem to gain some moisture probably due to subsurface flow from the surrounding level areas.

The findings by soil surface water content sampling agree only partially with soil water content observations from fall sampling (Figure 72). SH exhibited 0.5 g(100g soil)$^{-1}$ less soil water content in 1999, with similar SM contents in 0 - 30 cm depth throughout the field. In the dry year of 2000, BS showed the highest SM, whereas at LE and at SH positions the minimum amount was observed. This is different from the patterns seen for SSM on 12.05.2000 in

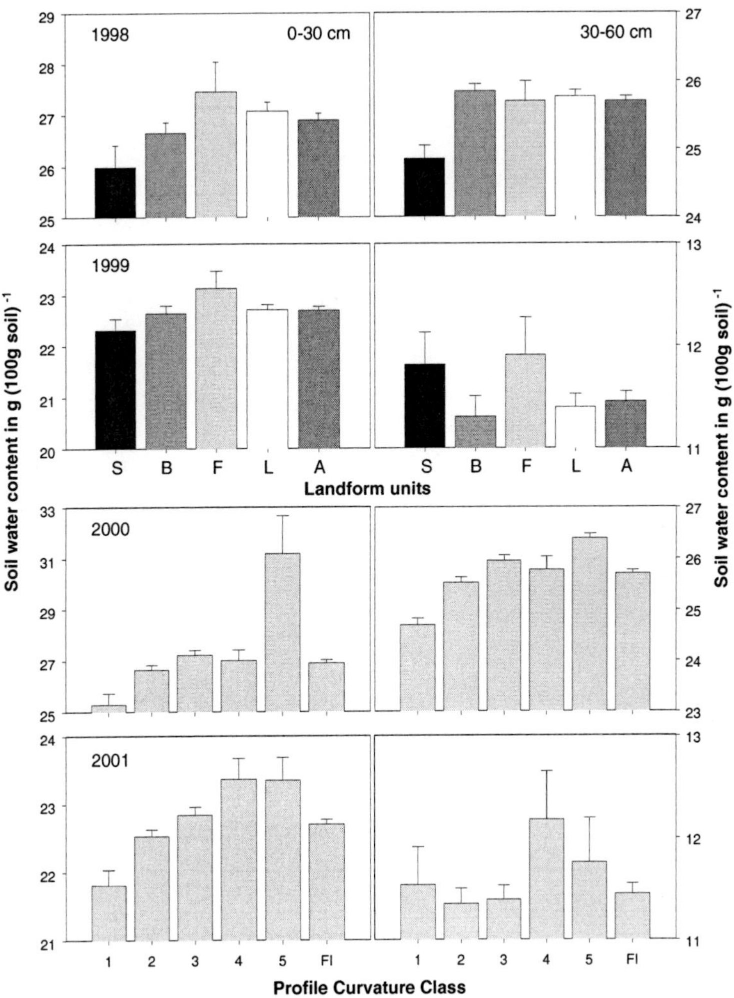

Figure 70. Gravimetric soil water content at 0 -10 cm depth (SSM) at the field site "Bei Lotte", stratified for landform classes (upper part) and curvature classes (lower part). Landform elements are Shoulder (S), Backslope (B), Footslope (F), Level (L) and the field average (A). Profile curvature classes are: -0.2- -0.1 as 1, -0.1-0 as 2, 0-0.1 as 3, 0.1-0.2 as 4, 0.3-0.4 as 5, Field average as FI . Bars show the standard error.

Figure 72. Results for the wet year of 2001 showed a similar distribution (Figure 72) for the SSM obtained on 12.05.2000 for 0 - 10 cm depth.

Post-harvest soil water content in 30 - 60 cm and 60 - 90 cm levels showed similar patterns in all three years, however differences were observed in 0 - 30 cm depth. In 1999, FS positions showed the minimum soil water content and in 2000 and 2001 SH and Level position contained the lowest soil water content.

Generally an increase in soil water content would be expected from SH-BS-FS as observed by Manning et al. (2001a). However, as shown in Figure 70 and Figure 71, differences from that pattern were observed for soil surface water content (0 - 10 cm depth) as well as for deeper soil water content distributions (30 - 60 cm). The differences observed in the amount of soil water content, may be due either to subsurface flow or as a side effect of different amounts of nitrogen application. Campbell et al. (1977), for example showed (I) a decrease in available water with increasing nitrogen uptake between tillering and anthesis and (II) that different amounts of nitrogen were taken up by the crop under wet conditions.

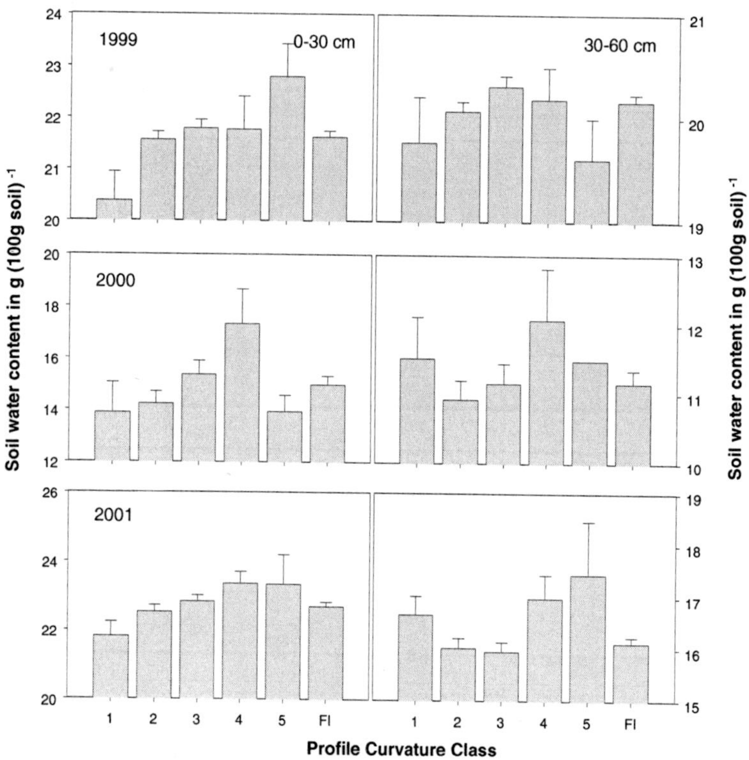

Figure 71. Gravimetric soil water content at the field site "Bei Lotte" for the years 1999, 2000 and 2001 for the depth 0-30 cm (left column) and 30-60 cm (right column), stratified for profile curvature. Profile curvature classes: -0.2- -0.1 as 1, -0.1-0 as 2, 0-0.1 as 3, 0.1-0.2 as 4, 0.3-0.4 as 5, Field average as FI. Bars show the standard error.

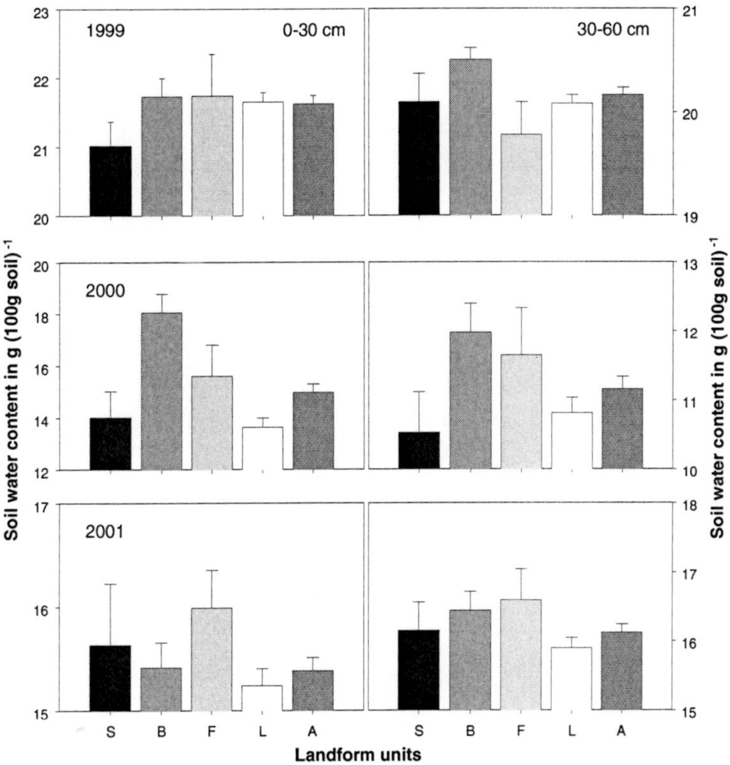

Figure 72. Gravimetric Soil Water content for the years 1999,2000 and 2001 at the field site "Bei Lotte" for the depth 0-30cm (left column) and 30-60cm (right column), separated for landform units. Shoulder (S), Backslope (B), Footslope (F), Level (L) and the field average (A). Bars show the standard error.

8.3.2. Soil mineralized Nitrogen versus profile curvature and landform elements

Mineralised soil nitrogen content is plotted in Figure 73 versus profile curvature for 0-30 cm and 30-60 cm depths. The highest values obtained for N_{min} regardless of depth were found in the profile curvature class 5 (0.3-0.4 located in the depression, see Figure 69) in all years and exceeded the field average by 17 kgha^{-1} in 0 - 30 cm depth (22 kgha^{-1}, 12 kgha^{-1}), 7 kgha^{-1} in 0 - 30 cm depth (15 kgha^{-1}, 7 kgha^{-1}) in 1999, 20 kgha^{-1} (0 kgha^{-1}, 8 kgha^{-1}) in 2000 and 70 kgha^{-1} in 0 - 30 cm depth(21 kgha^{-1}, 5 kgha^{-1}) in 2001. Numbers in parentheses are for 30 - 60 cm depth and 60 - 90 cm depth. This supports results obtained by Morgenstern (1986), who found that FS and lowlands positions contained higher total nitrogen content.

Planar areas (2, 3 in Figure 69) showed in 1999 (10 N_{min} kgha^{-1}) and 2000 (1 N_{min} kgha^{-1}) slightly larger values than convergent (1) and divergent (4) positions. For the years of 1998 and 2001, convergent positions (4 in Figure 69) showed similar values of N_{min} as that observed in planar sections.

The N_{min} content in 30-60 cm and 60-90 cm depth again showed the minimum nitrogen content found at divergent positions (1) for all given years, whereas all other profile curvature classes showed only slight differences. These results confirm findings by Aandahl (1948), who found less nitrogen at convex positions (ridge) compared to concave positions (FS). Already back in 1948, Aandahl (1948) recommended two strategies for investigations related to nitrogen management: (I) hold all slope properties constant, except those involved and (II) use different kinds of slopes as part of a comparison study.

Based on this, sampling on divergent positions (1) will lead to an underestimation between 3 and 13 kgha^{-1} N_{min} for the 0 - 30 cm depth (1-5 kgha^{-1} in 30 - 60 cm depth and 0-3 kgha^{-1} in 60 - 90 cm depth) for all 4 years compared to the field average. Sampling at strongly convergent positions (5) will even lead to larger differences, showing an overestimation of between 7 and 70 kgha^{-1} (8 - 23 kgha^{-1} in 30 - 60 cm and 0 - 12 kgha^{-1} in 60 - 90 cm) for all four years compared to the field average. Based on the observed differences, a practical application would be a „map-overlay“ map based on profile curvature, which could be used to limit the application of nitrogen fertilizer in a field (see Wenkel et al., 2001).

Residual N_{min} content after harvest separated for landform units is shown in Figure 74 for the years 1998-2001. In this representation differences between years are accentuated. SH positions for all years and almost all depths had the smallest nitrogen contents (between 0-12 kgha^{-1} less than the field average). The reason for these differences might be (I) favourable nutrient uptake, (II) erosion effects, (III) leaching into groundwater, (IV) lateral distribution to other landforms (BS, FS), and (V) and different amounts of soil organic matter.

Maximum N_{min} contents versus LF and years were observed either for BS (1x), FS (5x) and LE positions (6x). Differences between the maximum and minimum N_{min} per LF values were 4 kgha^{-1} in 1998 in the 0-30 cm depth (5 kgha^{-1}, 5 kgha^{-1}), 15 kgha^{-1} (5 kgha^{-1}, 2 kgha^{-1}) in 1999, 2 kgha^{-1} (2 kgha^{-1}, 2 kgha^{-1}) in 2000, and 15 kgha^{-1} (4 kgha^{-1}, 2 kgha^{-1}) in 2001. Numbers in parentheses are for 30-60 cm depth and 60-90 cm depths. Differences between LFs decreased in deeper soil layers, but still existed.

Applying a catena principle, the assumption can be made, that the values of N_{min} increase from SH to BS to FS. Such patterns were visible in 1998 (60-90 cm), 1999 (30-60 cm and 60-90 cm), 2000 (0-30 cm, 30-60 cm), and 2001 (0-30 cm) (Figure 74). However only small differences were visible (see additional Tables in Appendix Landforms).

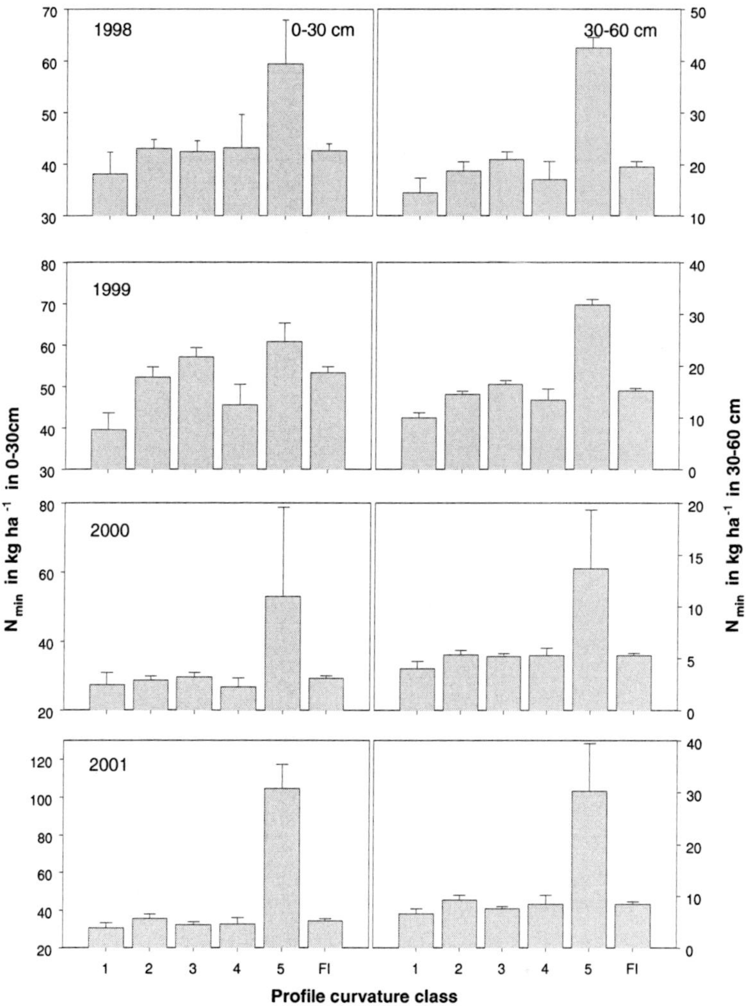

Figure 73. Residual mineralised soil Nitrogen (N_{min}) after harvest for four subsequent years (1998-2001) for the depths 0-30 cm (left column) and 30-60 cm (right column), stratified for different curvature classes for the field site "Bei Lotte". Profile curvature classes: -0.2- -0.1 as 1, -0.1-0 as 2, 0-0.1 as 3, 0.1-0.2 as 4 and 0.3-0.4 as 5, FI as field average. Bars show the standard error.

Observed results were in contrast to those of Manning et al. (2001a), who observed a decrease in residual nitrate for 0 - 90 cm depth over two years with increasing convergent character of the landscape (i.e. SH-BS-FS) in three of four fertilizer treatments applied

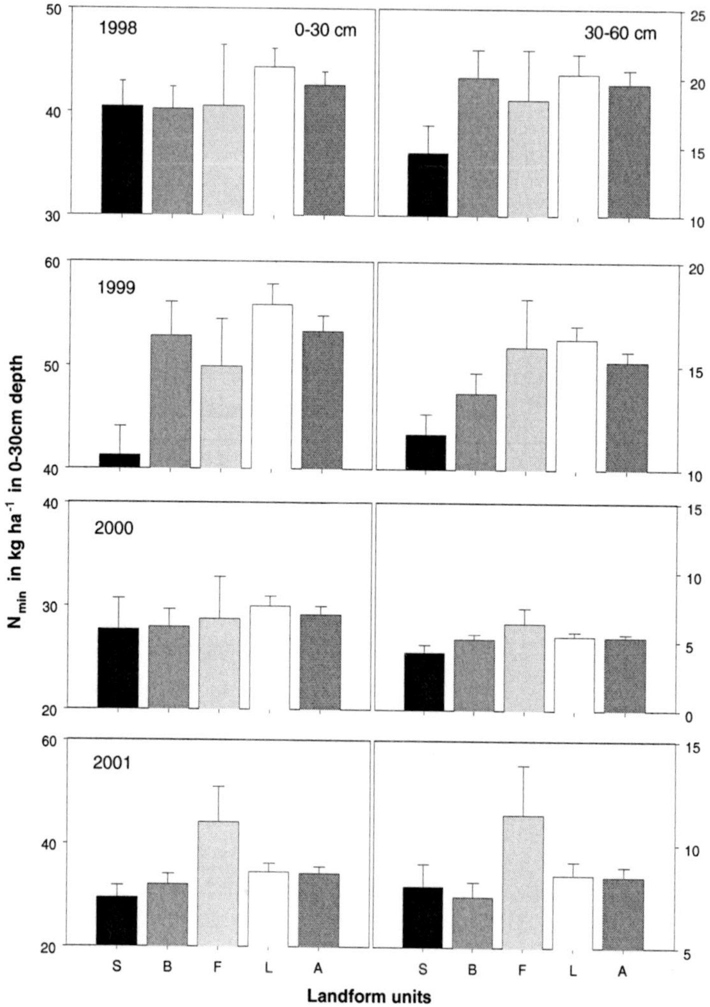

Figure 74. Residual mineralised soil Nitrogen (N_{min}) after harvest for four subsequent years (1998-2001) for the depths 0-30 cm (left column) and 30-60 cm (right column) at the field site "Bei Lotte", stratified for different landform elements. Shoulder (S), Backslope (B), Footslope (F), Level (L) and field average (A). Bars show the standard error.

8.3.3. Soil phosphorus and potassium content versus profile curvature and landform elements

Phosphorus contents for the years 1999-2001 were determined by EUF analysis for the 0-30 cm depth and plotted versus profile curvature class in Figure 75. The relationship between profile curvature classes and soil phosphorus was determined using the DL method in 1998 for the 0 - 30 cm depth and the 30 - 60 cm depth as seen in Figure 76.

Analysis of EUF-P_1 content versus profile curvature reveals, that divergent planar (2) and strong concave (4) areas contain larger amounts than convex (1) and convergent planar (3) areas. The same is valid for the EUF-P_2 fraction, but the differences were not as distinct. Maximum differences for EUF-P between LFs and years are for EUF-P_1 (0.33 in 1999, 0.18 in 2000, 0.12 in 2001, in mg(100g soil)$^{-1}$) and EUF-P_2 (0.14, 0.19, 0.8, in mg(100g soil)$^{-1}$). The phosphorus contents determined by the DL method showed only the divergent planar (2) class with elevated levels of phosphorus, whereas strong concave areas (4) show values similar to the rest of the field. Phosphorus content using the DL-method in the 30 - 60 cm showed a similar distribution (see Table 50 in Appendix Landform), however, with less distinction. Mulla (1993) observed an increase in phosphorus content with increasing SOC, and stated that the order of SOC (low, medium, high) closely resembles the order of landscape positions (top, back, toe). However, as SOC contents do not follow the order of landscape positions (Figure 80 in the field site "Bei Lotte", the observed results differed. Additional significant correlations are found for $P_{(DL)}$ and EUF-P to SOC in Appendix Landform.

The order of magnitude of phosphorus distribution across different landforms changed in different years. Phosphorus content at single LFs also differed between years. For the EUF P_1-fraction, the SH and FS positions contained the largest values as well as the largest standard errors (Figure 76), whereas BS and FS show the minimum values. Maximum differences for EUF-P_1 between landforms decreased from 1.62 mg(100g soil)$^{-1}$ in 1999, to 1.14 mg(100g soil)$^{-1}$ in 2000 and 0.88 mg(100g soil)$^{-1}$ in 2001. EUF-P_1-values observed at FS decreased from 1.67 mg(100g soil)$^{-1}$ in 1999, to 1.28 mg(100g soil)$^{-1}$ in 2000 and increased slightly to 1.33 mg(100g soil)$^{-1}$ in 2001. At SH positions EUF-P_1-values decreased from 1.24 mg(100g soil)$^{-1}$ to 1.09 mg(100g soil)$^{-1}$ and than to 1.06 mg(100g soil)$^{-1}$ for the years 1999-2001 (Table 55 in Appendix Landform)

SH and FS positions were determined to contain the largest values (DL-method) for 0 - 30 cm and 30 - 60 cm (Figure 76). In contrast to the results for the EUF P_1 fraction, the BS position contained the least phosphorus content for $P_{(DL)}$ in the 0 - 30 cm and 30-60 cm depths (Figure 76).

The EUF-P_2 content showed similar ranking for different landforms as EUF-P_1 and $P_{(DL)}$. The decrease at BS-positions as well as the elevated phosphorus content for SH and FS is visible in the EUF-P_2 fraction results.

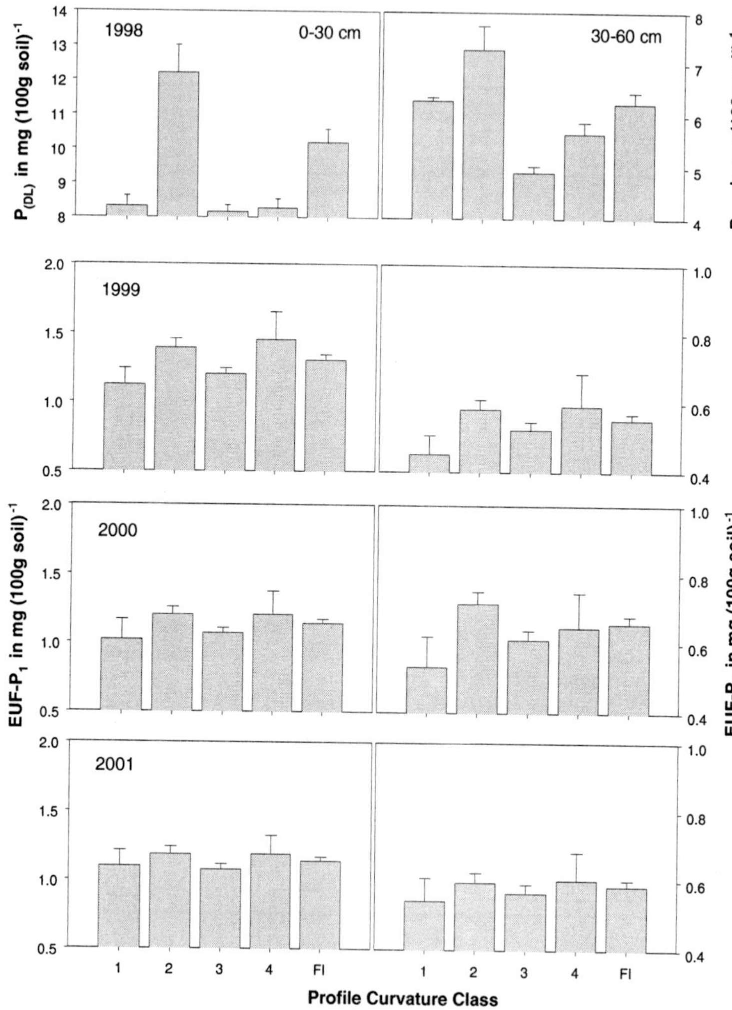

Figure 75. $P_{(DL)}$ in 0-30 cm and 30-60 cm depth (top row) and EUF-P_1-fractions (left column) and EUF-P_2-fraction (right column) in 0-30 cm depth after harvest for three subsequent years (1999-2001) stratified for profile curvature at the field site "Bei Lotte". Profile curvature classes: -0.2- -0.1 as 1, -0.1-0 as 2, 0-0.1 as 3, 0.1-0.2 as 4 and FI as field average. Bars show the standard error.

The higher phosphorus contents at SH positions was unexpected at first glance. However, the 1% greater clay content as well as 0.01% greater values of SOC than the field average (see Figure 81) explain the observed pattern (Scheffer and Schachtschabel, 1992).

The conclusion reached regarding sampling under uniform management for phosphorus at the field site "Bei Lotte" was two-fold. When trying to locate areas as a farmer likely to first show phosphorus deficiencies, the recommendation would be to sample BS positions. However, under legislative restrictions, areas of over-fertilization are of concern, therefore it would be more appropriate to sample the landform units shoulder and footslope in case of the field site "Bei Lotte".

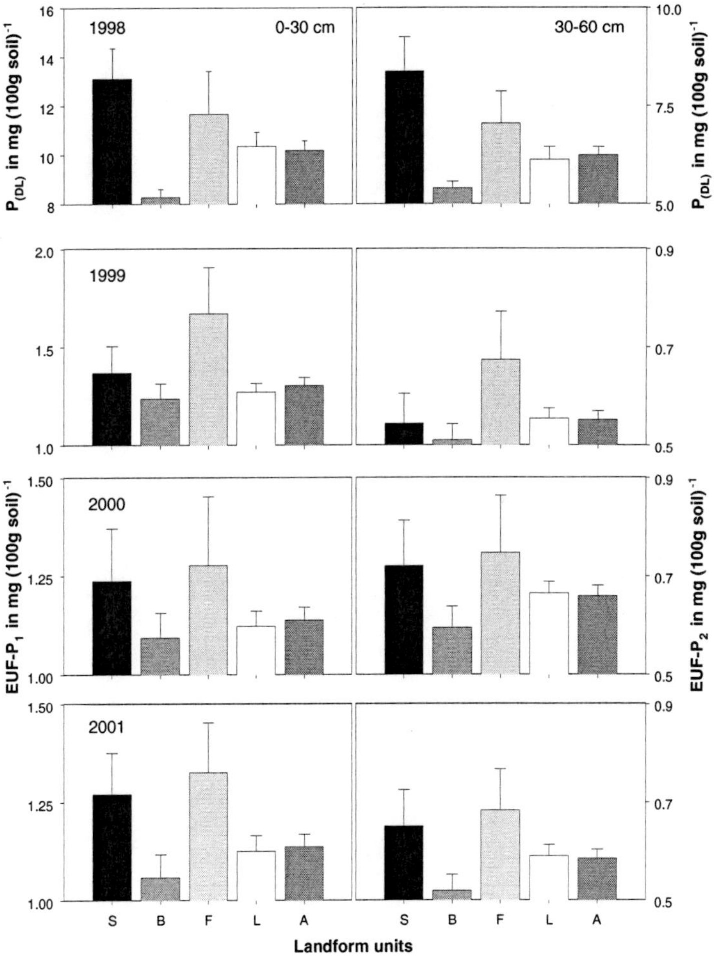

Figure 76. $P_{(DL)}$ for 0-30 cm (P30) and 30-60 cm (P60) and EUF-P-fraction (P_1 and P_2) for the Years 1999-2001 at the field site "Bei Lotte", separated for different landform units. Shoulder (S), Backslope (B), Footslope (F), Level (L) and field average (A). Bars show the standard error.

The potassium content distribution across profile curvature classes and for landforms at the field site „Bei Lotte" are presented in Figure 77 and Figure 78. The values are shown additionally in Table 49 and Table 50 in the Appendix Geostatistics.

Generally, differences between profile curvature classes diminish between years. Profile curvature classes 2 and 4 showed the highest values of EUF-K_1 and EUF-K_2 in 1999 and 2000, but not in 2001. Convergent planar areas (3) represented areas with the lowest EUF-K_1 and EUF-K_2 contents in all years. This would lead to the conclusion, that areas of potassium deficiency are located at convergent planar positions.

Manning et al. (2001a) found increasing potassium contents with increasing curvature character of the landscape. However, this was not the case for the field site "Bei Lotte". Scheffer and Schachtschabel (1992) state that potassium content is closely related to clay content. Clay content differed (see Figure 81) between LFs, and these differences of clay might explain difference in the observed results compared to the work of Manning et al. (2001a).

Potassium extracted using the $K_{(DL)}$ method shows the highest values at planar areas (2+3) for 0-30 cm and 30-60 cm. The maximum differences between curvature classes are 1.79 mg(100g soil)$^{-1}$ for the 0-30 cm and 1 mg(100g soil)$^{-1}$ for the 30-60 cm layer. A sampling scheme for uniform management would sample the profile curvature classes 1 and 4 (-0.2--0.1 1/100m and 0.1-0.2 1/100m), to locate potassium - deficient locations, rather than convergent planar locations as in the case for EUF-phosphorus sampling.

Potassium content was at minimum at BS and LE positions, whereas SH and FS contained approximately 1 mg(100g soil)$^{-1}$ greater potassium contents for the EUF-K_1-fraction. EUF-K_2 contents showed a similar distribution across LFs as EUF-K_1.

The content of EUF-K_1 (more readily plant available fraction) at the level landform in the dry year of 2000 is decreased (4.09) in relation to all other landforms and is higher than but still similar to values in 1999 (4.92) and 2001 (4.85). This would indicate a delivery from long-term source or potential resources such as clay and SOC.

Potassium is needed especially under drought conditions for plant respiration (Marschner, 1995). The decreases in EUF-K_1 were associated with the highest number of N_{Kernel}, N_{Spike} and the lowest TKM as compared to all other LF (see Figure 48). Note that the EUF-K_2 content for level LF was 2.35 mg(100g soil)$^{-1}$ in 1999, and subsequently increased to 2.44 mg(100g soil)$^{-1}$ in 2000 before returning to a value of 2.4 mg(100g soil)$^{-1}$ in 2001.

Generally, potassium content at LFs showed elevated levels at SH and FS positions compared to lower values at BS and LE positions in all three years. However, sampling during a dry year (e.g. in 2000), led to an underestimated of the EUF-K_1 compared to other years at the Level landform and indices based on the EUF-K_1/K_2 fractions may lead to a misinterpretation of the results.

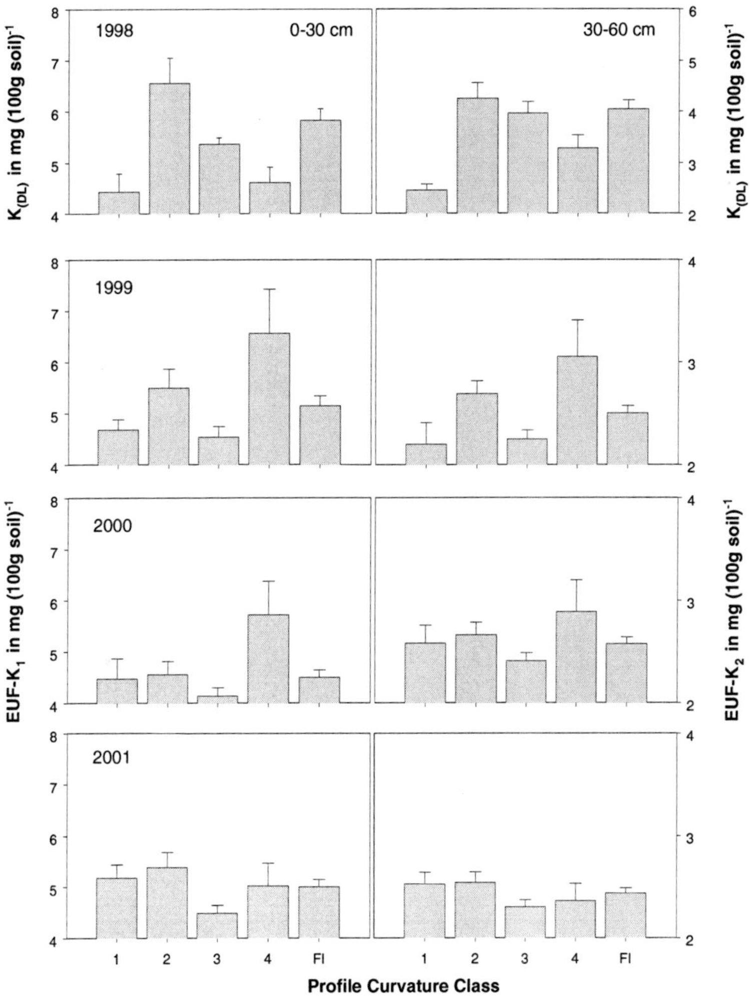

Figure 77. $K_{(DL)}$ in 0-30 cm and 30-60 cm depth (top row, samples taken in 1999) and EUF-K_1-(left column) and EUF-K_2-(right column) fractions for the Years 1999, 2000, and 2001 at the field site " Bei Lotte, stratified for profile curvature at the field site "Bei Lotte". Profile curvature classes: -0.2- -0.1 as 1, -0.1-0 as 2, 0-0.1 as 3, 0.1-0.2 as 4 and FI as field average. Bars show the standard error.

The landforms SH and FS were also found to contain elevated potassium contents for the 0-30 cm depth using the DL-method. Differences between LE (5.95 mg K (100g soil)$^{-1}$) and BS positions (5.32 mg K (100g soil)$^{-1}$) were larger compared to EUF-K_1 contents. Potassium contents at LE positions almost reached the values of SH and FS positions (6.2 mg K (100g

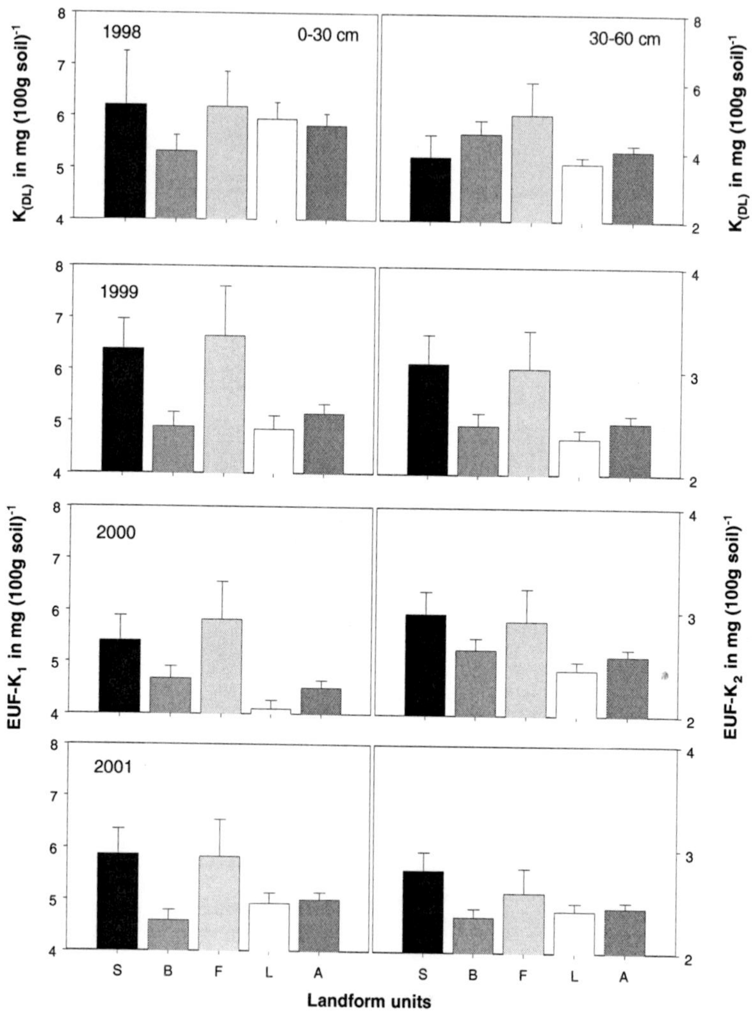

Figure 78. $K_{(DL)}$ in 0-30 cm and 30-60 cm (top row) and EUF-K-fractions (K_1 in the left column and K_2 in the right column) for the years 1999, 2000, and 2001 at the field site "Bei Lotte", separated for different landform units. Shoulder (S), Backslope (B), Footslope (F), Level (L) and field average (A). Bars show the standard error.

soil)$^{-1}$). For the 30-60 cm depth, an increasing trend in values was observed from SH (3.86 mg K (100g soil)$^{-1}$) over BS (4.94 mg K (100g soil)$^{-1}$) and FS (5.1 mg K (100g soil)$^{-1}$) positions, with LE positions showing values lower than all other LF (3.69 mg K (100g soil)$^{-1}$).

Conclusions drawn regarding sampling under uniform management were similar to phosphorus at the field site “Bei Lotte”. To locate areas with potential potassium deficiency, it would be advisable to sample BS positions, whereas to locate areas of over fertilization the landform units shoulder and footslope are the most appropriate landforms to sample.

Schmidt (1984) found, that plant nutrient contents on footslopes showed higher variability in areas of the Pleistocene lowlands, whereas BS and SH showed the least variability. Reasons given for these differences were processes like erosional deposition and air deficiency, which may increase variability (Schmidt, 1984). CV of EUF-nutrient sampling and $P_{(DL)}$ at the field site “Bei Lotte” were the largest at FS positions, results for $K_{(DL)}$ were largest at SH positions.

8.3.4. Soil texture, soil organic matter and pH versus profile curvature and landform elements

The distribution of soil texture at the field site “Bei Lotte” versus profile curvature is shown in Figure 81. The highest sand contents (7 $g(100g)^{-1}$ in 0-30 cm, 8 $g(100g)^{-1}$ in 30-60 cm, 9 $g(100g)^{-1}$ in 60-90 cm) as well as the lowest silt contents (76 $g(100g)^{-1}$, 75 $g(100g)^{-1}$, 74 $g(100g)^{-1}$) were found at the strong concave locations (5) as compared to all other LF. Besides these large differences, an increasing trend for convex, convex planar and concave planar profile curvature was observed in Figure 81 for sand in 0-30 cm (2.8 $g(100g)^{-1}$ - 3.2 $g(100g)^{-1}$ - 3.6 $g(100g)^{-1}$) and in 30-60 cm (2.3 $g(100g)^{-1}$ - 2.9 $g(100g)^{-1}$ - 3.3 $g(100g)^{-1}$) as well as for silt in all three depths (see Table 56 in Appendix Landform). The decrease in clay content at all three depth layers are therefore not surprising. Concave areas (4) showed values similar to planar convex (2) for silt and clay in 0-30 cm and 30-60 cm layers. The decrease in silt content and clay content agrees with the observations by Young and Hammer (2000), who showed that “silt“ content (clay was included in that value) in planar areas were 94.3 $g(100g)^{-1}$ and in concave areas 93 $g(100g)^{-1}$. Clay content varied from 28.5 $g(100g)^{-1}$ (planar) to 25.3 $g(100g)^{-1}$ (concave).

SOC content in Figure 80 increased in very strong concave positions (5) which contained up to 0.1% more SOC than the field average in the 0-30 cm, 0.43% in 30-60 cm and 0.4% in 60-90 cm depths. Apart from that, planar areas (2, 3 in Figure 80) contained slightly elevated values of SOC compared to convex (1) and concave areas (4),i.e, only 0.05% more in the 0-30 cm depth , 0.08% more in the 30-60 cm depth and 0.06% in the 60-90 cm depth. Differences appear to be very small. However, as discussed for nutrient distribution at different landforms, nutrient dynamics might be based strongly on these differences (see Stoot et al., 1989, Blake et al.,2000).

Soil pH values show a similar distribution as did SOC with the lowest pH values observed at divergent positions (0.22 less than field average for the 0-30 cm depth) and the largest pH at strong concave areas (0.3 more than field average for the 0-30 cm). A slight trend was observed between divergent and strong concave positions (Figure 80). Positions were resampled in the autumn of 2000 at fewer locations (n=15), and no relationship to the patterns in 1998 was observed (Figure 80). The increase in pH values with convergent landscape position supports observations by Brubaker et al. (1993), who additionally observed a marked

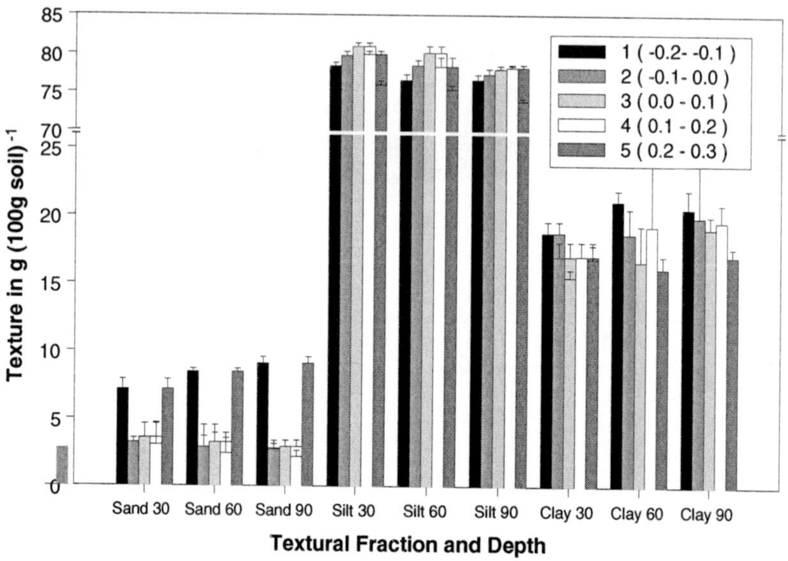

Figure 79. Texture for 0-30 cm depth (30), 30-60 cm depth (60) and 60-90 cm depth (90) for 64 sampling points at the field site "Bei Lotte", separated for profile curvature classes. Profile curvature classes: -0.2- -0.1 as 1, -0.1-0 as 2, 0-0.1 as 3, 0.1-0.2 as 4 and Fl as field average. Bars show the standard error.

drop in soil pH values at FS positions. Such a marked drop was not observed in the 0-30 cm soil layer, however it was found in the 30-60 cm depth.

Textural analysis of the landforms presented in Figure 81 showed that the FS positions contained approximately 2 $g(100g)^{-1}$ more sand for all depth layers compared to all other LF. This 2 $g(100g)^{-1}$ increase corresponded to a decrease of 2 $g(100g)^{-1}$ in silt content found at FS positions. Similar results were obtained by Young and Hammer (2000).

SH positions showed similar silt contents to that of FS positions. No increase in sand content for SH position was observed as shown in Figure 81. Clay content was found to be elevated compared to the field average by up to 1.64 $g(100g\ soil)^{-1}$ in the 0-30 cm, 1.77 $g(100g\ soil)^{-1}$ in 30-60 cm and 0.37 $g(100g\ soil)^{-1}$ in 60-90 cm depths. The highest silt contents were observed at BS and Level positions. Young and Hammer (2000) showed that footslope positions contained more clay and less silt than backslope positions in horizons below 30 cm. The same was observed during the investigation at the field site "Bei Lotte" for the 0 - 30 cm and the 30 - 60 cm depths, however not for deeper soil layers from 60-90 cm.

Different landforms showed different clay distributions with depth (Figure 81 and Table 55 in Appendix Landform). At SH and BS positions similar clay content differences between the 0-30 cm and the 30-60 cm (approximately 1 $g(100g\ soil)^{-1}$) were observed, whereas at FS and

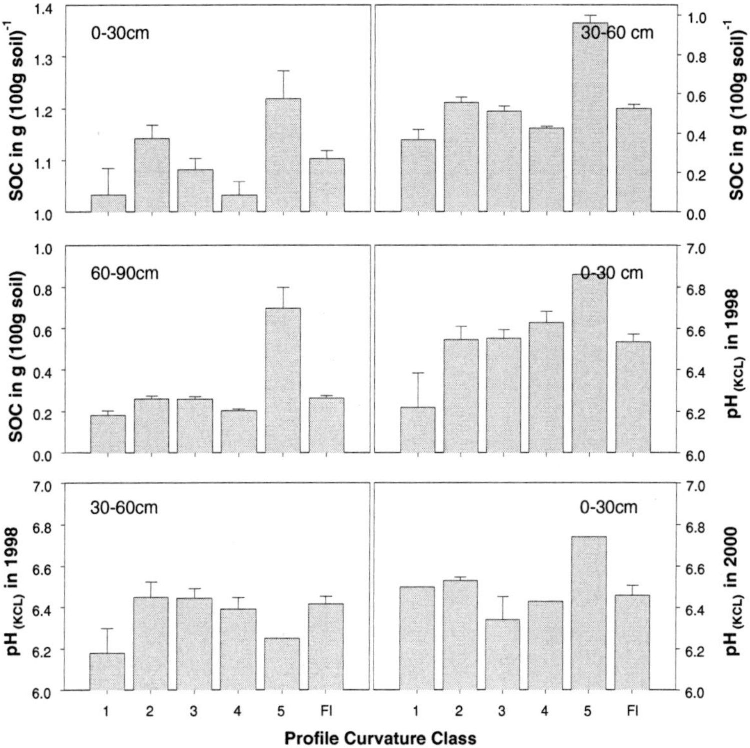

Figure 80. SOC in 1998, pH in 1998 (98) and 2000 (00) for 0-30 cm depth (30), 30-60 cm depth (60) and 60-90 cm depth (90) for 64 sampling points at the field site "Bei Lotte", separated for different profile curvature classes. (N=15 for pH in 2000). Profile curvature classes: -0.2- -0.1 as 1, -0.1-0 as 2, 0-0.1 as 3, 0.1-0.2 as 4 and FI as field average. Bars show the standard error.

Level positions clay content reached approximately 1.9 g(100g soil)$^{-1}$. However, the differences between the 30 - 60 cm and the 60 - 90 cm depth at BS positions showed a high difference of 2.4 g(100g soil)$^{-1}$, whereas SH (0.2), FS (0.4 and Level (0.9) showed quite lower values. An explanation for the higher clay content in the 60-90 cm soil layer would be lessivage. The question arises: what difference in environmental properties led to the increase in clay content only at the BS position. Scheffer and Schachtschabel (1992) state that for lessivage to occur a quick leachate is required. BS-positions provided elevated water content (indicating moisture transport processes in Figure 70 and Figure 71) and lower SOC contents (important for aggregate stability - Figure 82), therefore supporting this hypothesis. Clay content distribution versus landform were different to those observed by Morgenstern (1986) who found higher clay content at FS and lowlands positions on the Pleistocene lowlands.

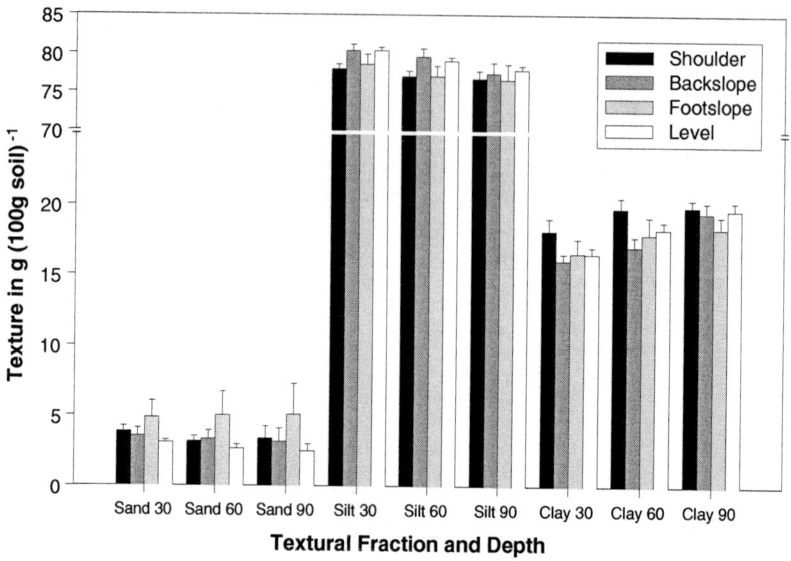

Figure 81. Texture for 0-30 cm depth (30), 30-60 cm depth (60) and 60-90 cm depth (90) for 64 sampling points at the field site "Bei Lotte", separated for different landform units. Shoulder (SH), Backslope (BS), Footslope (FS), and Level (LEVEL). Bars show the standard error.

Soil distribution processes led to different textural properties as shown above. In this context SOC was found to be largest at FS position in the 30 - 60 cm and the 60 - 90 cm depth (similar to results found by Morgenstern (1986)). The maximum SOC content was found in the 0 - 30 cm depth for all LFs representing similar values close to the field average (1.1%). Results are in contrast to Pennock and Corre (2001), who observed a decrease in SOC at SH positions either due to decreased organic matter input or higher mineralization. As Pennock and Corre (2001) also found horizontal truncation, the authors argue for physical removal and deposition due to tillage translocation. Such even distribution of SOC in the upper soil layer at the field site "Bei Lotte" would be an indicator of good practice under the no tillage system utilised at this site since 1992.

At SH positions a decrease could be observed in deeper soil layers (0.02% for 30-60 cm, 0.05% for 60-90 cm layer) in comparison to the field average. Generally, SOC was found to be homogeneously distributed except for SOC found at BS positions in 0-30 cm, SH in 60-90 cm and FS positions in all depth layers. The decrease in SOC at BS positions (see Figure 82) agrees partly with the results found by Young and Hammer (2000), who observed higher SOC in 0-15 cm depth, but less in 15-30 cm depth as compared to other LF positions. Additionally, only the deepest soil layer - the 60-90 cm depth showed an increase in SOC with the landform units shoulder - backslope - footslope as shown by Manning et al. (2001b).

Soil pH at SH positions was found lower than the field average for the 0-30 cm depth (6.53 versus 6.19) and 30-60 cm depth (6.42 versus 6.25) (Figure 82). For the 0-30 cm depth, an increasing trend in soil pH for the landform units SH-BS-FS was observed (6.19 - 6.58 - 6.72 in 0-30 cm depth); this trend was also distinct in the pH values obtained in 2000 (Figure 82). A similar increase in soil pH (from 6.2 - 6.3 - 6.6) with SH-BS-FS landforms was found by Manning et al. (2001b), who suggested that the higher pH values in the convergent LFs were related to (I) cultivation or (II) hydrological reversal (footslope landforms receive bases via capillary rise from a shallower, less dynamic water table). The deeper soil layer of 30-60 cm in 1998 does not show this trend. However, Young and Hammer (2000) showed a decrease soil pH for BS positions compared to other LFs up to 75 cm depth. In Figure 82 just the opposite is shown, which is probably related to the different amounts of SOC.

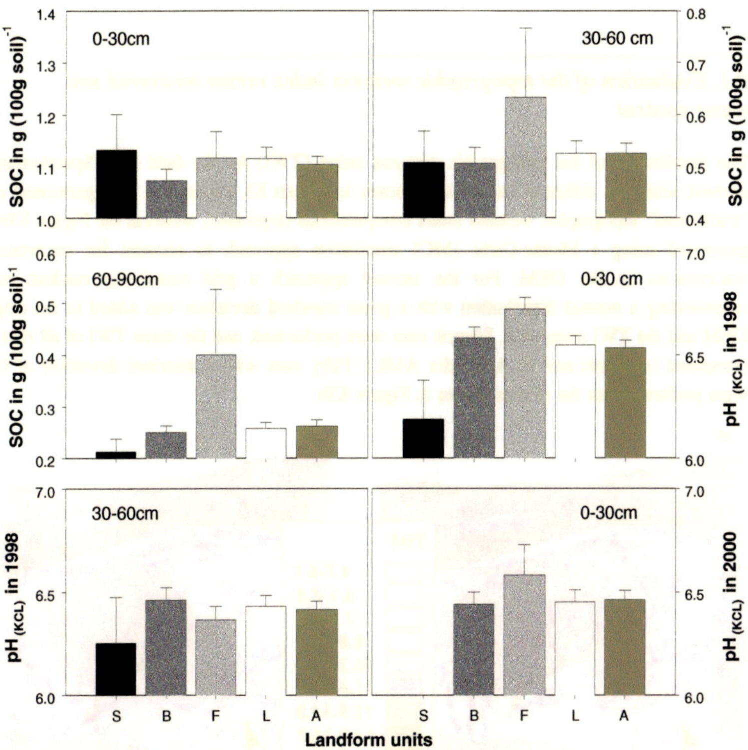

Figure 82. SOC in 1998, pH in 1998 (98) and 2000 (00) for 0-30 cm depth, 30-60 cm depth and 60-90 cm depth for 64 sampling points at the field site "Bei Lotte", separated for different landform units. (N=15 for pH in 2000). Shoulder (S), Backslope (B), Footslope (F), Level (L) and field average (A). Bars show the standard error.

CHAPTER 9

Results and Discussion for interactions between grain yield, soil components and relief properties

9.1. Evaluation of the topographic wetness index versus measured soil water content

The distribution of the topographic wetness index (TWI) for the field site "Sportkomplex", derived with two different methods is shown in Figure 83. Figure 83a was generated using "traditional" topographic wetness index computations (topo.aml), whereas the Figure 83b was generated using a Monte-Carlo (MC) simulation approach to account for uncertainties/ inaccuracies in the DEM. For the second approach a grid containing random values representing a normal distribution with a given standard deviation was added to the original DEM and the TWI computed. Several runs were performed, and the mean TWI of all runs was computed (montewi.aml in Appendix AML). Fifty runs with a standard deviation of 0.1 m were performed for the results shown in Figure 83b.

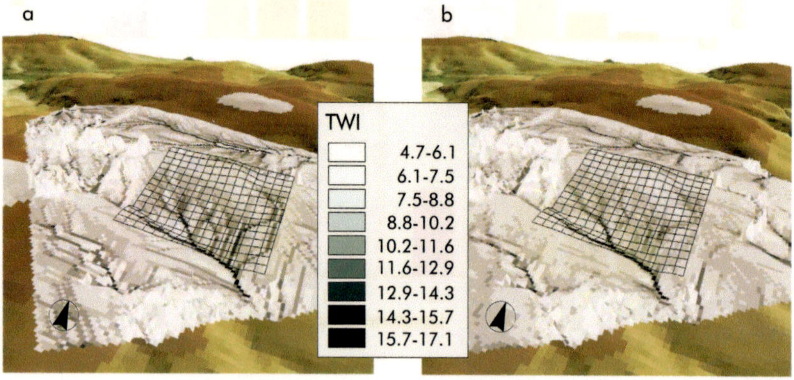

Figure 83. Three dimensional view of the Topographic Wetness Index (a) and the Topographic Wetness Index with Monte-Carlo-methods (b) at the field site "Sportkomplex" for a DEM resolution of 10 m by10 m.

A first visual assessment shows that local drainage pathways known from field experience, which are located 0.1 m below their surrounding, are preserved in the computed MC-TWI. Secondly, certain linear features (also known as zig-zag-flow) as seen the area south of the investiagtion raster in Figure 83a disappeared and are replaced by a smoother distribution with the MC-approach. Flow accumulation occurs with the MC-TWI-approach over several grid cells, not only along single cells. This results in a decrease in mean and standard deviation for MC-TWI, in combination with a smaller range.

Soil surface moisture (SSM) was classified for the field site "Sportkomplex", assigning a one if soil water content is above and a zero if below field average soil water content (Wendroth et al., 2001). This procedure was applied to 225 samples taken at two dates in 1998 and at two dates in 1999 for samples of the field site „Sportkomplex". The sum for the four dates of soil surface moisture was computed as an index and compared against Topographic Wetness Indices (Figure 84). If a position contained a zero, SSM - values were below field average at all four dates, if a three was computed, three of four dates showed SSM - values above field average and so on.

Relationships show an increase in soil water content index with increasing wetness index. A decrease in standard error of soil water content index could be observed by applying a MC - TWI computations, indicating a better representation of the soil water content distribution. Differences in standard deviation and mean of MC-TWI will decrease if a larger number of runs (n=50 in this case) and a different standard deviation (0.1 in this case) are provided as input parameter.

Two different aggregation methods were tested against soil surface moisture measurements at four different dates for the field sites "Sportkomplex" and "Bei Lotte" (Figure 84). Raster data from the 10 m by 10 m TWI-computation were then averaged to a 27 m by 27 m raster (signed 'raster' in Table 32) or the TWI value at the location of the SSM sampling location was determined directly from the underlying dataset (signed 'point' in Table 32).

For the field "Sportkomplex" MC-methods aggregated for a 27 m x 27 m raster provided largest correlation coefficients in three of four dates compared to all other methods. For the fourth date, the point – MC - TWI - data showed the highest correlation. On the other hand, TWI and SSM-results for the field site "Bei Lotte" did not show large differences. For two of four dates, results by MC-TWI show similar correlation coefficients than traditional TWI-computations. The other two dates show a slightly higher (up to 0.05) correlation of the MC-TWI. Based on the results from both field sites the conclusion can be drawn, that MC-TWI computations averaged to a 27 m x 27 m raster showed the largest correlations to measured soil water content conditions in 0-10 cm depth.

Table 32. Correlation coefficients between soil surface moisture at 0-10 cm (SSM) and Topographic Wetness Index (TWI) at four dates and two aggregation methods (point/raster) for the field sites "Sportkomplex" and "Bei Lotte". MC indicates that the results have been obtained using a Monte -Carlo - Simulation approach.

"Sportkomplex" Type / Date	02.04.1998	04.05.1998	25.03.1999	27.04.1999
TWI raster	0.35	0.22	0.40	0.37
TWI-MC raster	0.40	0.32	0.43	0.34
TWI point	0.29	0.24	0.33	0.31
TWI-MC point	0.34	0.29	0.42	0.40
"Bei Lotte" Type / Date	**25.03.1999**	**29.04.1999**	**10.05.1999**	**12.05.2000**
TWI raster	0.43	0.26	0.37	0.16
TWI-MC raster	0.45	0.26	0.37	0.23
TWI point	0.37	0.23	0.28	0.12
TWI-MC point	0.45	0.23	0.30	0.27

Differences between aggregation methods were observed based on the analysis shown above (Table 32). Therefore, the influence of different resolutions (1 m-50 m) and different quality (LS vs. TK 1:10.000) on the correlation coefficients was studied for the field site "Sportkomplex" (see Figure 85). The TK-DEM of 1 m by 1 m resolution was generated using

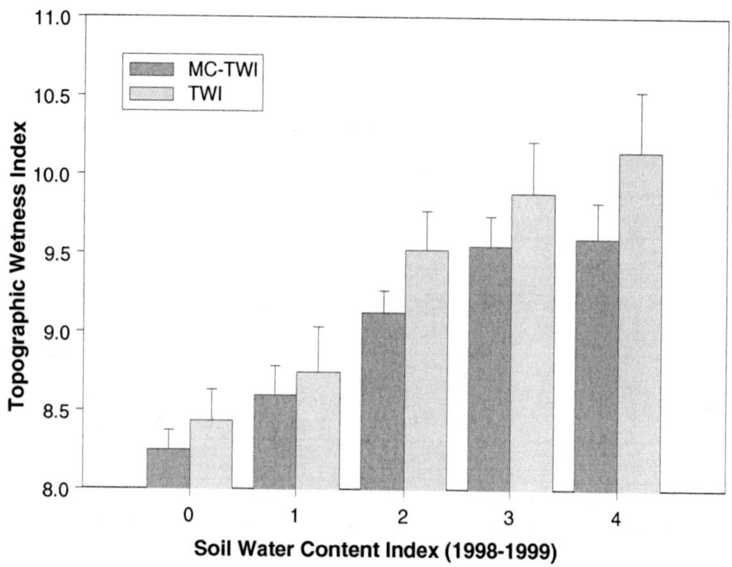

Figure 84. Comparison of Soil Water Content Index versus Topographic Wetness Index (TWI) and Monte-Carlo-(MC) Simulated TWI. Note the decreased standard errors for all five soil water content classes for the MC-TWI. Bars show the standard error.

the ArcInfo topogrid command. DEMs with different resolutions were generated using resampling of 1 m by 1 m DEM.

TK-based TWI – SSM correlation coefficients showed a slow increase up to a resolution of 25-30 m and a decrease afterwards (see Figure 85d). The highest correlation coefficients were observed at that resolution for the MC-TK-TWI, however results showed similar values over a larger resolution range of 5 – 30 m. Additionally a maximum can be observed at a resolution of 8 m. TWI-computations for LS datasets showed highest correlations with resolutions between 12 – 18 m for LS-TWI and LS-MC-TWI and again at a resolution of 30 m. These results indicate that two DEM-resolutions exist for TWI computations, which provide large correlation coefficients. The first represents a resolution of 30 m and can be found in LS and TK datasets. The second one shows a smaller resolution of 12 m for the much more precise LS dataset as well as a resolution of 8 m for the results of the MC-TK. The TWI is therefore able to represent SSM distributions, and vice versa SSM-measurements would also be possibly a procedure to determine topographic features.

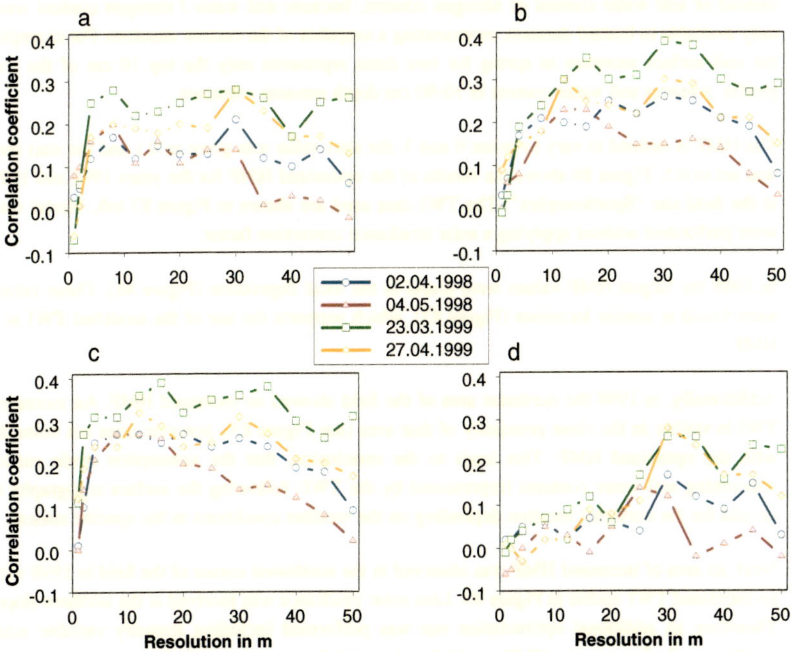

Figure 85. Correlation coefficients between topographic map (TK) based Monte-Carlo (MC) Topographic Wetness Index (TWI) (a); LaserScan (LS) based TWI (b), LS-MC-TWI (c) and TK-TWI (d) and four dates of soil surface moisture measurements. At the x-axis the respective DEM resolution for the LS and the TK 1:10.000 dataset is shown.

9.2. Evaluation of topographic wetness index as additional model parameter

A transformed TWI is used as a hydromorphic modification factor (HMF) for the field capacity parameter in the HERMES model (Kersebaum et al., 2002). A direct evaluation compares model runs with and without the HMF obtained from relief analysis with respect to the observed grain yield as done by Kersebaum et al. (2002). However, in the case presented, an indirect approach was chosen. The TWI based HMF (see 5.6 on page 37) was compared against an optimized HMF, which has been obtained using a nonlinear optimization technique. The HMF is a correction factor to the soil water capacity; therefore it has direct influence on air deficiency as well as moisture deficiency related processes.

The parameter optimization program PEST (Doherty, 2002) was used together with HERMES to minimize the difference between measured and simulated grain yield by searching for an optimum of HMF. Grain yield as an integrative parameter was chosen for an optimization instead of soil water content or nitrogen content, because soil water / nitrogen content were only available in limited datasets, representing a snapshot of the current situation. For example, the soil surface moisture in spring for two dates represents only the top 10 cm of the soil profile, whereas soil water content in 10-90 cm depth remains unknown.

The HMF is defined to vary between 0 and 3, the start value was given as 0.1 and the start step was set to 0.5. Figure 86 shows the results of the optimized HMF for the years 1998 and 1999 at the field site "Sportkomplex". The TWI data used are shown in Figure 83 left. Model runs were performed without applying a solar irradiance correction factor.

In 1998 the largest HMF-values were found in the local depression (Figure 86). These values were found at similar locations (Figure 83), which supports the use of the modified TWI as a HMF.

Additionally, in 1998 the northeast area of the field showed an increased HMF. An increased TWI is visible in the close proximity of that area (see Figure 83), however does not coincide with the optimized HMF. This leads to the conclusion, that the assumption made earlier concerning soil water contents (represented by the TWI) following the surface topography is invalid for the northeast location depending on the weather conditions in the specific season.

Next, an area of increased HMF was observed in the southwest corner of the field in 1998 with no increased TWI visible in Figure 83. Less solar irradiance was received at the northern slope. Therefore an additional optimization run was performed including spatially variable solar irradiance. A decrease in HMF could be observed between an optimization run including spatially solar irradiance (SR) and a model run using constant SR for four simulated positions in that area. The observed changes of optimized HMF are rather small compared to the larger differences in SR-input data (Results not shown).

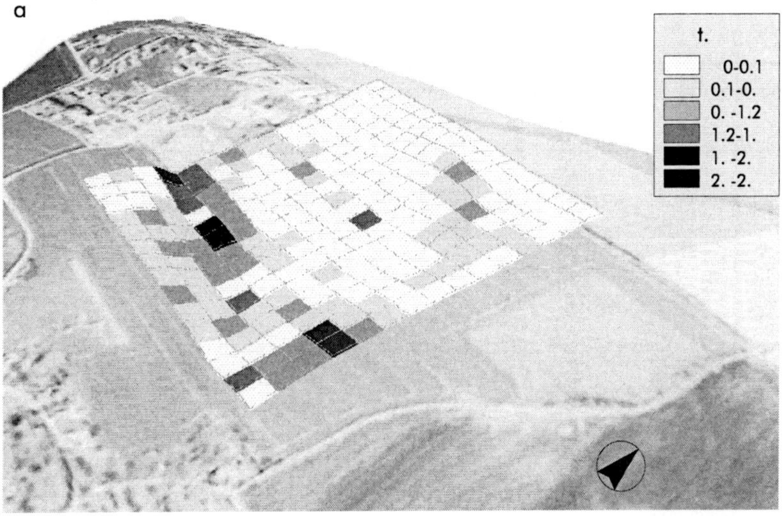

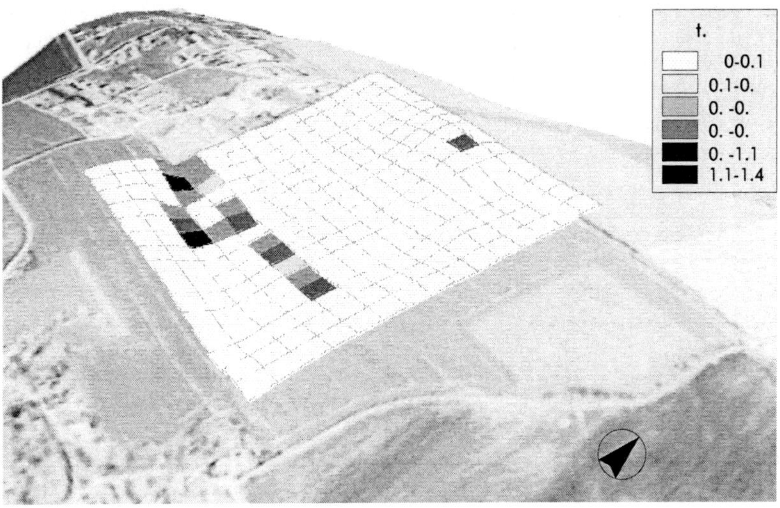

Figure 86. Optimized Hydromorphic Modification Factor (HMF) in respect to measured dry matter using PEST and HERMES for the field site Sportkomplex in 1998 (a) and 1999 (b). Model runs performed without spatially varying solar radiation.

The south-east part of the field in 1998 is another area studied. Again, no increase in TWI could be observed (Figure 83). The reasons for the increased HMF might be due to the former landuse. In the German soil appraisal (approximately 1935) that area of the field was mapped as a meadow. Additionally, higher soil penetration resistances in 30-60cm depth were observed in comparison to the rest of the field site "Sportkomplex".

For the year 1999 again the largest HMF values were found in the local depression, which validates the use of the TWI as a HMF (Figure 86). However, two additional locations showed an elevated HMF. The first is located at the middle top position of the field, closely to those areas found in 1998. The geological map shows a larger extended area at that location with a shallower loess thickness, than observed by optimizing HMF, which might have led to a decrease in yield.

The second location is a line running east-west in the southeast corner of the field showing neither elevated TWI values visible in Figure 83 nor optimized HMF values in 1998 in Figure 86. Other data sources (for example CIR-air photos) do not give any indication for this pattern. Unfortunately, no hand harvest was performed in 1998 and 1999, therefore the possibility of a combine harvester error exists. An indicator for such an error is the east west direction of the increased HMF-values, which coincides with the direction of combine harvest and the direction of tillage.

Generally, the use of TWI as a hydromorphic correction factor was found to be similar to the distribution of the obtained HMF-values obtained using nonlinear optimization techniques for depressions. Therefore, the use of the TWI as a hydromorphic correction factor can be confirmed. Unexpectedly, other areas with additional factors like geology influencing crop yield development could be identified. Therefore, a major advantage of using this technique is that areas can be identified where additional observations of soil and plant properties should be performed.

9.3. Evaluation of simulated solar irradiance against measured solar irradiance

Results for modelled solar irradiance (SR) for the location of REF1 compared against measured SR showed a close agreement between measured and simulated values with an R^2 of 0.9654. Simulated irradiance is overestimated at three months with low SR-values (Figure 87, lower left corner), either due to (I) errors in the parameter estimation for the SRAD model, (II) errors in measuring solar irradiance, or (III) errors in the simulation of SR. These months lead to the general overestimation of simulated SR at low values compared to measured values as observed by the regression equation with $SR_{simulated} = 0.9105\ SR_{measured} + 126.4$. The overestimation at low values as well as the underestimation at large values is consistent with observations by Wang et al. (2002).

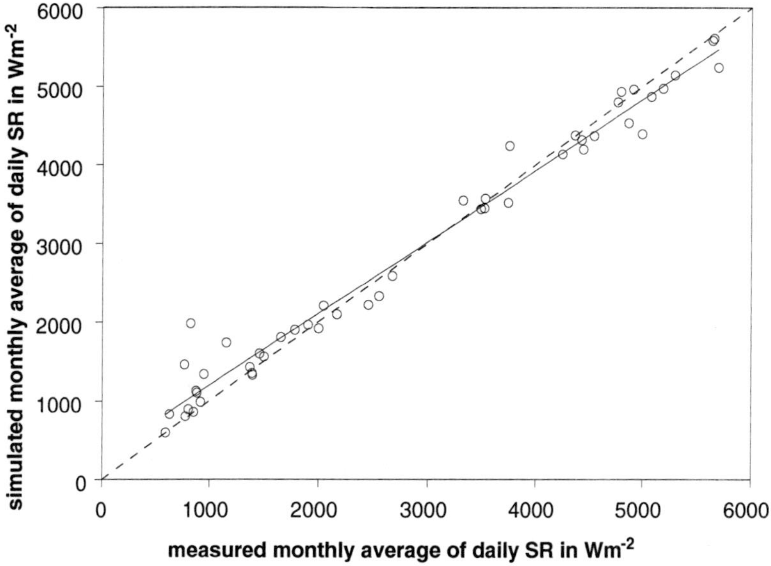

Figure 87. Measured monthly average of daily solar irradiance (SR) at REF1 compared against simulated average monthly daily irradiance at sloping surface. Solid line represents the least squares fit, the dashed line the 1:1 line between measured and simulated solar irradiance.

Solar irradiance measured at locations WS1 and WS2 for June 12, 2002 are shown in Figure 88 for 10 min intervals. During the first hours clear sky conditions are apparently indicated by the straight increase in SR, followed by some cloudy conditions for the time period between 8:00 am and 11:30 am. Both weather stations show similar reactions in measured SR. Later, clear sky conditions were observed. At both locations no significant differences in SR can be observed until 5:00 pm, and would not be expected based on the DEM as differences due to topographic shading would be minimal between WS1 and WS2. However, for the time period between 5:00 pm and 7:00 pm the influence of topographic shading on the amount of direct SR is clearly visible with differences up to 150 Wm^{-2} between WS1 and WS2 (Figure 88).

The monthly average of daily SISSR influenced by topography for the month of July and November 2002 is shown in Figure 89. Northern exposed areas in the left part of the Figure 89 receive less solar irradiance, whereas southern exposed positions receive larger amounts of solar irradiance indicated by darker colours. Elevation differences of up to 8 m between the middle northern (Location of WS1) and north-eastern part of the field site lead to slight differences in the July calculation results (Figure 89a), however appear even more pronounced for November (Figure 89b).

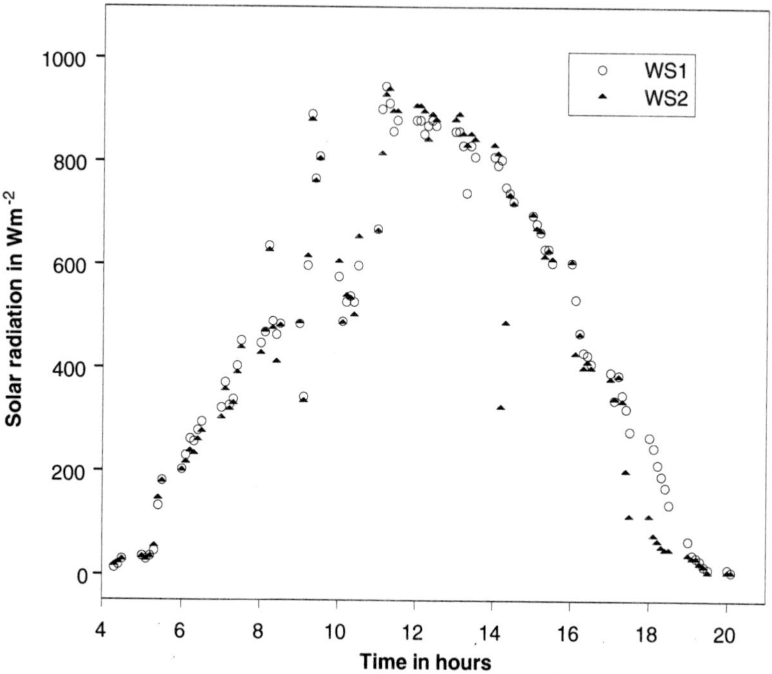

Figure 88. Solar irradiance measured at WS1 and WS2 on June, 12. 2003. Note the shading of WS2 after 5:00 pm with around 150 Wm^{-2} less solar irradiance.

The evaluation results of the accuracy of SISSR at the field scale for temporal and spatial distribution is shown in Table 33. Differences between WS1 and WS2 are well preserved for the summer period between measured and simulated values and similar to differences found in Geiger et al. (1961). However, simulated values in November show differences of up to 0.5 $MJm^{-2}d^{-1}$ between locations WS1 and WS2, which were not measured. Additional computations of the duration of direct solar irradiance using the solarflux.aml showed differences of up to 100 minutes for DOY 324 (Table 33). Therefore the measured results from the winter period are questionable. Reasons may be attributable to (I) the horizontal plane of measurements of the weather stations compared to the SISSR for sloping surface; (II) differences in sensor performance, which overcast the results due to the much lower SR values in winter (1/4 of SR observed in summer); (III) false parameterisation of the SRAD model; or (IV) due to some dust on top of the pyranometer.

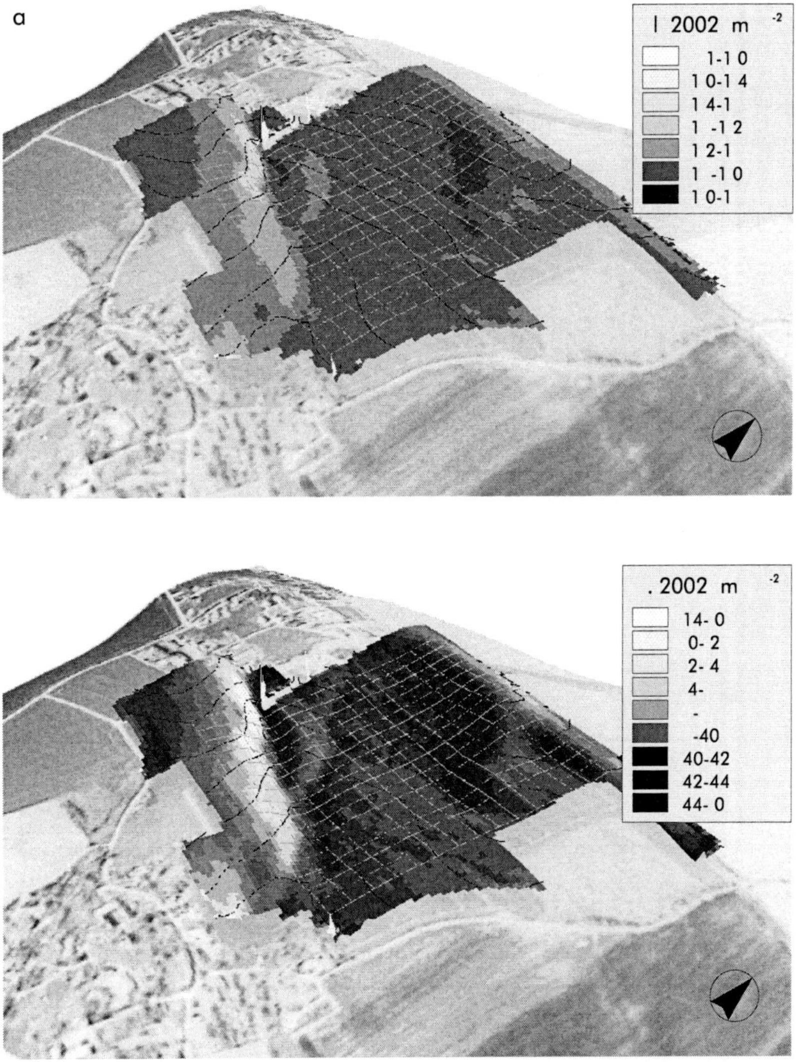

Figure 89. Simulated monthly average daily shortwave irradiance at sloping surface for July (a) and November (b) 2002 in Wm^{-2}. The grey lines represent the sampling grid, black lines the contours of the DEM. The backdrop is a colour infrared image of May, 30 1999.

Table 33. Average daily irradiances for a summer and a winter period. Values are given in MJ m^{-2} day^{-1}

Summer	**WS1**	**WS2**
Measured Solar Iradiance(11-22.07.2002)	13.809	13.238
Simulated Sloped Solar Irradiance (July 2002)	14.684	14.206
Simulated Day Length in h (DOY197)	15.60	14.92
Winter	**WS1**	**WS2**
Measured Solar Irradiance(9.-20.11.2002)	2.789	2.785
Simulated Sloped Solar Irradiance (November 2002)	3.101	2.607
Simulated Day Length in h (DOY 324)	7.99	6.41

The SRAD model was slightly altered to work with a high-resolution dataset and showed reliable SISSR-results in time and space. The evaluations performed for the field site „Sportkomplex“ add significantly to the scientific knowledge as results for SISSR and topographic shading are validated. Up to now a validation of the SRAD model was to the best of our knowledge, only published for the values of a single weather station by Moore et al. (1993b), McKenney et al. (1999) and Kumar and Skidmore (2000).

9.4. Evaluation of crop growth modelling including simulated solar irradiance influenced by topography

Solar irradiance is one parameter used in crop growth models, being a driving force for biomass production. Filzer (1939) observed lower dry grain yield with less light intensity. Hughes (1959) showed that duration and amount of grain development depended on light duration and intensity. However, little attention has been spent on how the spatial differentiation of solar irradiance can alter crop production in agricultural fields. Until now, a constant solar irradiance is assumed across a field site, even if under hummocky conditions the terrain affects the amount of incoming solar irradiance. The effect of shading during different stages of crop development is rather complex. This was already found quite early by Schoder (1932), who investigated the parameters influencing assimilation. Therefore, solar irradiance cannot simply be related to crop yield patterns, e.g., by empirical considerations. Instead modelling can help achieve a better understanding of these complex processes.

Results for the “HERMES” model runs including spatially varying solar irradiance for the years 1998 and 1999 are shown in Table 34 and the RSD-index in Figure 90 for the field site “Sportkomplex”. The HMF factor was held constant for these model runs. Average simulated grain yield show small differences between model runs and measured grain yield up to 0.5 tha^{-1}. Variances observed for the measured values (0.55 tha^{-1} in 1998 and 0.53 tha^{-1} in 1999) were not sufficiently represented by all four model runs. Slightly larger variances resulted from the R-model runs. Generally, the small span of simulated grain yield compared to measurements is disappointing in face of detailed model input provided. Lastly, the $Ln(\delta^2)$ criterion showed small improvements in model performance for the year 1998 and 1999 by using spatially variable solar irradiance.

Table 34. Model evaluation for model runs assuming spatially constant solar irradiance (P) and solar irradiance influenced by topography (R) and observed grain yield (MES) for the years 1998 and 1999 at the field site "Sportkomplex". Statistical parameters of simulated and measured dry matter are given in tha^{-1}.

DATASET	1998 R	1998 P	1998 MES	1999 R	1999 P	1999 MES
Mean	5.12	5.14	4.61	5.79	5.80	6.38
STE	0.03	0.03	0.05	0.04	0.04	0.05
Variance	0.24	0.24	0.55	0.31	0.32	0.53
Skewness	-6.12	-6.11	-0.82	-6.73	-6.71	0.05
Kurtosis	45.62	45.57	1.57	57.71	57.50	-0.25
Range	4.71	4.62	4.53	5.78	5.73	3.97
$Ln(\delta^2)$	-2.15	-2.13		-2.28	-2.27	

The spatial distribution of the model results and the measured yield is not shown in detail. Instead the RSD-index is used to evaluate model performance. If the RSD-index shown in Figure 90 is above zero, model runs including spatially variable solar radiation (R-run) perform better than model runs with spatially constant solar radiation (P-runs), each compared against the measured yield. In total, 121 of the 225 simulated positions for R-runs show an improvement in 1998 and 118 cases in 1999. At all other positions (104 in 1998; 107 in 1999) R-runs performed less sufficient than P-runs.

Generally, northern exposed slopes in the southwest area as well as in the northwest part of the field showed an improvement in 1998 and 1999 for the R-run. These areas of model improvements have been expected, as northern and southern exposed slopes are known to show different growth development (Beresneva and Popova, 1983, Wessolek et al., 1992).

However, certain areas exist where differences between measured and simulated grain yield development could not be explained by incorporation of spatially variable SR in the HERMES-model. First, the southeast area, which was a meadow approximately 40 years ago and has higher soil penetration resistances at 30 - 60 cm depth than the rest of the field, and secondly at locations related to the flow line at the valley bottom.

The pattern observed in 1999 is not as clear as in 1998. Possible reasons might be differences in yield development between spring barley in 1999 and winter rye in 1999 either (I) due to differences in radiation use efficiency (see Gallagher and Biscoe, 1978, Kiniry et al., 1989) or (II) due to differences in soil nitrogen supply (Weerakoon et al., 2000) as spring barley grown for malting in 1998 received significant less fertilizer than the winter rye grown in 1999.

Generally, the spatially structured areas with negative RSD values suggest that certain processes exists, which influence yield development at that field. These processes were (I) either not represented in the model, (II) not precisely measured, or (III) not precisely incorporated into the model. For an indepth discussion of these questions see the paper by van Ittersum and Donatelli (2003).

a

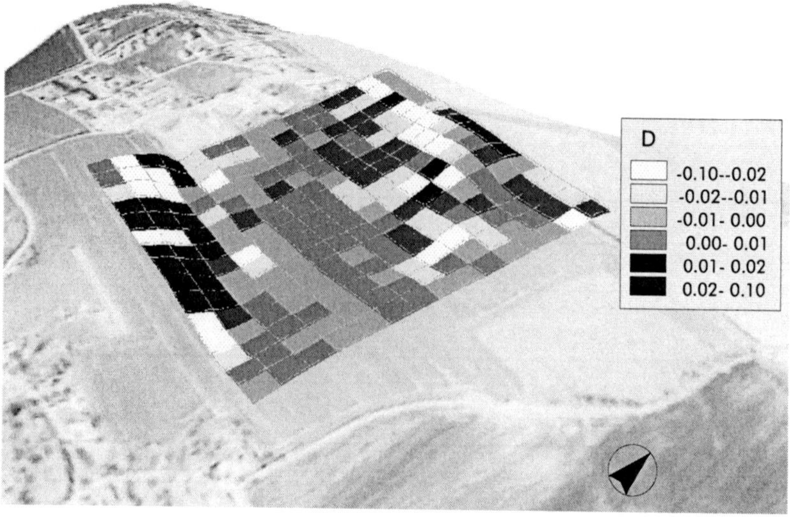

Figure 90. Evaluation of model runs using the RSD-Index for spring barley in 1998 (a) and winter rye in 1999 (b) at the field site "Sportkomplex".

The above ground biomass development at a daily time step was simulated for the years 1998 and 1999 at the locations of WS1 and WS2 and is shown in Figure 91 and Figure 91. R- and P-model runs show differences in crop biomass early in the development with little differences until June 12, 1998 for the spring barley, and increasing differences until harvest (up to 420 kg ha^{-1}). Differentiation between R- and P-model runs occurred earlier in April for the winter rye in 1999, and reached differences of up to 1000 kg ha^{-1} at harvest. The absolute differences in simulated above ground biomass may appear to be small, however are a first step on the understanding of differences of crop yield development influenced by differences in solar irradiance due to effects of topography.

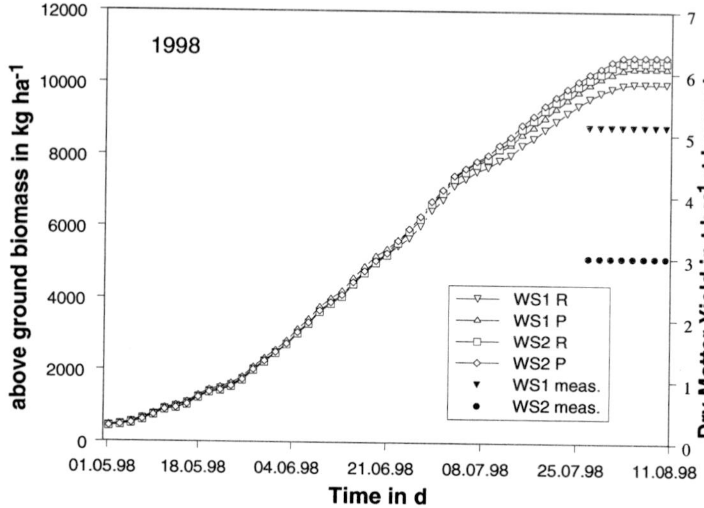

Figure 91. Example simulation of above ground biomass development for locations WS1 and WS2 in 1998 (spring barley) for model runs assuming spatially constant solar irradiance (P) and solar irradiance influenced by topography (R). Additionally, the dry matter yield obtained at location WS1 and WS2 is shown.

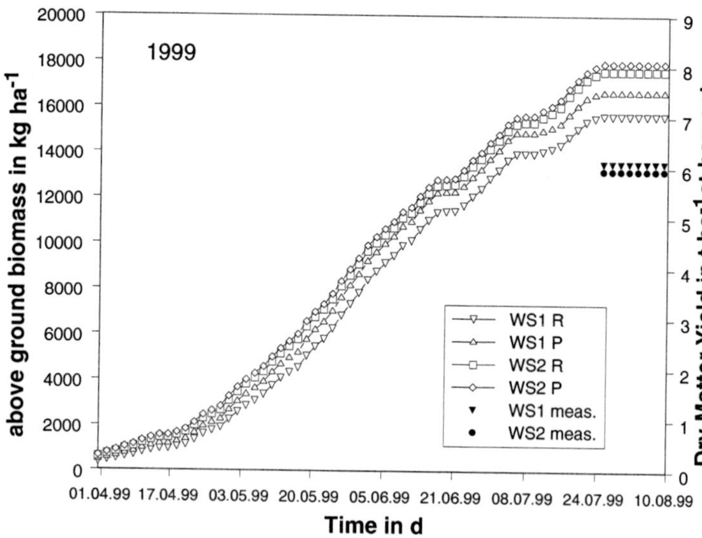

Figure 91. Example simulation of above ground biomass development for locations WS1 and WS2 in 1999 (winter rye) for model runs assuming spatially constant solar irradiance (P) and solar irradiance influenced by topographic shading (R). Additionally, the dry matter yield obtained at location WS1 and WS2 is shown.

The field “Sportkomplex” evaluated above contains different expositions as seen in Figure 89. In contrast the field “Bei Lotte” represents mainly north exposed positions (see Figure 14). The question arises, if spatially variable SR (R-model runs) is an improvement under these conditions. Results for the measured grain yield and the simulated grain yield including spatially variable solar irradiance (R-runs) and using constant solar irradiance (P-runs) are presented in Table 35 and evaluated using the RSD-Index in Figure 92.

The Ln(δ^2) criterion showed a decrease in model performance for the 1999 R-model run, whereas in 2000 and 2001 a very slight increase is visible, which might be based on the slightly extended standard deviation of the R-model runs.

The RSD-Index for the field ”Bei Lotte” for the time period 1999-2001 is shown in Figure 92. In 1999 the RSD-Index showed a model improvement at 59% of all cells, in 2000 at 52% and in 2001 only at 30% of all simulated 192 positions. All three years showed a model improvement for the depression in the northwest corner of the field (see Number 1 in Figure 92), which can be considered a positive result.

From the distribution of SISSR a positive RSD-index would have been expected at the south end of the field (shaded by topography and trees), which was not observed. Therefore, some other parameter must influence yield development in that area.

The Backslope position signed 2 in Figure 92 showed a different RSD-index for winter rye in 2000 and 2001. The R-model run improved the results for the Backslope in 2000 (RSD>0), whereas in 2001 the area shows a decrease (RSD<0). This is an indication of the influence of different meteorological conditions observed in both years, which are not reflected by the model.

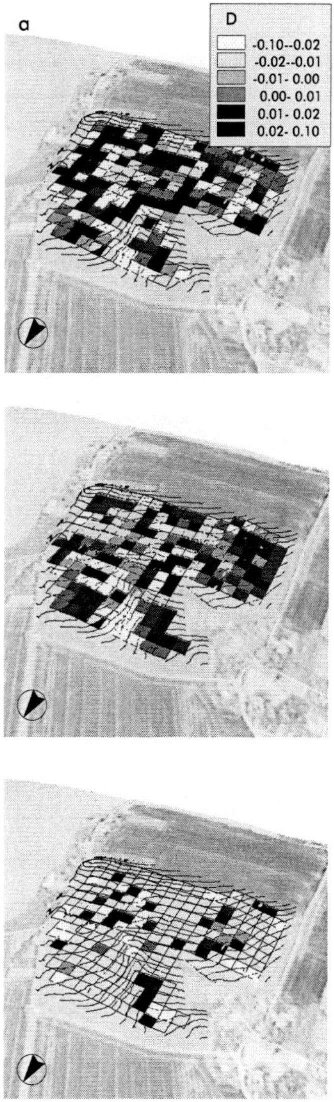

Figure 92. Evaluation of model runs using the RSD-Index for 1999- 2001 at the field site “Bei Lotte”. The number 1 shows the location of the depressional area, the number 2 the location of the Backslope area mentioned in the text.

As radiation penetrates the plant stands and reaches the soil surface, differences in temperature can be observed. In the approach shown here, the spatial differences in surface temperature, which can be additionally computed using the SRAD-model, has not been incorporated. Differences in mean annual ground surface temperatures were found using deep borehole temperature measurements up to 1.3°C due to slope orientation at a forest site (Safanda, 1999). Safanda (1999) concluded that differences in meadows may be even larger. Secondly, no reference has been made to the possible spatial variability of photosynthetically active radiation (PAR). PAR is defined as the radiation between the 400-700 nm wavelengths. Due to the sparse density of PAR measurements, researchers use either constant ratios related to broadband solar radiation, or calculated the ratios dynamically to local climatic conditions (Alados et al., 1996, Gonzalez and Calbo, 2002). Ehrhardt and Riedinger (1984) for example observed a decrease in PAR in shaded areas of their beech forest at a ratio of 2:1, whereas under sunlit conditions relations comparable to above canopy measurements were found.

Table 35. Model evaluation for model runs assuming spatially constant solar irradiance (P) and solar irradiance influenced by topography (R) and observed grain yield (MES) for the years 1998 and 1999 at the field site "Sportkomplex". Statistical parameters of simulated and measured dry matter are given in tha^{-1}.

	1999 MES	2000 MES	2001 MES	1999 P	2000 P	2001 P	1999 R	2000 R	2001 R
Mean	5.77	9.56	9.03	5.71	5.59	6.63	3.67	5.53	6.54
Median	5.75	9.47	9.02	5.67	5.44	6.63	3.63	5.38	6.53
STD	0.78	1.35	1.17	0.21	0.36	0.009	0.22	0.35	0.04
Variance	0.62	1.82	1.37	0.05	0.1307	0.00008	0.05	0.12	0.002
Skewness	0.047	0.176	0.359	0.253	2.230	-12.431	1.028	2.19	0.675
Curtosis	0.32	0.38	1.479	0.073	3.773	165.987	3.511	3.67	-0.922
Minimum	3.32	5.98	5.87	5.16	5.41	6.51	3.17	5.32	6.50
Maximum	8.22	13.40	14.09	6.24	6.69	6.64	4.84	6.66	6.65
Ln(δ)				-2.16	-2.23	-2.07	-2.12	-2.24	-2.08

CHAPTER 10

Conclusions

The aim of the investigations was to improve our understanding of spatial and temporal variability and development of soil properties and plant parameters in a loess landscape under special consideration of relief information.

The ArcInfo Macro Language was used to develop a suite of relief analysis tools. Using these tools, relief parameters can be computed and aggregated easily even by an inexperienced user. Procedures were designed as straight forward as possible. Novice users can work with the programs with minimum input, yet providing a variety of different features for an experienced user.

The landform classification routine by Pennock et al. (1987) was extended to include planar landforms and slightly different area-filtering algorithms. Differences between the original and the extended routines were evaluated.

Relief parameters for a 200 ha area in Saxony, south eastern Germany, were computed. The impact of different datasources - topographic maps with different scales and a digital elevation model (DEM) based on a Laserscan - on the results of the relief analysis were evaluated.

The impact of generalization of DEM - input data was analyzed for the landform classification procedure, showing that the occurrence of footslope and shoulder positions decreased significantly with increasing topographic map scale. Nonlinear parameter optimization was performed to overcome these problems and several parameter sets were identified for the agricultural landscape in the Luettewitz region for DEMs based on the topographic maps 1:10,000, 1:25,000 and 1:100,000.

To account for uncertainties / inaccuracies in the DEM a Monte-Carlo (MC) simulation approach was developed to compute the Topographic Wetness Index. Results for the fields "Bei Lotte" and "Sportkomplex" indicate a better representation of soil surface water content distribution using a Monte-Carlo-Topographic Wetness Index approach than a traditional Topographic Wetness Index. Additionally, the optimum DEM - resolution for a Laserscan (obtained at 1 m resolution, <5.7 m) and for topographic map data sources (1:10.000 <16 m, 1:25.000 <22.6 m and 1:100.000 < 32 m) were determined.

Conclusions drawn from the relief analysis were a first step to analyse the landscape. However, to understand the crop growth and soil properties inherent in the Luettewitz region, a raster-based sampling approach at several field sites was performed.

Grain yield properties were obtained by combine harvest and hand harvest. Differences between up to a maximum of 2 tha^{-1} in grain yield were observed in the worst case. Possible errors were (I) differences between sampling dates, (II) occurrence of lodging areas, (III) measurement errors in the processing of combine and harvest grain yields and (IV) small-scale heterogeneity in plant stands.

The grain yield's spatial variability was tested across different fields in different years. Semivariogram analyses for grain yields obtained at different fields showed various spatial correlation structures even for the same type of crop grown in one single year. Two cases, both based on the same meteorological conditions, influencing the different semivariogram parameters could be identified.

The first case included different spatial semivariogram ranges identified for two different field sites for grain yield in 1998-2001. A decrease in fitted semivariogram range could be observed with increasing precipitation sum between March and July for hand harvest grain yield and combine harvest grain yield.

The second case was related to landform elements. The assumption was made that each landform unit would reflect a characteristic crop yield development, based on the meteorological conditions and soil conditions observed. Therefore, the spatial semivariogram pattern would be expected and has been found to differ across landforms. Different semivariogram parameters were identified for landform units. Landform unit boundaries were used as faults in the kriging process in ISATIS. A larger number of outliers could be identified during a jackknifing process using landform elements as faults, compared to a jackknifing process without landform boundaries.

Several significant crosscorrelations were determined between grain yield / grain yield components and relief parameters, texture components, penetration resistance, NDVI and other parameters. However, the significance of crosscorrelations differed between years (i.e. clay and yield) and switched also their sign due to different meteorological conditions observed (i.e. yield and elevation). Differences in plant development were visible from NDVI images in every year. However significant crosscorrelations were not reflected in total grain yield in each year due to different processes influencing the final grain yield (measurement uncertainty by combine harvest, pests, heavy rainfall, lodging).

The landform classification procedure was used to stratify the field "Bei Lotte" into several landform units. Results for 192 sampling points in the time period 1999-2001 showed that single yield components and plant development reacted differently over three years. Similar results could be drawn from the relief parameter profile curvature. Plants compensated to some extent differences in yield development using various yield components, if environmental conditions were favourable for plant development. Especially important was the number of yielding spikes per area as a yield limiting factor in this loess landscape. Further research has to show how management can be changed to influence crop yield components during different time spans.

Beside the topography, soil properties are another factor influencing the observed crop yield and crop yield variability at the field sites "Bei Lotte" and "Sportkomplex". Therefore the spatial and temporal dynamics of selected soil properties were determined using a number of methods.

A homogeneous fertilizer distribution by the fertilizer spreader is assumed for any field research. To test if a systematic error is introduced by applying fertilizer, the mineralized nitrogen content was evaluated against the distance to tramlines with no systematic distribution of N_{min} in relation to distance to tramline observed.

Geostatistical analysis was performed for different soil properties using semivariogram and crosscorrelation analysis for the field site "Bei Lotte".

Crosscorrelation coefficients were determined between texture and penetration resistance versus relief parameters, NDVI and other parameters. No significant crosscorrelation coefficients could be observed between SOC, clay, silt, and sand. The relief parameter elevation showed the highest number of significant crosscorrelations for several soil properties like texture, nitrogen content, pH-value, penetration resistance and soil water content. Penetration resistance was found to be significantly crosscorrelated in different depth layers with the NDVI and the Plant-N-status in different years due to the observed environmental conditions.

Soil properties were found to be different across varying profile curvature classes and landforms. Differences could be identified for (I) strong profile curvature values and compared to the rest of the field and (II) between the planar profile curvature classes and divergent and convergent profile curvature classes. Similar conclusions can be drawn for landform elements, which have implications for soil nutrient sampling at different landforms or profile curvature classes. Additionally, outliers at different landform or profile curvature classes helped to identify areas where additional processes like subsurface flow occurred.

Up to this point the spatial and temporal variability of soil properties and grain yield components have been shown, especially under consideration of relief information. Still, to understand the observed differences, models are needed to simulate nonlinear processes. Relief parameters can be easily computed for large areas and therefore might be a cost effective way to improve model predictions and model results. Two relief parameters, the solar irradiance influenced by topography and the topographic wetness index were evaluated using the HERMES model.

First, the spatial and temporal distribution of solar irradiance in an agricultural landscape using the SRAD model was simulated and evaluated. The SRAD model was slightly altered to use higher resolution DEMs. Simulated solar irradiance was compared against three weather stations with different amounts of topographic shading. Good agreements were observed between simulated and measured shortwave solar irradiance.

Secondly, the spatially differing simulated solar irradiance was applied as a correction factor for the HERMES-model. Model results using spatially distributed solar irradiance improved in 54% of 225 simulated positions in 1998 and 52% in 1999 for the field site “Sportkomplex”. A spatially related pattern (north or south exposed positions) was observed. Results at the field site “Bei Lotte” for 192 simulated positions using simulated solar irradiance were not able to confirm these model improvements for three years.

Third, the transformed Topographic Wetness Index (TWI) was used as a hydromorphic modification factor (HMF) for the field capacity parameter in the HERMES model. Using an indirect approach the use of the transformed TWI as a HMF was compared against an optimized HMF, which has been obtained using a nonlinear optimization technique. Generally, the use of the TWI as a hydromorphic correction factor can be confirmed. The fast, easy and cheap computation the TWI appears to be one possibility to describe the processes connected to water flow in the hummocky landscape.

Both methods allowed to identify areas, where additional factors influenced crop yield development. To these belong geology, previous land use or water content distribution in flow lines. Another advantage of the indirect nonlinear optimization technique is to identify areas for additional observations.

The aim of the investigations, as stated in the beginning, was to improve our understanding of spatial and temporal variability and development of soil properties and plant parameters in a loess landscape. With the results shown for relief analysis, quantification of spatial and temporal dynamics of crop yield components and soil properties, and with the application of simulation models using selected relief parameters, one can conclude, that a first step has been taken to improve our understanding. Special attention has been taken in the thesis for the consideration of relief information. However, one can also state that many open questions are still waiting to be answered. Only two will be adressed.

One question would be related to the further development of the landform analysis, possibly combining best of both approaches, and even more important to port the codes into the GIS ArcView, so more users can access it.

The second question is related to the different development of yield properties inside a field. The understanding of these processes need to be improved to develop a kind of agricultural management model such as LISTOPAD (Ivanov et al. 1980) in Russia. However, instead of entire regions or fields, subsections of fields have to be managed and the interactions between different applications have to be evaluated carefully.

References

Aandahl,A.R., 1948. The Characterization of Slope Position and Their Influence on the Total Nitrogen Content of a Few Virgin Soils of Western Iowa. Soil Science Society of America Journal, 13: 449-454.

Adolf Thiess Gmbh &CO.KG, 1999. Dokumentation für Auftrag 992576.

AG BODEN, 1994. Bodenkundliche Kartieranleitung, 4.Auflage. Schweizerbart'sche Verlagsbuchhandlung,Stuttgart, 392 pp.

Agricon Precision Farming Company, 2000. Ein Schritt in die richtige Richtung. Jahna, Agricon.

Alados,I., Foyo-Moreno,I., and Alados-Arboledas,L., 1996. Photosynthetically active radiation: measurements and modelling. Agricultural and Forest Meteorology, 78: 121-131.

Anderl,A., Mangstl,A., and Reiner,L., 1981. Die Ertragsstruktur von Winterweizen, dargestellt an der Datenbasis ISPFLANZ. Bayrisches Landwirtschaftliches Jahrbuch, 58: 455-468.

Anthoni,P.M., Law,B.E., Unsworth,M.H., and Vong,R.J., 2000. Variation of net radiation over heterogeneous surfaces: measurements and simulation in a juniper-sagebrush ecosystem. Agricultural and Forest Meteorology, 102: 275-286.

Armstrong,M., 1998. Basic Linear Geostatistics. Springer, Berlin Heidelberg.

Auernhammer,H., Demmel,M., Murh,M., Rottmeier,J., and Wild,K., 1993. Yield Measurements on Combine Harvesters. ASAE Paper 931506, ASAE Winter meeting in Chicago.

Aufhammer,W., 1973. Untersuchungen zur Ertragsbildung und ihrer Beeinflussbarkeit durch physiologisch wirksame Substanzen bei der Sommergerste. PD-Thesis, University of Bonn.

Bakhsh,A., Colvin,T.S., Jaynes,D.B., Kanwar,R.S., and Tim,U.S., 2000. Using soil attributes and GIS for interpretation of spatial variability in yield. Transactions of the ASAE, 43(4): 819-828.

Bakhsh,A., Jaynes,D.B., Colvin,T.S., and Kanwar,R.S., 2000. Spatio-temporal analysis of yield variability for a corn-soybean field in Iowa. Transactions of the ASAE, 43(1): 31-38.

Baumann,H., 1949. Wetter und Ernteertrag. Deutscher Bauernverlag, Berlin.

Beckett,P.H.A. and Webster,R., 1971. Soil variability: A review. Soils and Fertilizers, 34: 1-15.

Beresneva,I.A. and Popova,E.N., 1983. Microclimatic variability of total solar radiation in the hilly relief of Badkhyz. Problems of desert development, 1: 76-81.

Bian,L. and Walsh,S.J., 1993. Scale Dependencies of Vegetation and Topography in a Mountainous Environment of Montana. Professional Geographher, 45(1): 1-11.

Bindlish,R. and Barros,A.P., 1996. Aggregation of Digital Terrain Data using a modified fractal interpolation scheme. Computers and Geoscience, 22(8): 97-917.

Blackmore,S., Godwin,R.J., Taylor,J.C., Cosser,N.D., Wood,G.A., Earl,R., and Knight,S., 1999. Understanding Variability in Four Fields in the United Kingdom. In: Robert, P. C., Rust, R. H., and Larson, W. E., Proceedings of the fourth international conference on precision agriculture, St.Paul, Minnesota, American Society of Agronomy, Crop Science Society of America, Soil Science Society of America.

Blake,L., Mercik,S., Koerschens,S., Moskal,S., Poulton,P.R., Goulding,K.W.T., Weigel,A., Powlson,D.S., 2000. Phosphorus content in soil, uptake by plants and balance in three European long-term field experiments.. Nutrient Cycling in Agroecosystems, 56: 263-275.

Bleines,C., Perseval,S., Rambert,F., Renard,D., and Touffait,Y., 2000. ISATIS Software Manual. GEOVARIANCES, Avon Cedex.

Bonner,J., 1962., Ideas of Biology, Mineola, N.Y. Dover Publications.

Böttcher,J., Strebel,O., 1988. Spatial variability of groundwater solute concentrations at the water table under arable land and coniferous forest. Part 2: Field data for arable land and statistical analysis. Zeitschrift Pflanzenernährung und Bodenkunde, 151: 191-195.

Boyer,D.G., Wright,R.J., Winant,W.M., and Perry,H.D., 1990. Soil Water Relations on a hilltop cornfield in Central Appalachia. Soil Science, 149(5): 383-392.

Breuer,B., 2001. Landform modelling of a research area in the Upper Palatinate, Germany, with the SARA software package (System for Automatic Relief Analysis). Zeitschrift für Geomorphologie, 45(1): 17-31.

Brown,D.G., Lusch,D.P., and Duda,K.A., 1998. Supervised classification of types of glaciated landscapes using digital elevation data. Geomorphology, 21: 233-250.

Brubaker,S.C., Jones,A.J., Lewis,D.T., and Frank,K., 1993. Soil Properties associated with Landscape Position. Soil Science Society of America Journal, 57: 235-239.

Burrough,P.A., van Gaans,P.F.M., and MacMillan,R.A., 2000. High -resolution landform classification using fuzzy k-means. Fuzzy Sets and Systems, 113: 37-52.

Cambardella,C.A. and Karlen,D.L., 1999. Spatial Analysis of Soil Fertility Parameters. Precision Agriculture, 1(1): 1-5.

Cambardella,C.A., Moorman,T.B., Novak,J.M., Parkin,T.B., Karlen,D.L., Turco, R.F., Konopka, A.E., 1994. Field scale variability of soil properties in central Iowa soil. Soil Science Society America Journal, 58: 1501-1511.

Campbell,C.A., Cameron,D.R., Nicholaichuk,W., and Davidson,H.R., 1977. Effects of fertilizer N and soil moisture on growth, N content and moisture by spring wheat. Canadian Journal of Soil Science, 57: 289-310.

Ciha,A.J., 1982. Slope Position and Grain Yield of Soft white Winter wheat. Agronomy Journal, 76: 193-196.

Collis-George,N. and Davey,B.G., 1960. The doubtfull utility of present day field experiments and other determinations involving soil-plant interactions. Soils & Fertility, 23: 307-310.

Conacher,A.J. and Dalrymple,J.B., 1977. The nine-unit landsurface model: an approach to pedogeomorphic research. Geoderma, 18: 1-154.

CSIRO, 2002. TOPOG V9.0. Canberra, CSIRO Land and Water, GPO Box 1666, Canberra, Australia, 2601.

Damisch,W. and Wiberg,A., 1977. Ergebnisse über den Einfluß der Temperatur auf Stoffzuwachs und Ertragsbildung bei Sommergerste. Archiv für Acker- und Pflanzenbau und Bodenkunde, 21(6): 485-494.

Damisch,W., 1971. Über einige Aspekte der Ertragsbildung bei Getreide, untersucht an Sommergerstensorten. Archiv für Acker- und Pflanzenbau und Bodenkunde, 15(11): 913-925.

Damisch,W., Müller,G., and Wiberg,A., 1984. Einfluß der Umweltfaktoren auf Entwicklungsgeschwindigkeit und Ertragsbildung bei Winterweizen. Archiv für Acker- und Pflanzenbau und Bodenkunde, 28(5): 305-312.

Dammer,K.-H., Wartenberg,G., Ehlert,D., Hammen,V., Reh,A., Wagner,U., and Dohmen,B., 2001. Recording of present plant parameters by pendulum sensor, remote sensing, and ground measurments, as fundamentals for site-specific fungicide application in winter wheat. In: Grenier, Gilbert and Blackmore, Simon. Third European Conference on Precision Agriculture, Montpellier, agro Montpellier, 2: 647-652

Darwinkel,A., 1980. Ear development and formation of grain yield in winter wheat. Netherlands Journal of Agricultural Science, 28: 156-163.

Dehn,M., Gärtner,H., and Dikau,R., 2001. Principles of semantic modeling of landform structures. Computers and Geoscience, 27: 1005-1010

DIN 19683 Blatt 1, 1973. Bestimmung der Korngrößenzusammensetzung durch Siebung.

DIN 19683 Blatt 2, 1973. Bestimmung der Korngrößenzusammensetzung nach Vorbehandlung durch Natriumpyrophosphat.

DIN ISO 10390, 1997. Bodenbeschaffenheit: Bestimmung des pH-Wertes.

DIN ISO 10693, 1994. Bodenbeschaffenheit: Bestimmung des Carbonatgehaltes mit einem volumetrischen Verfahren.

DIN ISO 10694, 1994. Bodenbeschaffenheit: Bestimmung des organischen Kohlenstoffgehaltes und des Gesamtkohlenstoffgehaltes nach trockener Verbrennung (Elementaranalyse).

DIN ISO 13878, 1994. Bodenbeschaffenheit: Bestimmung des Gesamtstickstoffgehaltes nach trockene Verbrennung (Elementaranalyse).

DIN ISO 11048, 1997. Bestimmung von wasser- und säurelöslichen Sulfat.

Doherty,J., 2002. PEST Model-Independent Parameter Estimation. 4. Version. Watermark Numerical Computing.

Dubsky,G., Wernecke,P., Müller,J., Eschenröder,A., Strutz,A., Pigla,U., and Diepenbrock,W., 1999. Soft- and Hardware-Components for Crop Growth Modelling. Schiefer, G., Helbig, R., and Rickert, U., Perspectives of modern information and communication systems in agriculture, food production and environmental control. University Bonn, Bonn, 735-741.

Eghball,B. and Varvel,G.E., 1997. Fractal analysis of temporal yield variability of crop sequences: implications of site-specific management. Agronomy Journal, 89(6): 851-855.

Eghball,B., Binford,G.D., Power,J.F., Baltensperger,D.D., and Anderson,F.N., 1995. Maize Temporal Yield Variability under Long-Term Manure and Fertilizer Application: Fractal Analysis. Soil Science Society of America Journal, 59: 1360-1364.

Ehrhardt,O. and Riedinger,F.P., 1984. Zum Strahlungshaushalt eines Buchenwaldes. International Symposium for Dr. Franz Sauberer, Universität für Bodenkultur, Wien. 125-128.

Eijkelkamp Agriresearch Equipment, 1999. Penetrologger. Eijkelkamp.

Entz,M.H. and Fowler,D.B., 1998. Critical stress periods affecting productivity of no-till winter wheat in western canada. Agronomy Journal, 80: 987-992.

Fiez,T.E., Miller,B.C., and Pan,W.L., 1994. Winter wheat yield and grain protein across varied landscape positions. Agronomy J, 86: 1026-1032.

Filzer,P., 1939. Lichtintensität und Trockensubstanzproduktion in Pflanzengesellschaften. Berichte Deutsche Botanische Gesellschaft, 57: 155.

Florinski,I., Eilers,R.G., Manning,G., and Fuller,L.G., 2002. Prediction of soil properties by digital terrain modelling. Environmental Modelling & Software, 17: 295-311.

Florinski,I.V., 1998. Accuracy of local topographic variables derived from digital elevation models. International Journal of Geographical Information Science, 12(1): 47-61.

Gallagher,J.N. and Biscoe,P.V., 1978. Radiation absorption, growth and yield of cereals. Journal Agriculture Science (Cambridge), 91: 47-60.

Gallant,J.C. and Hutchinson,M.F., 1996. Towards an Understanding of Landscape Scale and Structure. Third International Conference/Workshop on Integrating GIS and Environmental Modeling CD-ROM, Santa Fe, New Mexico, USA, National Center for Geographic Information and Analysis.

Gallant,J.C. and Hutchinson,M.F., 1997. Scale dependence in terrain analysis. Mathematics and Computers in Simulation, 43: 313-321.

Gallant,J.C. and Wilson,J.P., 1996. Tapes-G: a grid-based terrain analysis program for the environmental sciences. Computers and Geoscience, 22(7): 713-722.

Gates,D.M., 1980. Biophysical ecology. Springer, New York.

Geiger,R., Aron,R.H., and Todhunter,P., 1961. The climate near the ground. Vieweg, Braunschweig.

Geisler,G., 1983. Ertragsphysiologie von Kulturarten des gemäßigten Klimas. Paul Parey, Berlin,Hamburg.

Gitelson,A.A., Merzlyak,M.N., Zur,Y., Stark,R., and Gritz,U., 2001. Non-desdructive and remote sensing techniques for estimation of vegetation status. Grenier, G. and Blackmore, S., Montpellier, agro Montpellier. Third European Conference on Precision Agriculture. (1): 201-210.

Goel,P.K., Prasher,S.O., Landry,J.A., Patel,R.M., Bonnell,R.B., Viau,A.A., and Miller,J.R., 2003. Potential of airborne hyperspectral remote sensing to detect nitrogen deficiency and weed infestation in corn. Computers and Electronics in Agriculture, 38: 99-124.

Goodchild,M.F., 1986. Spatial autocorrelation. Catmog 47, Geo Books, Norwich

Gonzalez,J.A. and Calbo,J., 2002. Modelled and measured ratio of PAR to global radiation under cloudless skies. Agricultural and Forest Meteorology, 110: 319-325.

Grenzdörfer,G.J. and Gebbers,R.I.B., 2001. Seven years of yield mapping - Analysis and possibilities of multi year yield mapping data. Grenier, G. and Blackmore, S., Montpellier, agro Montpellier. ECPA 2001 Third European Conference on Precision Agriculture.

Grisso,R.D., Jasa,P.J., Schroeder,M.A., and Wilcox,J.C., 1999. Yield Monitor Accuracy: Sucessfull Farming Magazine Study. ASAE Paper 991047. Toronto.

Haque,I. and Lupwayi,N.Z., 2002. Landform and phosphorus effects on nitrogen fixed by annual clovers and its contribution to succeeding cereals in the Ethiopian highlands. Australian Journal of Agricultural Research, 50: 1393-1398.

Härtel,F., 1931. Erläuterungen zur Geologischen Karte von Sachsen im Maßstab 1:50000 - Nr.47 Blatt Lommatzsch. Finanzministerium, Leipzig.

Heisig,M., Giebel,A., Schwarz,J., 2001. Yield Data Preparation in the project MOSAIK.

Herbst,R., 2002. Bodenschätzung, geoelektrische Sondierung und pedostatistische Modellierungen als Basis von digitalen Hof-Bodenkarten im Präzisen Landbau. PhD-Thesis, Universität Kiel, Schriftenreihe des Instituts für Pflanzenernährung und Bodenkunde, Nr. 60.

Herbst,R., Lamp,J., and Reimer,G., 2001. Inventory and spatial modelling of soils on PA pilot fields in various landscapes of germany. Grenier, Gilbert and Blackmore, Simon, Montpellier, agro Montpellier. Third European Conference on Precision Agriculture. (1): 395-400.

Heuvelink,G.B.M. and Webster,R., 2001. Modelling soil variation: past, present and future. Geoderma 100, 269-301. 2001.

Holmes,K.W., Chadwick,O.A., and Kyriakidis,P.C., 2000. Error in USGS 30-Meter digital elevation model and its impact on terrain modeling. Journal of Hydrology, 233: 154-173.

Hugget,R.J., 1975. Soil Landscape Systems: A model of soil genesis. Geoderma, 13: 1-22.

Hughes,A.P., 1959. Plant growth in controlled environments as an adjunct to field studies. Journal agriculture Science, 53: 247-259.

Huising,E.J. and Gomes Pereira,L.M., 1998. Errors and accuracy estimates of laser data acquired by various laser scanning systems for topographic applications. ISPRS Journal of Photogrammery & Remote Sensing, 53: 245-261.

Hutchinson,M.F., 1996. A locally adaptive approach to the interpolation of digital elevation models. NCGIA. Proceedings of the Third International Conference Integrating GIS and Environmental Modelling. Santa Barbara,CA, University of California, National Center for Geographic Information and Analysis.

HYDRO, 2002. Hydro N-Tester. HYDRO [Düngung aktuell].

HYDRO, 2000. PRECISE Der sechste Sinn für die Landwirtschaft.

Idso,S.B., 1969. Atmospheric attenuation of solar radiation. Journal of Athmospheric Sciences, 26: 1088-1095.

Irrgang,A., Johnen,T., and Gebbers,R.I.B., 2001, Evaluation of yield maps by low-cost remote sensing. Grenier, G. and Blackmore, S., Montpellier, agro Montpellier. Third European Conference on Precision Agriculture. (1) :217-222.

Irvin,B.J., Ventura,S.J., and SLater,B.K., 1997. Fuzzy and isodata classification of landform elements from digital terrain data in Pleasant Valley, Wisconsin. Geoderma, 77: 137-154.

Ivanov,A.F., Klimov,A.A., Listopad,G.E., and Ustenko,G.P., 1980. Ertragsprogrammierung - Wesen der Methode. VEB Deutscher Landwirtschaftsverlag, Berlin.

Ivanov,V.Y., Vivoni,E.R., Bras,R.L., and Entkhabi,D., 2003. Coupling hydrological processes within the framework of a tin-based distributed model and experience in simulating long-term catchment dynamics. 5, No.04431. Nice, EGS. Geophysical Research Abstracts.

Jaynes,D.B. and Colvin,T.S., 1997. Spatiotemporal Variability of Corn and Soybean Yield. Journal of Production Agriculture, 89: 30-37.

Jensen,T., Kelly,R., and Strong,W., 2001. Using protein mapping to define yield-limiting factors. Grenier, G. and Blackmore, S., Montpellier, agro Montpellier. Third European Conference on Precision Agriculture. (2) 797-802.

Jowkin,V. and Schoenau,J.J., 1998. Impact of tillage and landscape position on nitrogen availability and yield of spring wheat in the Brown soil zone in southwestern Saskatchewan. Canadian Journal of Soil Science, 78(3): 563-572.

Jürschik,P., 2003. Accuracy of velocity sensors of combine harvest. personal communication

Jürschik,P. and Flemming-Fischer,E., 1995. Fliegend Ackerflächen erfassen. LANDTECHNIK, 95(2): 70-71.

Jürschik,P., Wendroth,O., and Giebel,A., 1999. Processing of point data from combine harvesters for precision farming. Stafford, J. V., Odense, Denmark, Proceedings of the Second European Conference of Precision Agriculture. 297-307.

Kerry,R. and Oliver,M.A., 2001. Comparing spatial structure in soil properties and ancillary data by using variograms. Grenier, G. and Blackmore, S., Montpellier, agro Montpellier, Third European Conference on Precision Agriculture, (1)413-418.

Kerschberger,M., Hege,U., and Jungk,A., 1997. Phosphordüngung nach Bodenuntersuchung und Pflanzenbedarf. http://www.vdlufa.de/vd_00.htm.

Kersebaum,K.C. and Beblik,A.J., 2001. Performance of a nitrogen dynamics model applied to evaluate agricultural management practices. In: M.Shaffer, L.Ma, and S.Hansen (Editors), Modeling carbon and nitrogen dynamics for soil management. CRC Press, Boca Raton, pp. 551-571.

Kersebaum,K.C., 1995. Application of a simple management model to simulate water and nitrogen dynamics. Ecological Modelling, 81: 145-156.

Kersebaum,K.C., Lorenz,K., Reuter,H.I., and Wendroth,O., 2002. Modelling Crop Growth and Nitrogen Dynamics for Advisory Purposes Regarding Spatial Variability. In: L.R.Ahuja, L.Ma, and T.A.Howell (Editors), Agricultural system models in field research and technology transfer. CEC Press LLC, Boca Raton, pp. 230-251.

Kiniry,J.R., Jones,C.A., O'Toole,J.C., Blanchet,R., Cabelguenne,M., and Spanbel,D.A., 1989. Radiation-use efficiency in biomass accumulation prior to grain-filling for five grain-crop species. Field Crops Research, 20: 51-64.

Kler,D.S., 1987. Bi-directional sowing on use of solar radiation and sarson yield. Transactions. Indian Soc. Deser. Technol. Univ. Cent. Desert. Stud. Jodhpur: The Society, 12(2): 55-61.

Kraus,K., 1997. Photogrammetrie -Verfeinerte Methoden und Anwendungen. Dümmler, Bonn.

Kravchenko,A.N., Bullock,D.G., and Boast,C.W., 2000. Joint multifractal Analysis of Crop Yield and Terrain Slope. Journal of Production Agriculture, 92: 1279-1290.

Krebs,C.J., 1998. Ecological Methodology. Addison-Wesley.

Kumar,L. and Skidmore,A.K., 2000. Radiation-Vegetation Relationships in a Eucalyptus Forest. Photogrammetric Engineering & Remote Sensing, 66(2): 193-204.

Lamp,J., Herbst,R., and Reimer,G., 2001. Precise and efficient soil surveys as a basis for application maps in precision agriculture. Grenier, G. and Blackmore, S.. Montpellier, agro Montpellier. Third European Conference on Precision Agriculture. (1): 49-54.

Lark,R.M. and Webster,R., 1999. Analysis and elucidation of soil variation using wavelets. European Journal of Soil sciene, 50: 185-206.

Lark,R.M., 1999. Soil-Landform relationships at within-field scales: an investigation using continuous classification. Geoderma, 92: 141-165.

Lee,J. and Marion,L.K., 1994. Analysis of spatial autocorrelation of U.S.G.S 1:250,000 Digital Elevation Models. http://spatialodyssey.ursus.maine.edu/gisweb/spatdb/gis-lis/gi94064.htm, 504-513.

Li,H., Lascano,R.J., Booker,J., Wilson,L.T., and Bronson,K.F., 2001. Cotton lint variability in a heterogeneous soil at a landscape scale. Soil & Tillage Research, 58: 245-258.

Lieberoth,I., Schmidt,H., and Cronewitz,E., 1985. Auswertung der Mittelmaßstäbigen Standortkartierung (MMK) - Flächennachweis auf Bezirksebene. Müncheberg, Akademie der Landwirtschaftswissenschaften der DDR, FZB Müncheberg, Bereich Bodenkunde / Fernerkundung Eberswalde. pp81.

Löpmeier,F., 1999. Die agrarmeteorologische Situation. KSB 1999. Offenbach, Deutscher Wetterdienst, Referat Grundlagen der Klimaüberwachung. Klimastatusbericht.

Löpmeier,F., 2000. Die agrarmeteorologische Situation. KSB 2000. Offenbach, Deutscher Wetterdienst, Referat Grundlagen der Klimaüberwachung. Klimastatusbericht.

Löpmeier,F., 2001. Die agrarmeteorologische Situation. KSB 2001. Offenbach, Deutscher Wetterdienst, Referat Grundlagen der Klimaüberwachung. Klimastatusbericht.

Lozan,J.L. and Kausch,H., 1998. Angewandte Statistik für Naturwissenschaftler. Parey Buchverlag, Berlin.

Machet,J.M., Beaudoin,N., Mary,B., Boffety,D., and Bernard,M., 2001. Characterization of the Variability in Grain Production and Quality within a winter wheat field. Grenier, G. and Blackmore, S.. Montpellier, agro Montpellier. ECPA 2001 Third European Conference on Precision Agriculture.

MacMillan,R.A. and Pettapiece,W.W., 1997. Soil Landscape Models: Automated landscape characterization and generation of soil-landscape models. Research Report No. 1E.1997. Lethbridge, Agriculture and Agri-Food Canada, Research Branch.

MacMillan,R.A., Pettapiece,W.W., Nolan,S.C., and Goddard,T.W., 2000. A generic procedure for automatically segmenting landforms into landform elements using DEMs, heuristic rules and fuzzy logic. Fuzzy Sets and Systems, 113: 81-109.

Manning,G., Fuller,L.G., Eilers,R.G., and Florinsky,I., 2001a. Soil moisture and nutrient variation within an undulating Manitoba landscape. Canadian Journal of Soil Science, 81(3): 449-458.

Manning,G., Fuller,L.G., Eilers,R.G., and Florinsky,I., 2001b. Topographic influence on the variability of soil properties within an undulating Manitoba landscape. Canadian Journal of Soil Science, 81(3): 439-447.

Manning,G., Fuller,L.G., Flaten,D.N., and Eilers,R.G., 2001c. Wheat yield and grain protein variation within an undulating soil landscape. Canadian Journal of Soil Science, 81(3): 459-467.

Marschner,H., 1995. Mineral Nutrition of higher plants. Academic Press Limited, Cambridge.

Maul,A. and Richter,M.K.A., 2001. Statistisches Jahrbuch über Ernährung, Landwirtschaft und Forsten der Bundesrepublik Deutschland. Landwirtschaftsverlag, Münster.

McBratney,A.B., de Gruijter,J.J., and Brus,D.J., 1992. Spatial prediction and mapping of continuous soil classes. Geoderma, 54: 39-64.

McKenney,D.W., Mackey,B.G., and Zavitz,B.L., 1999. Calibration and sensitivity analysis of a spatially-distributed solar radiation model. International Journal of Geographical Information Science 13[1], 49-65.

Michael,G., 1990. Vorstellungen über die Regulation der Wurzelhaarbildung. Kali-Briefe, 20: 411-429.

Miller,M.P., Singer,M.J., and Nielsen,D.R., 1988. Spatial variability of wheat yield and soil properties on complex hills. Soil Science Society of America Journal, 52: 1133-1141.

Milne,G., 1936. A provisional soil map of east africa. Eastern Africa Agriculture Research Station, Amanin Memoirs, 1-34.

Moore,I.D., Gessler,P.E., Nielsen,G.A., and Peterson,G.A., 1993a. Soil Attribute Prediction Using Terrain Analysis. Soil Science Society of America Journal, 57: 443-452.

Moore,I.D., Nortin,T.W., and Williams,J.E., 1993b. Modelling environmental heterogeneity in forested landscapes. Journal of Hydrology, 150: 717-747.

Moore,I.D., Grayson,R.B., and Ladson,A.R., 1991. Digital Terrain Modelling: A Review of Hydrological, Geomorphological and Biological Applications. Hydrological Processes, 5: 3-30.

Morgenstern,H., 1986. Zur arealen Verteilung der Bodenfruchtbarkeitskennziffern der reliefierten Grundmoräne im Junpleistozän der DDR. Archiv für Acker- und Pflanzenbau und Bodenkunde, 30(4): 187-193.

Mulla,D.J., 1993. Mapping and managing spatial patterns in soil fertility and crop yield. In: P.C.Robert, R.H.Rust, and W.E.Larson (Editors), Proceedings of First Workshop, Soil Specific crop management. ASA,CSSA,SSSA, Madison,CA, pp. 15-26.

Mulla,D.J., Bhatti,A.U., Hammond,M.W., and Benson,J.A., 1992. A comparison of winter wheat yield and quality under uniform versus spatially variable fertilizer management. Agriculture Ecosystems & Environment, 38: 301-311.

Nemeth,K., 1982. Electro-ultrafiltration of aqueous soil suspension with simultaneously varying temperature and voltage. Plant and Soil, 64: 7-23.

Neuhof, Schiele, Schützenmeister, and Schmidt,R., 1999. MMK 25 [TK4845]. Freiberg, Sächsiches Landesamt für Umwelt und Geologie, Bereich Boden und Geologie Freiberg, Halsbrücker Str. 31a, 09599 Freiberg.

Nielsen,D.R. and Wendroth,O., 2003. Spatial and temporal statistics. CATENA, Reiskirchen.

Panten,K., Haneklaus,S., and Schnug,E., 2002. Spatial accuracy of online yield mapping. Landbauforschung Völkenrode, 52(4): 205-209.

Park,S.J., McSweeney,K., and Lowery,B., 2001. Identification of the spatial distribution of soils using a process-based terrain characterisation. Geoderma, 103: 249-272.

Pattey,E., Strachean,I.B., Boisvert,J.B., Desjardins,R.L., and McLaughlin,N.B., 2001. Detecting effects of nitrogen rate and weather on corn growth using micrometeorological and hyperspectral reflectance measurements. Agricultural and Forest Meteorology, 108: 85-99.

Penning deVries,F.W.T. and van Laar,H.H., 1982. Simulation of plant growth and crop production. Centre for Agricultural Production and Documentation (Pudoc), Wageningen.

Pennock,D.J. and Corre,M.D., 2001. Development and application of landform segmentation procedures. Soil & Tillage Research, 58: 151-162.

Pennock,D.J., Anderson,D.W., and De Jong,E., 1994. Landscape-scale changes in indicators of soil quality due to cultivation in Saskatchewan, Canada. Geoderma, 64: 1-19.

Pennock,D.J., Walley,F., Solohub,M., Si,B., and Hnatowich,G., 2001. Topographically controlled yield response of Canola to nitrogen fertilizer. Soil Science Society of America Journal, 65(6): 1838-1845.

Pennock,D.J., Zebarth,B.J., and De Jong,E., 1987. Landform Classification and Soil Distribution in Hummocky Terrain, Saskatchewan, Canada. Geoderma, 40: 297-315.

Pollmer,W.G., 1957. Untersuchungen zur Ertragsbildung bei Sommerweizen. Zeitschrift Pflanzenzüchtung, 37: 231-262.

Porter,P.M., Laurer,J.G., Huggins,D.R., Oplinger,E.S., and Crookston,R.K., 1998. Assessing Spatial and Temporal Variability of Corn and Soybean Yields. Journal of Production Agriculture, 11: 359-363.

Primost,E., 1977. Wirkung von Stabilan (Chlorcholinchlorid, CCC) auf die Länge der einzelnen Internodien von Wintergerste im Feldversuch. Zeitschrift für Pflanzenernährung und Bodenkunde, 140: 387-396.

Ratnam,B.P. and Goudreddy,B.S., 1977. Effect of solar radiation on crop yield. Agriculture and agro-industries journal, 10(10): 19-20.

Reyniers,M., Maertens,K., Reyns,P., and De Baerdemaeker,J., 2001. Management of combine harvester precision farming data to make usefull maps. Grenier, G. and Blackmore, S., Montpellier, agro Montpellier. Third European Conference on Precision Agriculture. (1):91-96

Rich,P.M., Hetrick,W.A., and Saving,S.C., 2002. Modeling Topographic Influences on Solar Radiation: a Manual for the SOLARFLUX Model. LA-12989-M. Los Alamos National Laboratory, Los Alamos, New Mexico 87545.

Roth,D., Bergmann,H., Paul,R., John,K., and Autorenkollektiv, 1988. Naturwissenschaftliche Grundlagen und Lösungen für eine hohe Ertragswirksamkeit des natürlichen Wasserdargebotes und von Zusatzwasser in der Pflanzenproduktion. 26/5. Berlin, Institut für Landwirtschaftliche Information und Dokumentation Berlin. Fortschrittsberichte für die Landwirtschaft und Nahrungsgüterwirtschaft.

Roth,D., Kachel,K., Teichardt,R., Schwarz,K., and Berger,W., 1980. Ergebnisse und Empfehlungen zur Beregnung von Schwarzerdestandorten. Nr.180. Berlin, Akademie der Landwirtschaftswissenschaften der DDR. 33-45

Roth,H.A., 1956. Untersuchungen über die Beziehungen zwischen den von der Bodenschätzung erfaßten natürlichen Ertragsbedingungen und den Ernteerträgen des Ackerlandes. Akademie - Verlag Berlin, Berlin.

Rouse,J.W., Haas,R.H., Schell,J.A., Deering,D.W., and Harlan,J.C., 1974. Monitoring the vernal advancement of retrogradation of natural vegetation. Final Report Type III. 1974. Greenbelt,MD, NASA Goddard Space Flight Centre.

Röver,M. and Kaiser,E.-A., 1999. Spatial heterogeneity within the plough layer: low and moderate variability of soil properties. Soil biology and biochemistry, 31: 175-187.

Ruhe,R.V., 1960. Elements of the soil landscape. Madison. 165-170

Ruhe,R.V., 1956. Geomorphic Surfaces and the nature of soils. Soil Science, 82(6): 441-455.

Safanda,J., 1999. Ground surface temperature as function of slope angle and slope orientation and its effect on the subsurface temperature field. Tectonophysics, 306: 367-375.

Saldana,A., Stein,A., and Zinck,J.A., 1998. Spatial variability of soil properties at different scaleswithin three terraces of the Henares River (Spain). Catena, 33: 139-153.

Schaab,G., 2000. Modellierung und Visualisierung der räumlichen und zeitlichen Variabilität der Einstrahlungsstärke mittels eines Geo-Informationssystems. Ph.D., Technische Universität Dresden.

Scheffer,F. and Schachtschabel,P., 1992. Lehrbuch der Bodenkunde. Enke, Stuttgart.

Schmidt,R., 1984. Zur Arealvariabilität von Bodenparametern und Böden im pleistozänen Tiefland der DDR. Archiv für Acker- und Pflanzenbau und Bodenkunde.

Schoder,A., 1932. Über die Beziehung des Tagesganges der Kohlensäureassimilation von Freilandpflanzen zu den Außenfaktoren. Jahrbuch wissenschaftlicher Botaniker, 76: 441.

Schwarz,J., Giebel,A., Heisig,M., Kersebaum,K.C., Reuter,H.I., and Wendroth,O., 2000. Teilflächenspezifische Düngung unter Berücksichtigung der räumlichen und zeitlichen Stickstoffvariabilität. VDI-MEG. 307-312.

Simmons,F.W., Cassel,D.K., and Daniels,R.B., 1989. Landscape and Soil property Effects on Corn Grain Yield Response to Tillage. Soil Science Society of America Journal, 53: 534-539.

Sinai,G., Zaslavsky,D., and Golany,P., 1981. The effect of soil surface curvature on moisture and yield - beer sheba observation. Soil Science, 132(5): 367-375.

Smit,B., Stachiowiak,M., and van Volkenburgh,E., 1989. Cellular processes limiting leaf growth in plants under hypoxic stress. Journal Experimental Botany, 40: 89-94.

Stenger,R., Priesack,E., and Beese,F., 1998. Distribution of inorganic nitrogen in agricultural soils at different dates and scales. Nutrient cycling in Agroecosystems, 50: 291-297.

Stevenson,F.C. and vanKessel,C., 1996. A landscape-scale assessment of the nitrogen and non-nitrogen rotation benefits of pea. Soil Science Society of America Journal, 60(6): 1797-1805.

Stevenson,F.C., Knight,J.D., Wendroth,O., van Kessel,C., and Nielsen,D.R., 2001. a comparison of two methods to predict the landscape-scale variation of crop yield. Soil & Tillage Research, 58: 163-181.

Stone,R.C., Hammer,G.L., and Marcussen,T., 1996. Prediction of global rainfall probabilities using phases of the Southern Oscillation Index. Nature, 384: 252-255.

Stoot,D.E., Martin,J.P., 1989. Organic Matter Decomposition and Retention in Arid Soils. Arid Soil Research and Rehabilitation, 3: 115-148.

Südzucker, 2000. Betriebsspiegel - Hofgut Lüttewitz., Südzucker AG.

Thieken,A.H., Lucke,A., Diekruger,B., and Richter,O., 1999. Scaling input data by GIS for hydrological modeling. Hydrological Processes, 13: 611-630.

Thun,R., Herrmann,R., Knickmann,E., and Hoffmann,G., 1991. Methodenbuch Band 1; Die Untersuchung von Böden. 4.Auflage. 1991. Darmstadt, VDLUFA-Verlag.

Thylen,L. and Algerbo,P.A., 2001. Development of a protein sensor for combine harvesters. Grenier, Gilbert and Blackmore, Simon. Montpellier, agro Montpellier. Third European Conference on Precision Agriculture. (2) 869-873.

Thylen,L., Algerbo,P.A., and Petterson,C.G., 1999. Grain Quality variations within fields of malting barley. Stafford, J. V., Odense, SCI Sheffield Academic Press. Precision Agriculture '99. (1) 287-296.

Timlin,D.J., Pachepsky,Ya., Snyder,V.A., and Bryant,R.B., 1998. Spatial and Temporal Variability of Corn Grain Yield on a Hillslope. Soil Science Society of America Journal, 62: 764-773.

Tomer,M.D. and Anderson,J.L., 1995. Variation of Soil Water Storage across a Sand Plain Hillslope. Soil Science Society of America Journal, 59: 1091-1100.

Tsegaye,T. and Hill,R.L., 1998. Intensive Tillage effects on spatial variability of soil zesz, plant growth, and nutrient uptake measurements. Soil Science, 163(2): 155-165.

Vachaud,G. and Chen,T., 2002. Sensitivity of computed values of water balance and nitrate leaching to within soil class variability of transport parameters. Journal of Hydrology, 264: 87-100.

van Ittersum,M.K., and Donatelli,M., 2003. Modelling cropping systems - highlights of the symposium and preface to the special issues. European Journal of Agronomy, 18: 187-197.

Wang,S., Chen,W., and Cihlar,J., 2002. New calculation methods of diurnal distribution of solar radiation and its interception by canopy over complex terrain. Ecological Modelling, 155: 191-204.

Webster,R., Oliver, M.A., 1990. Statistical Methods in Soil and Land Resource Survey. Oxford University Press.

Weerakoon,W.M.W., Ingram,K.T., and Moss,D.D., 2000. Atmospheric carbon dioxide and fertilizer nitrogen effects on radiation interception by rice. Plant and Soil, 220: 99-106.

Weltzien,C., 2002. Technikbetreung und -vergleich. In: Precision Agriculture Herausforderung an integrative Forschung, Entwicklung und Anwendung in der Praxis. KTBL, Darmstadt, pp. 153-167.

Wendroth,O., Jürschik,P., Kersebaum,K.C., Reuter,H.I., van Kessel,C., and Nielsen,D.R., 2001. Identifying, understanding, and describing spatial processes in agricultural landscapes - four case studies. Soil & Tillage Research, 58: 113-127.

Wenkel,K.-O., Brozio,S., Gebbers,R.I.B., Kersebaum,K.C., Lorenz,K., 2001. Development and Evaluation of different methods for site-specific nitrogen fertilization of winter wheat. Grenier, G. and Blackmore, S.. Montpellier, agro Montpellier. Third European Conference on Precision Agriculture. (1):91-96.

Wernecke,P., Buck-Sorlin,G., and Diepenbrock,W., 1999. Virtual Canopies: Combining Ontogenesis- with Architectural Models. Second European Conference of EFITA, LLeida - Spain, European Society of Agronomy. 123-124.

Wessolek,G., König,R., and Renger,M., 1992. Entwicklung und Anwendung von Photosynthesemodellen für Hangstandorte. Bork, H.-R., Renger, M., Alaily, F., Roth, C., and Wessolek, G., No.8. Berlin, Institut für Ökologie, TU Berlin. Bodenökologie und Bodengenese.

Western,A.W., Grayson,R.B., Blöschl,G., Willgose,G.R., and McMahon,T.A., 1999. Observed spatial organization of soil moisture and its relation to terrain indices. Water Resources Research, 35(3): 797-810.

Willgose,G., Bras,R.L., and Rodriguez-Iturbe,I., 1991a. A coupled Channel Network Growth and Hillslope Evolution Model 1 Theory. Water Resources Research, 27(7): 1671-1684.

Willgose,G., Bras,R.L., and Rodriguez-Iturbe,I., 1991b. Results for a new model of river basin evolution. Earth Surface Processes & Landforms, 16: 237-254.

Wilson,J.P. and Gallant,J.C., 2001. Terrain Analysis - Principles and Applications. John Wiley & Sons,Inc., New York.

Wilson,J.P., Spangrud,D.J., Nielsen,G.A., Jacobson,J.S., and Tyler,D.A., 1998. Global Positioning System Sampling Intensity and Pattern Effects on Computed Topographic Attributes. Soil Science Society of America Journal, 62: 1410-1417.

Wolock,D.M. and McCabe,G.J., 2000. Differences in topographic characteristics computed from 100- and 1000-m resolution digital elevation model data. Hydrological Processes, 14: 987-1002.

Wu,J.G. and Hobbs,R., 2002. Key issues and research priorities in landscape ecology: An idiosyncratic synthesis. Landscape Ecology, 17(4): 355-365.

Yang,C., Peterson,C.L., Shropshire,G.J., and Otawa,T., 1998. Spatial Variability of field Topography and wheat yield in the Palouse region of the Pacific Northwest. Transactions of the ASAE, 41(1): 17-27.

Young,A., 1972. Slopes. Edingburgh.

Young,F.J. and Hammer,R.D., 2000. Soil-landform relationships on a loess-mantled upland landscape in Missouri. Soil Science Society of America Journal, 64(4): 1443-1454.

Zarco-Tejada,P.J., Miller,J.R., Mohammed,G.H., Noland,T.L., and Sampson,P.H., 2001. Estimation of chlorophyll fluorescence under natural illumination from hyperspectral data. International Journal of Applied Earth Observation and Geoinformation, 3(4): 321-327.

Zevenbergen,L.W. and Thorne,C.R., 1987. Quantitative Analysis of land surface topography. Earth Surface Processes & Landforms, 12: 47-56.

Appendix Abbreviations

Table 36. List of Abbreviations.

Abbreviation / Description		Abbreviation / Description	
α	cloud transmittance	ACT	boardcomputer for precision farming - company AGROCOM
ALS	agriculture area of the laserscan	AML	Arc Macro Language
ArcInfo	GIS Software	asl	above sea level
ASP	Exposition to north	AT	atmospheric transmittance in the context of SR-modelling, also the Abbreviation of the fieldsite „Am Teich" in the context of the site description
ATB	Institute for Agricultural Engineering Potsdam-Bornim	BD	Bulk density
BH	Bauhof	BI	Biomass Index
BL	Fieldsite Bei Lotte	BS	Backslope
CBS	**C**onvergent **B**ack**S**lope	CC	Cross correlation
CFS	**C**onvergent **F**oot**S**lope	CH	Combine harvest
CIR	Colour-Infra-Red	CIRC	Circumsolar coefficient
CSH	**C**onvergent **SH**oulder	CURV	Curvature (based on plan and profile curvature)
CV	Coefficient of variation	DBS	**D**ivergent **B**ack**S**lope
DEM	Digital Elevation Model	DFS	**D**ivergent **F**oot**S**lope
DGPS	Differential GPS	DGW	Depth of ground water table
DL	theorethical hours of sunshine	DM	Dry mass
DOY	Day of Year	DSH	**D**ivergent **SH**oulder
ECa	Soil conductiviy measured with EM38	EM	Soil conductiviy measured with VERIS31000
EUF	Electro-Ultra-Filtration	EUF-K_1	first fraction of potassium using EUF-analysis
EUF-K_2	second fraction of potassium using EUF-Analysis	EUF-P_1	first fraction of phosphorus using EUF-analysis
EUF-P_2	second fraction of phosphors using EUF-Analysis	FI	Field
FLACC	Flow accumulation	FS	Footslope
GH	Fieldsite Gasthof	GIS	Geographic Information System
GK	Gauss Krüger Reference System	GPS	Global Positioning System
GSA	German Soil Appraisal	HCL	**H**igh **C**atchement **L**evel
HERMES	Crop growth and Nitrogen model	HH	Hand Harvest
HMF	hydromorphic modification factor	ISATIS	Geostatistical Software
ISRM	Incoming Solar radiation model	$K_{(DL)}$	Potassium content using the DL method
KB	Fieldsite Kleinbahn	KpS	Kernel per Spike

Table 36. List of Abbreviations.

Abbreviation / Description		Abbreviation / Description	
α	cloud transmittance	ACT	boardcomputer for precision farming - company AGROCOM
LAI	Leaf Area Index	LCL	**L**ow **C**atchment **L**evel
LEC	Landform Element Complex	Level	Level Areas
LF	Landform	LFC	Landforms (area filtered)
LS	Laserscan	LS10	Laserscan resmapled to a resolution of 10 m by 10 m
MC	Monte-Carlo-Simulation	MMK	Medium Scale Soil Map based on the Results of the GSA for Eastern Germany
$Moist_{meas}$	Grain Yield moisture measured at combine	MW	Mean value
N	Number of Samples/Iterations	N-Sensor	Fertilizer Application Sensor - company HYDRO
NDVI	Normalized Difference Vegetation Index	NIR	Near infrared image
N_{Kernel}	Number of Kernels	N_{min}	mineralized Nitrogen
NS-ratio	Nugget to Sill ratio	N_{Spike}	Number of Spikes
$P_{(DL)}$	Phosphorus content using the DL method	PBS	**P**lanar **B**ack**S**lope
Pen.Res.	Penetration Resistance	PEST	Nonlinear Parameter Optimization software
PF	Precision Farming	PFS	**P**lanar **F**oot**S**lope
PLAN	Planform Curvature	proc	processed yield data
PROF	Profile curvature	Prot	Protein Content
PSH	**P**lanar **SH**oulder	REF1	Weatherstation at farm house
REF2	Weatherstation at field site „Bei Lotte“	RMSS	Root Mean Square Slope
RSD	Relative Standardized Difference	RTK	Real Time Kinematik
SB	Spring barley	SCA	Specific Catchment Area
SCD	Significant Crosscorrelation Distance	SD	Standard Deviation
sf	Sunshine fraction	SH	Shoulder
SISSR	Simulated incoming sloped SR	SK	Fieldsite Sportkomplex
SLP	Slope	SM	Soil moisture
SOC	Soil organic carbon	SOM	Soil organic matter
SPI	Stream Power Index	SPSS	Statistic software
SR	Solar irradiance	SRAD	Model to simulate solar radiation and temperature across a landscape
SRCF	solar radiation correction factor	SSM	Soil surface moisture
STC	Sediment Transport Capacity	TK	Topographic map sheet
TK10	DEM based on contour lines of the TK 1:10.000	TK100	DEM based on contour lines of the TK1:100.000
TK25	DEM based on contour lines of the TK 1:25.000	TKM	Thousand Kernel Mass
TWI	Topographic Wetness Index	WGS84	World Geodetic Reference System 1984
WR	Winter rye	WS1	Pyranometer without topographic shading
WS2	Pyranometer with topographic shading	WW	Winter wheat
Yie_{moist}	Grain yield at field moisture	Yie_{std}	Grain yield at specific moisture
YpS	Yield per Spike	ZALF	Center for Agriculture Landscape and Landuse Research Müncheberg

Appendix Site Description

Table 37. Depth profiles of the TK4845 from Härtel (1931) and own investigations.

Lommatzsch (Brickworks of K.Hohnstein) Nr. 22

Depth in m	< 2mm (Gravel)	2- 0.05 mm (Sand)	< 0.05 mm (Clay+Silt)
0.25-0.3	0.0	11.4	88.6
1	0.0	10.7	89.3
2	0.0	9.2	90.8
3	0.0	9.8	90.2
4	0.0	4.7	95.3
5	0.0	8.6	91.4
6	0.1	10.5	89.4

SW of Munschwitz Nr. 26

Depth in m	< 2mm (Gravel)	2- 0.05 mm (Sand)	< 0.05 mm (Clay)
0-0.25	0.0	15.2	84.8
0.30-0.4	0.0	11.1	88.9
0.5-0.6	0.0	9.0	91.0

Field "Bei Lotte" Point G9

Depth in m	Sand	Silt	Clay
0-0.3	2.2	78.1	19.6
0.3-0.6	2.0	77.0	21.0
0.6-0.9	1.9	78.3	19.8
2.4-2.5	7.8	67.2	25.0

Table 38. Average monthly precipitation in mm from Härtel(1931) and own calculations from data of the meterological service of the GDR.

Month	Geol.Map	RÜSEINA	MOCHAU
January	36	50	45
February	43	37	36
March	47	44	41
April	44	54	47
May	68	63	56
June	87	67	56
July	89	60	49
August	72	80	57
September	51	47	42
October	53	45	43
November	46	53	45
December	51	62	60
Location/Period	unknown / 1864-1923	see text / 1969-1989	see text / 1969-1989

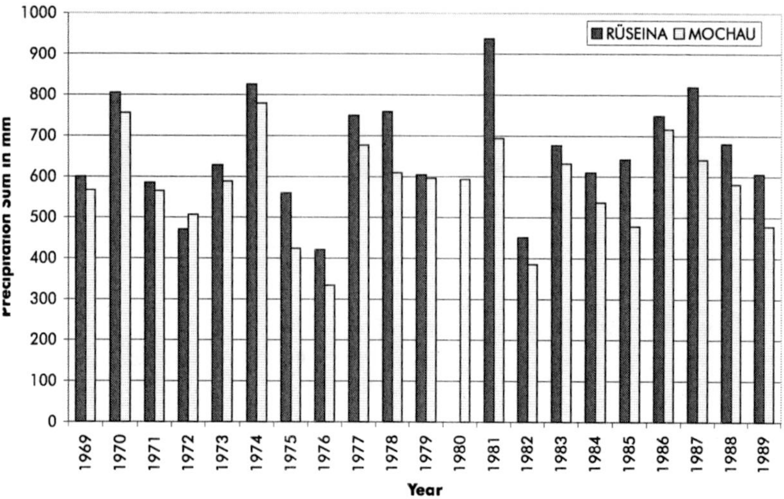

Figure 93. Annual precipitation for the time period 1969 -1989 for the precipitation stations Rüseina and Mochau.

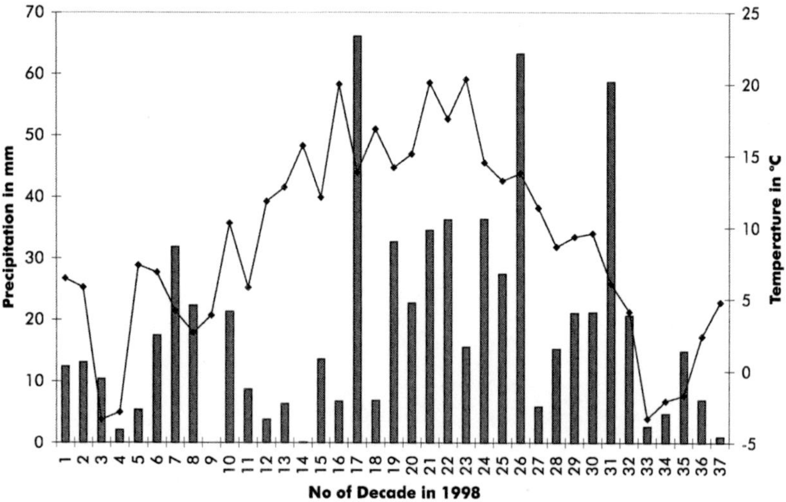

Figure 94. Precipitation Sum in mm and Average Temperature in degree C for decades in 1998.

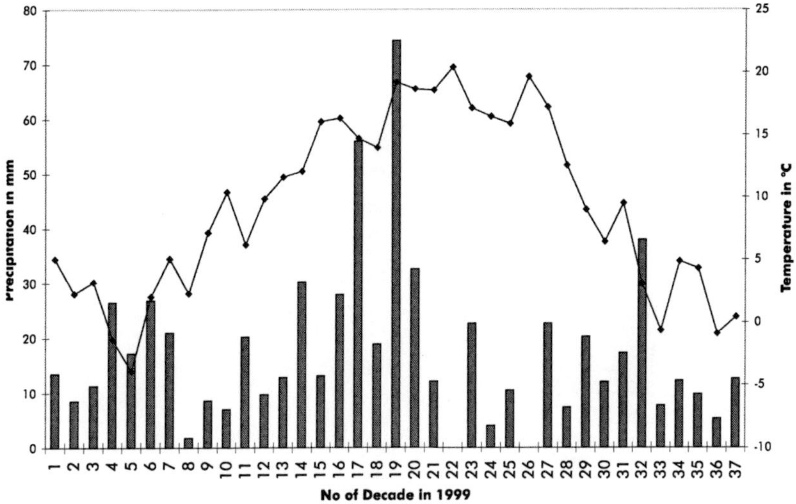

Figure 95. Precipitation Sum in mm and Average Temperature in degree C for decades in 1999.

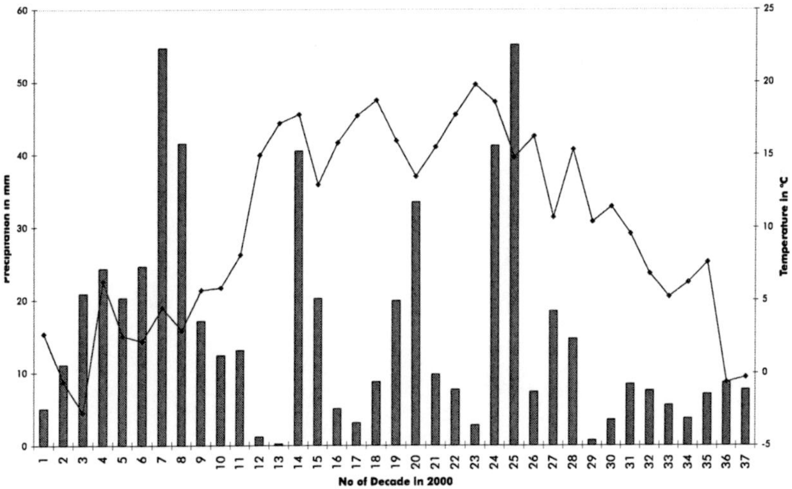

Figure 96. Precipitation Sum in mm and Average Temperature in degree C for decades in 2000.

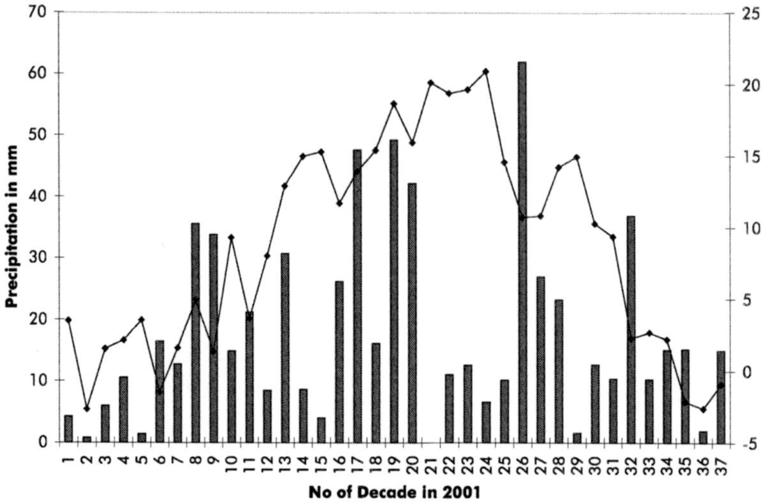

Figure 97. Precipitation Sum in mm and Average Temperature in degree C for decades in 2001.

Appendix Statistics

Table 39. Descriptive parameters of Spring Soil Moisture in $g(100g\ soil)^{-1}$ for the field site "Bei Lotte".

Depth/Date	N	Minimum	Maximum	Mean	Standarddeviation
0-30 cm 100599	12	19.67	23.06	21.1254	0.8667
30-60 cm 100599	12	19.9	21.67	21.07	0.5962
60-90 cm 100599	12	19.75	22.86	21.675	1.0332
0-30 cm 270499	12	22.15	25.25	23.82	0.8895
30-60 cm 270499	12	20.88	23.72	21.8358	0.8216
60-90 cm 270499	12	19	23.79	21.9837	1.3523
0-30 cm 080399	12	22.51	26.29	24.1723	1.2112

Table 39. Descriptive parameters of Spring Soil Moisture in $g(100g\ soil)^{-1}$ for the field site "Bei Lotte".

30-60 cm 080399	12	17.9	19.85	18.7106	0.5876
60-90 cm 080399	12	17.74	19.44	18.6902	0.5613
0-30 cm 270598	2	13.02	13.33	13.175	0.2192
30-60 cm 270598	2	13.8	16.88	15.34	2.1779
60-90 cm 270598	2	18.7	19.59	19.145	0.6293
0-30 cm 220300	12	25.37	30.56	27.3454	1.8146
30-60 cm 220300	12	19.67	25.99	23.6279	1.7302
60-90 cm 220300	12	16.74	24.91	22.9271	2.1738
0-30 cm 180400	12	22.78	27.91	24.8629	1.7395
30-60 cm 180400	12	20.73	24.46	22.2313	1.1193
60-90 cm 180400	12	20.45	23.65	22.4608	0.9243
0-30 cm 030500	12	15.49	20.86	16.8846	1.4717
30-60 cm 030500	12	17.76	22.13	19.49	1.1057
60-90 cm 030500	12	15.78	21.89	20.6367	1.6674
0-30 cm 160301	12	24.92	27.35	25.7566	0.727
30-60 cm 160301	12	21.01	23.97	22.5852	0.9846
60-90 cm 160301	12	18.62	24.14	22.4736	1.5424
0-30 cm 180401	12	24.91	28.52	26.7532	1.168
30-60 cm 180401	12	22.95	25.72	24.1198	0.9349
60-90 cm 180401	12	19.33	24.16	22.5952	1.4581
0-30 cm 080501	12	24.72	29.33	26.6891	1.3808
30-60 cm 080501	12	21.32	24.42	22.6094	0.9225
60-90 cm 080501	12	20.35	24.3	22.44	1.0996

Table 40. Descriptive parameters of Spring N_{min} in $kgha^{-1}$ at the field site "Bei Lotte".

Depth/Date	N	Minimum	Maximum	Mean	Standarddeviation
0-30 cm 270598	2	30.15	46.8	38.475	11.7733
30-60 cm 270598	2	7.65	9	8.325	0.9546
60-90 cm 270598	2	11.7	13.95	12.825	1.591
0-30 cm 080399	12	24.52	49.67	36.2492	7.5507
30-60 cm 080399	12	13.97	40.01	25.0145	7.5676
60-90 cm 080399	12	8.52	22.25	13.3117	4.1861
0-30 cm 270499	12	93.15	133.2	111.75	13.7236
30-60 cm 270499	12	20.25	47.25	34.5	8.1121
60-90 cm 270499	12	10.35	29.7	18.825	6.5291
0-30 cm 100599	12	66.15	149.85	95.25	24.1749
30-60 cm 100599	12	16.65	58.5	34.5375	11.9076
60-90 cm 100599	12	9.45	32.4	22.05	7.7895
0-30 cm 220300	12	12.24	38.93	17.9676	7.5116
30-60 cm 220300	12	5.61	14.38	8.2441	2.4226
60-90 cm 220300	12	2.92	13.61	6.5304	2.8798
0-30 cm 180400	12	17.47	33.55	22.9921	5.4508
30-60 cm 180400	12	1.98	10.59	5.6687	2.4706
60-90 cm 180400	12	0.53	9.96	3.019	2.7404
0-30 cm 030500	12	9.64	21.48	14.7716	4.1088
30-60 cm 030500	12	0.8	6.44	2.8793	2.1665
60-90 cm 030500	12	0.52	2.5	1.0975	0.666
0-30 cm 160301	12	8.83	15.36	11.8526	2.6359
30-60 cm 160301	12	2.05	7.06	3.5609	1.4297
60-90 cm 160301	12	0.83	2.36	1.5001	0.5357
0-30 cm180401	12	1.5	42.74	18.8655	10.5766

Table 40. Descriptive parameters of Spring N_{min} in $kgha^{-1}$ at the field site "Bei Lotte".

30-60 cm 180401	12	2.29	17.42	7.1867	4.6165
60-90 cm 180401	12	1.02	14.07	4.3286	3.3133
0-30 cm 080501	12	17.29	68.65	33.0061	14.65
30-60 cm 080501	12	5.96	18.41	9.2922	3.7977
60-90 cm 080501	12	0.94	8.76	4.3336	2.4695

Table 41. Descriptive parameters for soil surface water content (SSM) and soil water content (SM) in $g(100g\ soil)^{-1}$ for the field site „Bei Lotte“. (Values given in parentheses for soil water content in 2000 show results from the first sampling date).

Property	Depth in cm	Date	Count	Minimum	Maximum	Mean	CV
SSM	0-10	25.03.1999	192	23.28	35.73	26.9039	4.66
SSM	0-10	29.04.1999	192	23.44	28.45	25.6991	11.91
SSM	0-10	10.05.1999	192	19.67	26.29	22.7006	6.87
SSM	0-10	12.05.2000	192	7.2055	16.1996	11.4537	3.79
SM	0-30	10.08.1998	192	16.75	28.55	21.71	13.15
SM	30-60	10.08.1998	192	11.64	22.47	16.77	12.63
SM	60-90	10.08.1998	192	8.80	21.43	16.72	11.99
SM	0-30	16.08.1999	192	16.12	27.35	21.62	7.46
SM	30-60	16.08.1999	192	17.70	24.34	20.17	4.71
SM	60-90	16.08.1999	192	14.68	22.92	20.34	6.04
SM	0-30	17.08.2000	192(138)	8.52(8.52)	26.06(16.39)	14.97(12.4)	31.03(11.14)
SM	30-60	17.08.2000	192(138)	6.44(6.44)	24.36(14.01)	11.16(10.4)	23.05(10.31)
SM	60-90	17.08.2000	192(138)	7.70(7.70)	21.00(14.01)	12.44(11.98)	15.97(8.79)
SM	0-30	13.08.2001	191	10.69	22.86	15.38	11.24
SM	30-60	13.08.2001	191	10.27	21.77	16.12	10.56
SM	60-90	13.08.2001	191	8.87	23.30	17.90	8.78

Table 42. Descriptive parameters for soil organic carbon (SOC) and soil texture in $g(100g\ soil)^{-1}$ for the field site „Bei Lotte“.

Property	Depth in cm	N	Min	Max	Mean	CV
SOC	0-30	64	0.82	1.47	1.10	11.13
SOC	30-60	64	0.25	1.00	0.53	29.96
SOC	60-90	64	0.13	0.80	0.26	39.26
Sand	0-30	63	1.84	9.03	3.45	41.03
Sand	30-60	63	1.73	11.08	3.13	62.45
Sand	60-90	62	1.23	13.65	3.00	70.15
Silt	0-30	64	68.06	85.22	80.02	3.77
Silt	30-60	64	66.08	90.12	78.86	4.41
Silt	60-90	64	53.01	87.68	77.49	5.64
Clay	0-30	64	11.42	20.80	16.53	15.27
Clay	30-60	64	7.16	22.51	18.01	16.86
Clay	60-90	64	9.34	26.35	19.51	15.85

Table 43. Descriptive parameters for phosphorus in mg(100g soil)$^{-1}$, potassium in mg(100g soil)$^{-1}$, pH, EUF-P_1 in mg(100g soil)$^{-1}$, EUF-P_2 in mg(100g soil)$^{-1}$, EUF-K_1 in mg(100g soil)$^{-1}$ and EUF-K_2 in mg(100g soil)$^{-1}$ for the field site „Bei Lotte".

Property	Depth in cm	Year	N	Mean	Min	Max	CV
$P_{(DL)}$	0-30	1998	49	29.09	5.72	37.06	52.92
$P_{(DL)}$	30-60	1998	49	9.04	3.25	21.92	48.16
$K_{(DL)}$	0-30	1998	49	9.29	3.21	19.91	52.31
$K_{(DL)}$	30-60	1998	49	5.74	1.87	11.95	59.21
pH	0-30	1998	49	6.53	5.70	7.08	4.46
pH	30-60	1998	49	6.42	5.70	7.01	4.83
pH	0-30	2000	15	6.47	5.92	6.74	2.79
P_1	0-30	1999	192	0.50	3.78	1.31	41.89
P_2	0-30	1999	192	0.16	1.45	0.55	44.77
K_1	0-30	1999	192	1.97	25.23	5.15	52.51
K_2	0-30	1999	192	0.53	7.66	2.50	40.69
P_1	0-30	2000	192	0.37	3.05	1.14	40.29
P_2	0-30	2000	192	0.17	1.81	0.66	46.04
K_1	0-30	2000	192	0.65	16.19	4.50	43.56
K_2	0-30	2000	192	0.30	7.33	2.57	35.62
P_1	0-30	2001	191	0.41	3.57	1.14	37.91
P_2	0-30	2001	191	0.18	1.88	0.59	43.30
K_1	0-30	2001	191	2.16	21.31	5.00	41.37
K_2	0-30	2001	191	0.99	7.73	2.44	31.48

Table 44. Descriptive parameters for residual N_{min} in kgha^{-1} for the field site „Bei Lotte" for the time period 1998-2001.

Property	Depth in cm	Date	N	Min	Max	Mean	CV
N_{min}	0-30	10.08.1998	192	13.95	130.95	42.65	41.98
N_{min}	30-60	10.08.1998	192	1.35	98.55	19.59	74.43
N_{min}	60-90	10.08.1998	192	1.34	50.49	8.41	83.06
N_{min}	0-30	16.08.1999	192	15.50	120.32	53.30	39.28
N_{min}	30-60	16.08.1999	192	2.25	37.42	15.21	45.75
N_{min}	60-90	16.08.1999	192	1.11	16.68	6.39	48.69
N_{min}	0-30	17.08.2000	192	2.82	78.78	29.17	38.58
N_{min}	30-60	17.08.2000	192	0.94	23.19	5.28	61.27
N_{min}	60-90	17.08.2000	190	0.46	34.97	3.07	91.65
N_{min}	0-30	13.08.2001	191	8.38	157.91	34.28	50.91
N_{min}	30-60	13.08.2001	191	1.33	52.17	8.44	78.87
N_{min}	60-90	13.08.2001	187	0.46	17.76	2.95	78.31

Appendix AML

All AML-scripts provided here are found at http://arcscripts.esri.com
LOGOFF.AML

```
/* NAME: logoff.aml
/* WHAT: Turn off grid log file writing
/* INPUT: OFF/ON
/* DEVELOPER: hannes isaak reuter, zalf muencheberg, germany
/* CONTACT: hreuter@zalf.de / gisxperts@web.de
/* CHANGES: HIR 23/01/2002 developing
/*    HIR 23/06/2003 cleaning for arcsripts upload....
/* COMMENTS
/*          only for unix workstations
/*         Workaround for BUG CQ00045482
/*         LOGFILE OFF is not working in GRID
/*         Improvements: extend for windows NT
/*         could implement a unique filename
/* USAGE: &r logoff  <ON/OFF>
&sev &error &routine BailOut
&args status
/* copy workspace log to log.save
/*
/* logoff <OFF/ON>
&sv status = [upcase %status%]
&if %status% = OFF &then
&do
&if [exists log.save -file] &then &return ---> Saved LOG (log.save) already exists, quitting
&if [exists log -file] &then
&do
&sys cp log log.save
&type LOG OFF
&end
&else &return ---> LOG does not exists
&end
&else
&do
&if [exists log.save -file] &then
&do
&sys mv log.save log
&type LOG ON
&end
&else &return ---> Saved LOG does not exist, Do nothing
&end
/* end of aml
&return

/* ########################
&routine BailOut
&sev &error &fail
&type \--> AML ERROR: %aml$errorfile%
&type --> LINE : %aml$errorline%
&type --> ERROR: %aml$message%
```

```
&return
```

KILLGRIDS.AML

```
/* NAME:  killgrids.aml
/* WHAT:  killing many similar named grids
/* INPUT: DEM's with wildcard
/* DEVELOPER: hannes isaak reuter, zalf muencheberg, germany
/* CONTACT:  hreuter@zalf.de / gisxperts@web.de
/* CHANGES: HIR 10/06/2001 developing based on killcover -script
/*                         by  (SB, VCGI, July 1995)
/*                   HIR 23/06/2003 cleaning for arcsripts upload....
/*  COMMENTS:
/* USAGE: &r killgrids <DEM with wildcard>
&sev &error &routine BailOut
&args wildcard option
&if [null %wildcard%] &then
  &return --> USAGE: killgrids <wildcard> {ALL | ARC | INFO}
&if not [exists %option% -dir] &then
  &s option = ALL
&else
  &s option = [translate %option%]
&if not %option% in {'ALL','ARC','INFO'} &then
  &return --> USAGE: killgrids <wildcard> {ALL | ARC | INFO}
&s covcount = [filelist %wildcard% xxkillgrids.tmp -grid]
&if %covcount% = 0 &then
  &type \--> No grids match specification %wildcard%
&else &do
  &s file = [open xxkillgrids.tmp openstat -r]
  &s cov = [read %file% readstat]
  &do &while %readstat% = 0
   kill %cov% %option%
   &s cov = [read %file% readstat]
  &end
  &s close = [close %file%]
&end
&if [exist xxkillgrids.tmp -file] &then &s del [delete xxkillgrids.tmp -file]
&return

/* ################
&routine BailOut
&sev &error &fail
&type \--> AML ERROR: %aml$errorfile%
&type --> LINE : %aml$errorline%
&type --> ERROR: %aml$message%
&return
```

TOPO.AML

```
/* NAME: TOPO.AML
/* WHAT: calculates several relief indices for a DEM - grid
/* INPUT: DEM, flow threshold, stream cover
/* DEVELOPER: hannes isaak reuter, zalf muencheberg, germany
/* CONTACT:  hreuter@zalf.de / gisxperts@web.de
/* CHANGES:  HIR 1999-2001 developing
/*         HIR 23/06/2003 cleaning for arcsripts upload....
/* IMPROVEMENTS: clean up code, more relief parameter possible like distance to stream etc...
/* COMMENTS: base relief parameters are computed
/*         if stream is provided, topo expects a cover
/* !!!!!!!!!!!!!!!!!!!! BE CAREFUL !!!!!
/* CHECK WITH YOUR ARC VERSION HOW THE CURVATURE PARAMETERS ARE HANDLED
/* SOMEONE AT ESRI CHANGED THE SIGN OF THE CURVATURE DURING DEVELPOMENT
```

```
/* IMPORTANT  IMPORTANT  IMPORTANT
/* USAGE: topo <DEM> {streamflow threshold} {streamcover}
&sev &error &routine BailOut
&args file flow stream
&term 9999
&if [null %file%] &then &return USAGE: topo <DEM> {streamflow threshold} {streamcover}
&if [null %flow%] &then
&do
&s flow = 100
&type
&type Set streamflow threshold to 100
&type
&end
&else type Set streamflow threshold to %flow%
&if not [exists %file% -grid] &then &return Input DEM does not exists
display 0  /* faster....
/* logfile and grid testing
&sv programold = %:program%
&if %:program% = ARC &then &do
&r logoff off
GRID
&end
&else
&r logoff OFF
/* logfile off end
/* setup
&describe %file%
&sv cell_size = %GRD$DX%
setcell %cell_size%
/* clear old files
&if [exists %file%ASP -grid] &then kill %file%ASP all
&if [exists  %file%bas -grid] &then kill  %file%BAS all
&if [exists  %file%basa -grid] &then kill  %file%BASa all
&if [exists  %file%cur -grid] &then kill  %file%CUR all
&if [exists  %file%flacc -grid] &then kill %file%FLACC all
&if [exists  %file%fldir -grid] &then kill %file%FLDIR all
&if [exists  %file%plan -grid] &then kill  %file%PLAN all
&if [exists  %file%prof -grid] &then kill  %file%PROF all
&if [exists  %file%slp -grid] &then kill  %file%SLP all
&if [exists  %file%strlnk -grid] &then kill  %file%STRLNK all
&if [exists  %file%strnet -grid] &then kill   %file%STRNET all
&if [exists %file%strord -grid] &then kill   %file%STRORD all
&if [exists %file%wshd -grid] &then kill %file%wshd all
&if [exists %file%WI -grid] &then kill   %file%WI all
&if [exists %file%SPI -grid] &then kill   %file%SPI all
&if [exists %file%wsha -grid] &then kill   %file%wsha all
&if [exists %file%STC -grid] &then kill   %file%STC all
&if [exists %file%fl1 -grid] &then kill   %file%fl1 all
&if [exists %file%lf -grid] &then kill %file%lf
&if [exists %file%lfr -grid] &then kill %file%lfr
&if [exists tmp12345 -grid] &then kill tmp12345
&if [exists %file%pctg1 -grid] &then kill %file%pctg1
&if [exists %file%rdg -grid] &then kill %file%rdg
/* computing 1 -5
&type
&type Compute general indices
&type
%file%cur = curvature ( %file%, %file%prof, %file%plan, %file%slp, %file%asp )
/* computing 6
%file%fldir =  flowdirection ( %file% )
%file%flacc = flowaccumulation ( %file%fldir )
/* convert
&sv cellsqr = %cell_size% * %cell_size%
```

```
%file%fl1 = con ( %file%flacc == 0, 1 * %cellsqr%, %file%flacc * %cellsqr% )
/* create ridge grid
%file%rdg = con ( %file%flacc == 0, 1, 0 )
/* streamnet delineating
/* change the value in here.. 100, 250, watershadsize should be by around ??? )
%file%strnet = con (%file%flacc > %flow%, 1)
%file%strord = streamorder ( %file%strnet , %file%fldir )
%file%strlnk = streamlink ( %file%strnet , %file%fldir )
/* get into watersheed computations
&if [exists tmp12345 -grid] &then kill tmp12345
&if [exists tmp12346 -grid] &then kill tmp12346
%file%bas = basin( %file%fldir )
&if not [null %stream%] &then
%file%wshd = watershed ( %file%fldir , %stream% )
&else
&do
/* %file%wshd = watershed ( %file%fldir , %file%strnet )
/* found now algorithm do define watershed on www
/* 02.04.2002 www.ce.utexas.edu/prof/maidment/gishyd97/class/wshed/wshed.htm
tmp12345 = zonalmax ( %file%strlnk, %file%flacc )
tmp12346 = con ( tmp12345 == %file%flacc, %file%strlnk )
%file%wshd = watershed ( %file%fldir , tmp12346 )
/* commented deleting tmp12346 off - need it for pctg1
&if [exists tmp12345 -grid] &then kill tmp12345
/* &if [exists tmp12346 -grid] &then kill tmp12346
&end
&type
&type COMPUTING INDICES NOW
&type
/* indices from i.d.moore 1993
/* computing WETNESS and Stream POWER index
/* xx has to be upslope area
/* prepare wsha to set NODATA at least to 1 m due to depressionless sinks
/* &type replace nodata in watershad area with a 1 for landform classification
&if [exists tmp12345 -grid] &then kill tmp12345
tmp12345 = zonalarea ( %file%wshd )
%file%wsha = con ( isnull ( tmp12345), 1, tmp12345 )
/* buildvat %file%wsha  /* remove 16052003 due to error
&if [exists tmp12345 -grid] &then kill tmp12345
tmp12345 = zonalarea ( %file%bas )
%file%basa = con ( isnull (tmp12345), 1, tmp12345 )
&if [exists tmp12345 -grid] &then kill tmp12345
/* buildvat %file%basa /* remove 16052003 due to error
&type computing Wetness index
&type
%file%slp1 = con ( %file%slp == 0, 0.0001, %file%slp )
%file%WI = ln ( %file%fl1 / tan ( %file%slp1 div DEG ) )
kill %file%slp1
&type
&type computing stream power index
&type
%file%SPI = %file%fl1 * tan ( %file%slp div DEG )
&type
&type computing Sediment transport index
&type
%file%STC = pow ( ( %file%fl1 / 22.13 ), 0.6 ) *  pow ( ( sin ( %file%slp div DEG ) / 0.0896 ) , 1.3 )
&type
&type WI, SPI and STC are done
&type
/* &echo &on
/* implemented 22.04.2002
/* relief slope position algorithm based on alberta soil landscape algorithm
/* tmp12346 contains point of lowest elevation of
```

```
&type
&type Compute Relative Relief Position
&type
&if [exists tmp12345 -grid] &then kill tmp12345
&if [exists tmp12346 -grid] &then kill tmp12346
/* zonalamax for watershed
tmp12345 = zonalmax ( %file%wshd, %file% )
/* zonalmin for watershed
tmp12346 = zonalmin ( %file%wshd, %file% )
/* pctg1 ( !!! pctg is used by elevres.aml !!! )
%file%pctg1 = (( %file% - tmp12346 ) / ( tmp12345 - tmp12346 ) ) * 100
/* ##############################
/* some indices are still missing
/* ##############################
&type
&type Ready for landform classification
&type
&if %programold% = GRID &then
&r LOGOFF ON
&else
&do
&r logoff ON
q
&end
/* add aml call to log file
LOG * add
topo %file% %flow% %stream%
&messages &on
&return

/* #######################################################
/* routine bailout start
/* #######################################################
&routine BailOut
&sev &error &fail
&type \--> AML ERROR: %aml$errorfile%
&type --> LINE : %aml$errorline%
&type --> ERROR: %aml$message%
&r logoff on
&messages &on
&return
/* #######################################################
/* routine bailout end
/* #######################################################
```

MONTEWI.AML

```
/* NAME: montewi.aml
/* WHAT: calculates monte carlo topographic wetness index for a DEM - grid
/* INPUT: DEM, outgrid name, expected error of height, number of runs/or stop threshold
/* DEVELOPER: hannes isaak reuter, zalf muencheberg, germany
/* CONTACT: hreuter@zalf.de / gisxperts@web.de
/* CHANGES: HIR 24/02/2002 developing based on aml for viewshed analysis by anonymous
/*        HIR 23/06/2003 cleaning for arcsripts upload....
/* IMPROVEMENTS: extend to allow for different relief parameters
/* COMMENTS:
/* Citation Reuter,H.I. 2002 AML zur Berechnung des
/*          Topograhischen Wetness Indexes
/*          unter Berücksichtigung der Höhenfehler
/*          des DGM mittels Monte Carlo Simulation,
/* USAGE: &r montewi <inputDEM> <outputdem> <standard deviation> <number of iterations> {break}
/* ####################################################
```

```
/* Requirements: logoff.aml ( in GRID  LOG FILE turn off )
/* killgrids.aml ( delete many grids )
/* ##########################################################
&sev &error &routine BailOut
&args  .dgm .out .std .n .break
&sv .s 0
&r logoff OFF
&call error
&call calculation
&call statistic
&r logoff ON
LOG * add
montewi %file% %.out% %j1% %.n% %.std% %.break%
&return

/* ############################################
/* ############ setup ##############
/* ############################################
  &routine error
 /* check grid
   &if [show program] ne GRID &then grid
 /* Input Parameter check
   &do I &list %.dgm% %.std% %.n% %.out%
    &if [null %I%] &then
      &do
       &type Usage: MONTEWI <dem> <outgrid> <stdev> <n steps> {break}
       &return; &return
      &end
    &end
 /* Input data check
&if [exists %.dgm% -grid] eq .false. &then
    &do
     &type Input-DEM does not exist!
     &return; &return
    &end
   &if [exists %.out% -grid] eq .true. &then
    &do
     &type Output-Grid already exists!
     &return; &return
    &end
   /* compute optimum break off points by dividing std / number of runs
   &if [null %.break%] &then &do
   &sv .break = [calc %.std% / %.n% ]
   &type
   &type Break occurs at %.break%, if you want more iterations, set break to a smaller value
   &type
   &end
 /* löschen aller tempgrids
   &r killgrids temp*
   &r killgrids sum*
/* create file for saving
&sv stdfile = %.out%std.txt
&setvar file_unit = [open %stdfile% openstatus -write]
&return

/* ################################################
/* ############## calculation ####################
/* ################################################
&routine calculation
 &messages &off
 &sv j = 1
 &sv di = 0
  &do &until %j% > %.n%
```

```
/* create DEM + normal distribution DEM
   &if [exists temp -grid] eq .true. &then kill temp all
   &ty Computing tempdem ...
/* may be using rand() ?? also possible ??
   temp = %.dgm% + normal () * %.std%
   /* moran temp
/* temp ist elevation dgm to compute
/* tempwis is result of TWI

 &call wetness

    &if [exists sumvis%j% -grid] eq .true. &then kill sumvis%j% all
    &if %j% eq 1 &then sumvis1 = tempvis  /*Anlegen des Summengrid
    &else
      &do
       sumvis%j% = sumvis%di% + tempvis /* Summengrid = Summengrid_j-1 + Visabil._j
       kill sumvis%di% all  /*Löschen von sumvis(j -1)
      &end
   /* looking for optimum
&sv j1 = %j% /* need that for division
   &if [exists tmp -grid] &then kill tmp
   &if %j% > 1 &then &do
    tmp = sumvis%j% / %j%
    &describe tmp
    /* &sv mean = %GRD$MEAN%
    &sv std = %GRD$STDV%
    &if %j% > 2 &then &do
     /* break off if optimum is reached
/* &messages &on
/* &echo &on
     &sv test = [ABS [calc %stdo% - %std%]]
     /* &type %test%
     &if ( %test% < %.break% ) &then &do
&sv j1 = %j% /* need that for division
&sv j = %.n% /*need that for break off
&end
 &end
&sv stdo = %std%
&sv record = %j1%' '%std%
&sv a = [write %file_unit% %record%]
 kill tmp
   &end
   /* optimum end
   &type done %j1% of %.n% loops ...
   &sv j %j% + 1
   &sv di %di% + 1
  &end
/* &sv j = %j% - 1
/*   &type %j%
  &type
  %.out% = sumvis%di% / %j1% /* Summengrids auf out bringen und durch anzahl der läufe dividieren
  &type The Stddev of loop %j1% showed only a difference of %test% to the Stddev of the run before
  &type therefore MC is stopped here.
  kill tempvis all
/* delete all tempgrids
&r killgrids tmp*
&r killgrids sum*
/* close std recording file
&sv a = [close %file_unit%]
&messages &on
&return
/* ######################################
/* ################ Statistik##############
```

```
/* ######################################
  &routine statistic
      &describe %.out% /* get variables
    &format 4
     &ty  statistics for %.out%
     &ty ____________________________________
     &ty
     &ty [format 'minimum value: %1%' %grd$zmin%]
     &ty [format 'maximum value: %1%' %grd$zmax%]
     &ty [format 'mean value: %1%' %grd$mean%]
     &ty [format 'standard deviation: %1%' %grd$stdv%]
     &ty
&return
/* ##############################################
/* ############ TOPOGRAPHIC WETNESS-INDEX
/* ##############################################
&routine wetness
&type computing Wetness index
/* &messages &off
&if [exists tempvis -grid] &then kill tempvis
&describe temp
&sv cell_size = %GRD$DX%
&sv file temp
&sv cellsqr = %cell_size% * %cell_size%
%file%fldir = flowdirection ( %file% )
&sys echo -n „."
%file%flacc = flowaccumulation ( %file%fldir )
&sys echo -n „."
%file%slp = slope ( %file% )
&sys echo -n „."
%file%fl1 = con ( %file%flacc == 0, 1 * %cellsqr%, %file%flacc * %cellsqr% )
&sys echo -n „."
%file%slp1 = con ( %file%slp == 0, 0.0001, %file%slp )
&sys echo -n „."
tempvis = ln ( %file%fl1 / tan ( %file%slp1 div DEG ) )
&sys echo -n „."
&ty
kill %file%slp1
kill %file%slp
kill %file%flacc
kill %file%fl1
kill %file%fldir
kill %file%
&return
/* ########################
/* routine wetness index end
/* ########################

/* ######################################################
/* routine bailout start
/* ######################################################
&routine BailOut
&sev &error &fail
&type \--> AML ERROR: %aml$errorfile%
&type --> LINE : %aml$errorline%
&type --> ERROR: %aml$message%
&sv a = [close %file_unit%]
&r logoff on
&messages &on
&return
/* ######################################################
/* routine bailout end
/* ######################################################
```

ELEVRES.AML

```
/* NAME: elevres.aml
/* WHAT:  DEM elevation residual analysis
/* INPUT: DEM,  filter size
/* DEVELOPER: hannes isaak reuter, zalf muencheberg, germany
/* CONTACT:  hreuter@zalf.de / gisxperts@web.de
/* CHANGES: HIR 10/06/2001 developing
/*    HIR 23/06/2003 cleaning for arcsripts upload....
/* COMMENTS:
/*       based on work published in wilson & gallant 2000
/* size should be the scale to be investigates, e.g. hillslope length
/* ls.aml should be helpfull to determine the hillslope length,
/* or looking at flowaccumulation
/* USAGE: &r elevres <DEM> {window size in cells}
&args dgm size
&if [null %dgm%] &then &return USAGE: elevres <DEM> {window size in cells}
&if [null %size%] &then &sv size = 200
&type Window size is set to %size%
&type
&type BE CAREFULL - LARGE WINDOW SIZE NEED HUGE COMPUTING TIMES
&type
&if [exists %dgm%mean -grid] &then kill %dgm%mean
&if [exists %dgm%diff -grid] &then kill %dgm%diff
&if [exists %dgm%sd -grid] &then kill %dgm%sd
&if [exists %dgm%rang -grid] &then kill %dgm%rang
&if [exists %dgm%dev -grid] &then kill %dgm%dev
&if [exists %dgm%pctg -grid] &then kill %dgm%pctg
&if [exists %dgm%min -grid] &then kill %dgm%min
&if [exists %dgm%max -grid] &then kill %dgm%max
&if [exists tmp12345 -grid] &then kill tmp12345
&if %:program% = ARC &then grid
/* mean elevation
/* smoothed DEM
%dgm%mean = focalmean (%dgm%, circle, %size% )
&type done with mean
/* diff from mean elevation
/* shows relative topographic position - lightning strikes
%dgm%diff = %dgm% - %dgm%mean
&type done with diff
/* std of elevation
/* provides measure of local relief -
/* if using large windows- roughtness of landscape
%dgm%sd = focalstd ( %dgm%, circle, %size% )
&type done with SD
/* elevation range
/* similar to SD, disadvantage -> sometimes changing abruptly from place to place
%dgm%range = focalrange ( %dgm%, circle, %size% )
&type done with range
/* deviation from mean elevation
%dgm%dev = %dgm%diff / %dgm%sd
&type done with deviation
/* percentile as percentage of elevation range
&type compute min and max - needed parameter for percentage
%dgm%min = focalmin (%dgm%, circle, %size% )
%dgm%max = focalmax (%dgm%, circle, %size% )
%dgm%pctg = 100 * ((%dgm% - %dgm%min) / (%dgm%max - %dgm%min ))
&type done with perecentage
&type
&type performed elevation residual analysis
&type
&return
```

LANDFORM.AML

```
/* NAME: landform.aml (previous name pennock94/pennock95.aml)
/* WHAT: landform classification analysis
/* INPUT: DEM, running topo.aml as prequisite,
/*          default parameter for 10 m x 10 m DEM are set
/* DEVELOPER: hannes isaak reuter, zalf muencheberg, germany
/* CONTACT: hreuter@zalf.de / gisxperts@web.de
/* CHANGES: HIR 1999-2002
/*    HIR 23/06/2003 cleaning for arcsripts upload....
/* COMMENTS:
/* does landform classification after original paper by pennock 1994
/* and further literature sources
/* does extended landform classification for planar landforms,
/* preserves shape of each landform, not only SH,BS,FS, and Level
/* implements an filtering algorithm to reduce the number of
/*  misclassified relief units
/* method is the 8 or 11 relief classes , default is set to 11 classes
/* pennock is the area threshold, default is set to 5
/* !!!!!!!!!!!!!!!!!!! BE CAREFUL !!!!!
/* CHECK WITH YOUR ARC VERSION HOW THE CURVATURE PARAMETERS ARE HANDLED
/* SOMEONE AT ESRI CHANGED THE SIGN OF THE CURVATURE DURING DEVELPOMENT
/* IMPORTANT  IMPORTANT  IMPORTANT
/* IMPROVEMENTS: clean code for easier readibility,
/* include several other LF models which are already coded in aml
/* but not included in here.
/* Citation:
/* 1) Reuter,H.I. 2002
/* Computing Landforms using a Digital Elevation Model inside ArcINFO
/* 2) used and described in paper:
/*     MOSAIC:Crop yield observation - can landform stratification improve
/* our understanding of crop yield variability.
/*4.European Conference on Precision Agriculture
/* Berlin, 2003, ISBN 9076998213, page 579-584
/* USAGE: &r landform landform <DEM> <OUTDEM> {method (11 or 8- number of LF units} {threshold for area filtering
algorithm} {profile curvature threshold} {planform curvature threshold} {slope threshold} {threshold for level areas} {all/
original (my approach or pennocks approach)} {graphik y/n (draw graphik, default no)}
&sev &error &routine BailOut
&args file out method pennock curv plan slope warea org gra
/* input parameter area
&term 9999
&if [null %file%] &then
&return USAGE: &r landform <DEM> <OUTDEM> {method} {threshold} {profile} {planform} {slope} {watershedarea}
{all/original} {graphik y/n}
&if not [exists %file%slp -grid] &then &return --> Input DEM SLOPE does not exist
&if not [exists %file%flacc -grid] &then &return --> Input DEM Watershed Area (flacc) does not exist
&if not [exists %file%plan -grid] &then &return --> Input DEM PLAN does not exist
&if not [exists %file%prof -grid] &then &return --> Input DEM PROFILE does not exist
&if [null %out%] &then &return ---> PLease provide a output file name
&if [length %out%] > 14  &then &return ---> Output limit is 14 chars
&if [exists %out% -grid] &then &return ---> Output LF DEM already exists
&if [exists %out% -cover] &then &return ---> Output Name exists as Cover
&if [exists %out%c -grid] &then &return ---> Output Name+c DEM already exists
&if [exists %out%c -cover] &then &return ---> Output Name+c exists as Cover
&if [null %method%] &then &sv method = 11
&if [null %pennock%] &then &sv pennock = 5
&if [null %curv%] &then &sv curv = 0.1
&if [null %plan%] AND %method% = 8  &then &sv plan = 0.0
&else
&if [null %plan%] AND %method% = 11 &then &sv plan = 0.1
&if [null %warea%] &then &sv warea = 500
&if [null %slope%] &then &sv slope = 3
```

```
&if [null %org%] &then &sv org = all
&if [null %gra%] &then &sv gra = y
&if (%org% = #) &then &sv org = all
&if (%warea% = #) &then &sv warea = 500
&if (%slope% = #) &then &sv slope = 3
&if (%curv% = #) &then &sv curv = 0.1
&if (%plan% = #) &then &sv plan = 0.1
&sv gra = [upcase %gra%]
&type
&type Delete old temporary files
&type
&r killgrids tmp1234*
/* logfile and grid testing
&sv programold = %:program%
&if %:program% = ARC &then &do
&r logoff off
GRID
&end
&else
&r logoff OFF
&if (gra = Y) &then
&do
display 9999 1
mape %file%
&end
&else display 0
setwindow %file%
setmask %file%
&describe %file%
&sv cell_size = %GRD$DX%
&messages &off
&type
&type Process DEM %file% with Parameters SLOPE %slope%, PROFILE CURVATURE %curv% and PLANFORM
CUVRATURE %plan%.
&type Classification threshold is set to %pennock% cells
&type Classify will run on %org% method
&type
/* used cellsize * cellsite due to some problems with some arc versions...
&sv thrshold = [calc %cell_size% * %cell_size% * %pennock%]
&sv warea = [calc %warea% / ( %cell_size% * %cell_size% )]
/* &lv
&if [exists %file%lf%method% -grid] &then kill %file%lf%method%
&if [exists %file%lfr%method% -grid] &then kill %file%lfr%method%
&if [exists work -grid] &then kill work
&if [exists %file%lfc%method% -grid] &then kill %file%lfc%method%
&if %method% = 8 &then
&DO
/* 8 classes
DOCELL
if (%file%slp > 0 & %file%prof > %curv% & %file%plan >= %plan%) %file%lf%method% = 3
else if (%file%slp > 0.0 & %file%prof > %curv% & %file%plan <= %plan%) %file%lf%method% = 4
else if (%file%slp > %slope% & %file%prof >= -%curv% & %file%prof <= %curv% & %file%plan > %plan%)
%file%lf%method% = 1
else if (%file%slp > %slope% & %file%prof >= -%curv% & %file%prof <= %curv% & %file%plan < %plan%)
%file%lf%method% = 2
else if (%file%slp > 0.0 & %file%prof < -%curv% & %file%plan >= %plan%) %file%lf%method% = 5
else if (%file%slp > 0.0 & %file%prof < -%curv% & %file%plan < %plan%) %file%lf%method% = 6
if (%file%slp <= %slope% & %file%prof >= -%curv% & %file%cur <= %curv% & %file%flacc <= %warea%)
%file%lf%method% = 7
if (%file%slp <= %slope% & %file%prof >= -%curv% & %file%cur <= %curv% & %file%flacc > %warea%)
%file%lf%method% = 8
END
/* reclass for main 3 intersections
```

```
/* shoulder, backslope, footslope, and level
DOCELL
if ( %file%lf%method% <= 2 ) %file%lfr%method% = 1
else if ( %file%lf%method% <= 4 ) %file%lfr%method% = 2
else if ( %file%lf%method% <= 6 ) %file%lfr%method% = 3
else if ( %file%lf%method% <= 8 ) %file%lfr%method% = 4
END
&END
&ELSE
&DO
/* 11 classes
/* extended by planar landforms
DOCELL
if (%file%slp > 0.0 & %file%prof > %curv%  & %file%plan >= %plan%)              %file%lf%method% = 7
else if (%file%slp > 0.0 & %file%prof > %curv%  & %file%plan > -%plan% &  %file%plan < %plan%) %file%lf%method%
= 8
else if (%file%slp > 0.0 & %file%prof > %curv%  & %file%plan <= -%plan%)             %file%lf%method% = 9
else if (%file%slp > 0.0 & %file%prof < -%curv%  & %file%plan >= %plan%)             %file%lf%method% = 1
else if (%file%slp > 0.0 & %file%prof < -%curv%  & %file%plan > -%plan% &  %file%plan < %plan%) %file%lf%method%
= 2
else if (%file%slp > 0.0 & %file%prof < -%curv%  & %file%plan <= -%plan%)             %file%lf%method% = 3
else if (%file%slp > %slope% & %file%prof >= -%curv% & %file%prof <= %curv% & %file%plan >= %plan%)
%file%lf%method% = 4
else if (%file%slp > %slope% & %file%prof >= -%curv% & %file%prof <= %curv% & %file%plan > -%plan% &
%file%plan < %plan%) %file%lf%method% = 5
else if (%file%slp > %slope% & %file%prof >= -%curv% & %file%prof <= %curv% & %file%plan <= -%plan%)
%file%lf%method% = 6
else if (%file%slp <= %slope% & %file%prof >= -%curv% & %file%prof <= %curv% & %file%flacc <= %warea%)
%file%lf%method% = 10
else if (%file%slp <= %slope% & %file%prof >= -%curv% & %file%prof <= %curv% & %file%flacc > %warea%)
%file%lf%method% = 11
else %file%lf%method% = 12
END
/* &return
/* reclass for main 3 intersections
/* shoulder, backslope, footslope, and level
DOCELL
if ( %file%lf%method% <= 3 ) %file%lfr%method% = 1
else if ( %file%lf%method% <= 6 ) %file%lfr%method% = 2
else if ( %file%lf%method% <= 9 ) %file%lfr%method% = 3
else if ( %file%lf%method% <= 11 ) %file%lfr%method% = 4
END
&END
/* determine iteraschleife
/*  mask grid erstellen auf zellen unter 5 zelleinheiten
&messages &off &all
/* decision to use pennocks major lf approach or using all avaaiable lf's
copy %file%lf%method% %out%
&if %org% = all &then
copy %file%lf%method% work
&else
copy %file%lfr%method% work
/* test value set empirically
setmask %file%
&sv test = 1000
&sv count = 3
/* convergenz criteria ?
&DO &UNTIL %leaveIT% = 0
/*setup
/* &r killgrids tmp1234*
&type iterative count %count%
&type %test%
tmp12345 = regiongroup ( work, # , four )
```

```
/* setcell %cell_size%
tmp12346 = zonalarea ( tmp12345 )
kill tmp12345
tmp12345 = select ( tmp12346, [quote value < %thrshold%])
&sv count = %count% + 1
/* copy tmp12345 thr%count%
&if gra = Y &then
&do
mape tmp12345
gridpaint tmp12345
&end
&sv testold = %test%
&describe tmp12345
&sv test = %GRD$MEAN%
kill tmp12345
&call smoothing
&sv leaveIT = %testold% - %test%
&type %leaveIT%
&end
&type
&type Done with Iteration
&type
&type Looking for Tie cells
&type
&messages &on
/* remove the last few remaining cells from the grid
/* test value set empirically
&sv test = 200
&sv testold = 0
&sv count1 = 20
&sv count = 1
&DO &UNTIL %leaveIT% = 0
&messages &off &all
&r killgrids tmp1234*
&sv count = [calc %count% + 2]
/* &sv count1 = [calc %count1% + 1]
&type %test%
tmp12345 = regiongroup ( work, # , EIGHT )
tmp12346 = zonalarea ( tmp12345 )
kill tmp12345
tmp12345 = select ( tmp12346, [quote value < %thrshold%])
&sv testold = %test%
&describe tmp12345
&sv test = %GRD$STDV%
kill tmp12345
kill tmp12346
&type
&type Filling Count %count%
&call filling
&sv leaveIT = %testold% - %test%
&type %leaveIT%
&if %count% = 21 &then &s leaveit = 0
&END
&type Done with filling
&type
&type Final Fill /* just to make sure everything is filled
&sv count = [calc %count% + 2]
tmp12345 = regiongroup ( work, # , EIGHT )
tmp12346 = zonalarea ( tmp12345 )
kill tmp12345
tmp12348 = setnull ( tmp12346 < %thrshold%, work)
tmp12349 = focalmajority (tmp12348 , rectangle, %count%, %count% )
&type
```

```
&type count %count%
work1 = con ( isnull(tmp12348), tmp12349, work )
kill work
/* final cleanup and setup
rename work1 work
&r killgrids tmp1234*
copy work %out%c
kill work
&messages &on
&if %programold% = GRID &then
&r LOGOFF ON
&else
&do
&r logoff ON
q
&end
LOG * add
landform %file% %out% %method% %PENNOCK% %CURV% %PLAN% %slope% %warea%
&return
&return
/* ############################################################
/* routine smoothing start
/* ############################################################
&routine smoothing
/* create grid where below threshold data are NODATA values
/* for masking the data
tmp12348 = setnull ( tmp12346 < %thrshold%, work)
/* perform data analysis to select next cluster which is above area threshold
/* initially thought need DOCELL, but focalmaj does the job
tmp12349 = focalmajority (tmp12348 , rectangle, 3, 3 )
/* replace all data below threshold with focalmajority results
tmp12345 = con ( isnull(tmp12348), tmp12349, work )
/* all data with nodata, replace with iter.inputwork
work1 = con ( isnull(tmp12345), work, tmp12345 )
kill work
rename work1 work
/* cleaning
&r killgrids tmp1234*
&messages &off
&return
/* ############################################################
/* routine smoothing end
/* ############################################################

/* ############################################################
/* routine filling start
/* ############################################################
&routine filling
&messages &off &all
tmp12345 = regiongroup ( work, # , four )
tmp12346 = zonalarea ( tmp12345 )
kill tmp12345
tmp12348 = setnull ( tmp12346 < %thrshold%, work)
tmp12349 = focalmajority (tmp12348 , rectangle, %count%, %count% )
work1 = con ( isnull(tmp12348), tmp12349, work )
kill work
rename work1 work
&messages &off
&return
/* ############################################################
/* routine filling end
/* ############################################################
```

```
/* ####################################################
/* routine bailout start
/* ####################################################
&routine BailOut
&sev &error &fail
&type \--> AML ERROR: %aml$errorfile%
&type --> LINE : %aml$errorline%
&type --> ERROR: %aml$message%
&r logoff ON
&messages &on
&return
/* ####################################################
/* routine bailout end
/* ####################################################
```

RASTERGRD2.AML

```
/* WHAT: aggregateds many DEM's for a given polygon cover and joins it
/* INPUT: several DEM with the same name basis, a polygon cover
/* DEVELOPER: hannes isaak reuter, zalf muencheberg, germany
/* CONTACT: hreuter@zalf.de / gisxperts@web.de
/* CHANGES: HIR 1999-2002
/*    HIR 23/06/2003 cleaning for arcsripts upload....
/* COMMENTS:
/* USAGE: &r rastergrd2 <DEM> <JOINCOVER> <CELL SIZE FOR AGGREGATIONS> <STATISTICAL TYPE OF
AGGREGATION ( mean, std, range...)>
&sev &error &routine BailOut
&args file joinfile size type
/* &r rastergrd2 sktrd sk54 1 mean
&term 9999
&echo &off
&messages &off &all
&type working ....
/* logfile and grid testing
&sv programold = %:program%
&if %:program% = GRID &then &do
&r logoff off
q
&end
&else
 &r logoff OFF
/* log setup end
&if [null %file%] &then
&s file = [getchoice n sk bl -other -prompt 'Select file extension']
&if [null %size%] &then &sv size = 10
&if [null %type%] &then &sv type = mean
&if [NULL %joinfile%] &then &s joinfile = [getcover -other]
&if [exists %joinfile%g -grid] &then kill %joinfile%g
&sv joinfilefull = %joinfile%
&call getname
polygrid %joinfile% %join%g %join%-id
%size%
y
~
&sv raster = %join%g
grid
/* kill stats files
&sv x = [length %file%]
&s dateien = [listfile %file%* -grid]
&if [null %dateien%] &then &return No grids exist.
&s num := [token %dateien% -count]
&do I := 1 &to %num%
```

```
&type Creating stats for [extract %i% %dateien%]
&sv name = [extract %i% %dateien%]
&if [exists %name%_s -info] &then &sv ha =[delete %name%_s -info]
%name%_s = zonalstats(%raster%, %name% , %type%)
&end
/* go to arc prompt
q
&sv x = 2
&sv type1 = [substr %type% 1 3]
/* change names for joining (e.g.blvalues-id )
&do I := 1 &to %num%
&type Changing attributes names for [extract %i% %dateien%]
&sv name = [extract %i% %dateien%]
&sv ext = [substr %name% %x%]
tables
sel %name%_s
alter
%type%
%type1%%ext%
~
~
~
~
~
alter
value
%join%-id
~
~
~
~
~
q
&end
&type JOIN DATA
&do I := 1 &to %num%
&sv name = [extract %i% %dateien%]
joinitem %joinfilefull%.pat %name%_s %joinfilefull%.pat %join%-id
&end
kill %join%g
&messages &on
/* setup finish
&if %programold% = GRID &then
&do
&r LOGOFF ON
GRID
&end
&else
&r logoff ON
LOG * add
rastergrd2 for %file%, joind on  %joinfile% %size%  %type%
/* setup finish
&return
/* ##########################
&routine getname
/* strips unneccessary rest
&sv x = 1
&sv text = %joinfile%
&do &until %x% = 1
&sv x = [calc [search %text% /] + 1]
&sv y = [length %text%]
&sv z = [calc %y% - %x% + 1]
&sv text = [substr %text% %x% %z%]
```

```
&end
&sv join = %text%
&return
/* ########################################################
/* routine bailout start
/* ########################################################
&routine BailOut
&sev &error &fail
&type \--> AML ERROR: %aml$errorfile%
&type --> LINE : %aml$errorline%
&type --> ERROR: %aml$message%
&r logoff on
&messages &on
&return
/* ########################################################
/* routine bailout end
/* ########################################################
```

ARCISA2.AML

```
/* WHAT : import/process arc pointcovers to isatis
/* INPUT: Cover with point data, outfile and directiry, missing value convention
/* DEVELOPER: hannes isaak reuter, zalf muencheberg, germany
/* CONTACT: hreuter@zalf.de / gisxperts@web.de
/* CHANGES: HIR 1999-2002
/* COMMENTS:
/* will bail if umlaute (äüö) will be provided
/* will bail if variable names contain to many charachters.... ;-) 800 charachters- short variable names please
/* will bail if no ISATIS can't be started
/* improvements: for bailing cases
/* use gtx server instead of dummy template like done here ;-)
/* NEEDS: ARCISA.AML ARCISA.TMPL
/* USAGE: &r arcisa <COVER> <PATH/FILENAME> {missing value}
&args incover outfile missing
/* create ascii file
/* create variables
&if [null %incover% ] &then &return ---> Please provide an INPUTCOVER
&if [null %outfile%] &then &return ---> PLease provide an ISATIS study/file name
&sv study = [ before %outfile% /]
&sv filename = [ after %outfile% /]
&if [length %filename%] < 1 &then &return ---> Please provide an ISATIS file name
&if [null %missing% ] &then &sv missing = -9999
&type
&type data processing is started
&type
&r arcisa %incover% %missing%
&type
&type First step done -starting import
&type
&sv incover = [locase %incover%]
/* replace template with needed data
&if [exists tmp1234567890 -file] &then [delete tmp1234567890 -file]
&sys cat $AMLPATH/arcisa2.txt | sed [quote s/XXXXX/%incover%/g] | sed [quote s/YYYYY/%study%/g] | sed [quote s/
ZZZZZ/%filename%/g] > tmp1234567890
/* import file into isatis file system
&sys isatis -batch tmp1234567890
&type done
&type
&messages &on
&return
```

ARCISA.AML

```
/* create ascii file for isatis
/* takes argument filename
/* single items can be implemented - but need some time
/* if no items are provided, than use all
/* hir 8/99
/* hir 2/00
/* hir 8/01 added NODATA support for -9999.0
/* hir need to check for length of x and y ?
/* hir need to check for üäö (umlaut) in *.hd file
&sev &error &routine BailOut
&args file missing
&r LOGFILE OFF
&type create ascii table for import in isatis
&type written by hannes isaak reuter / februar 2000
&type no special characters e.g. äüö  are supported by isatis..
&type does not support polygon files at the moment
&type supports only files up to 800 signs wide
&type change your data- otherwise no import.
&type
/* units for x and y coordinates
&if not [exists %file% -cover] or [Null %file%] &then
&return &type Cover does not exist OR Provide a valid cover name
/*
/* &messages &off
&if %:PROGRAM% <> ARC &then q
/* setup
&messages &off
&sv file = [locase %file%]
&sv outfile = %file%.hd
&if [exists %outfile% -file] &then &sv ttt = [delete %outfile%]
/* &term 9999
/* &if [null %item% ] &then &s item = yield
&if [null %file% ] &then &s file = [getcover]
&type processing ... %file%
&if [exists tmp12345 -cover] &then kill tmp12345
copy %file% tmp12345
&sv output = tmp54321
/* check for file ending
/* et it for now to PAT
&sv ende = PAT
&sv type = point
&if [null %missing%] &then &do
&ty Assuming N/A as missing value
&sv missing = [quote N/A]
&end
&else
&ty Assuming %missing% as missing value
&sv count = 1
&lv
/*
/* the follwing puts x-coord and y-coord in first place
/* if xccoord and y coord already exits
/* first delete them
&if [iteminfo tmp12345 -point x-coord -exists] &then dropitem tmp12345.pat tmp12345.pat x-coord
&if [iteminfo tmp12345 -point y-coord -exists] &then dropitem tmp12345.pat tmp12345.pat y-coord
&sys echo # >> %outfile%
&sys echo #  FILE SAVING: Directory:   File: >> %outfile%
&sys echo # >> %outfile%
&sys echo # structure=free , x_unit=m , y_unit=m >> %outfile%
&sys echo # >> %outfile%
```

```
&describe tmp12345
&if %DSC$PRECISION% = SINGLE &then
&do
additem tmp12345.pat tmp12345.pat x-coord 4 12 F 3 $RECNO
additem tmp12345.pat tmp12345.pat y-coord 4 12 F 3 x-coord
&sys echo # field=1 , type=xg , name=X-Coord~
   , f_type=decimal , f_length=12 ,~
  f_digits=3  >> %outfile%
&sys echo # >> %outfile%
&sys echo # field=2 , type=yg , name=Y-Coord~
   , f_type=decimal , f_length=12 ,~
  f_digits=3  >> %outfile%
&sys echo # >> %outfile%
&sv t = #++++++++++++-----------
&end
&else
&do
additem tmp12345.pat tmp12345.pat x-coord 8 18 F 5 $RECNO
additem tmp12345.pat tmp12345.pat y-coord 8 18 F 5 x-coord
&sys echo # field=1 , type=xg , name=X-Coord~
    , bitlength = 64 , f_type=decimal , f_length=18 ,~
  f_digits=5  >> %outfile%
&sys echo # >> %outfile%
 &sys echo # field=2 , type=yg , name=Y-Coord~
    , bitlength = 64 , f_type=decimal , f_length=18 ,~
  f_digits=5  >> %outfile%
 &sys echo # >> %outfile%
&sv t = #++++++++++++++++++-----------------
&end
addxy tmp12345
&sv count = 0
&sv sign = plus
&sv tmp = ''
/* set t variables to nothing
/* this version support 800 charachters !!! LIMITATION !!!!
&sv t1 =
&sv t2 =
&sv t3 =
&sv t4 =
&sv t5 =
&sv t6 =
/* get all items for this object in liste %itemnames%
&sv itemnumber = 2
&sv itemnames = [listitem tmp12345 -%type% -all itemfile]
  &type create item definitions for %file%
&sv ifile = [open itemfile openstat -read]
&sv item = [read %ifile% readstat]
&sv item = [read %ifile% readstat]
&do y = 3 &to %itemnames%
&sv item = [read %ifile% readstat]
&sv %item% = [iteminfo tmp12345 -%type% %item% -definition]
 &sv itemn = %item%
&sv item = [value %item%]
&sv itemnumber = [calc %itemnumber% + 1]
 /* output to file
 /* &type '#   ' %itemnumber% > %output%
 /* cut the itemdefinition in pieces
 &sv first = [before %item% ',']
 &sv kuerze1 = [calc [length %first%] + 2]
 &sv item1 = [substr %item% %kuerze1%]
 &sv second = [before %item1% ',']
 &sv kuerze1 = [calc [length %second%] + 1 + %kuerze1%]
 &sv item1 = [substr %item% %kuerze1%]
```

```
 &sv third  = [before %item1% ',']
 &sv kuerze1 = [calc 2 + %kuerze1% ]
 &sv fourth = [substr %item% %kuerze1%]
/* do checking of name for length of variable for writing out record length
/* variable longer than 1024(800) will result in problems
&if [length %t%] > 800 &then
&do
&sv t%count% = %t%
&sv count = %count% + 1
&sv t =
&return --> Variable Description to long
&end
&type [length %t%]
/* translate it into variablen defintion and write it to output file
&select %third%
&when B
&do
/* INTEGER
&sys echo # field=%itemnumber% , type=numeric , name=[quote %itemn%]~
 , ffff=%missing% , bitlength=32 , f_type=integer ,~
 f_length=%second% >> %outfile%
&sys echo # >> %outfile%
&call descript
&end
/*
&when N
&do
/* REAL
&sys echo # field=%itemnumber% , type=numeric , name=[quote %itemn%]~
 , ffff=%missing% , bitlength=32 , f_type=decimal ,~
 f_length=%second% , f_digits=%fourth%  >> %outfile%
&sys echo # >> %outfile%
&call descript
&end
&when F
&do
/* REAL2
&sys echo # field=%itemnumber% , type=numeric , name=[quote %itemn%]~
,ffff=%missing% , bitlength=32 , f_type=Decimal, f_length=%second% ,~
 f_digits=%fourth%  >> %outfile%
&sys echo # >> %outfile%
&call descript
&end
&when C
&do
/* CHAR
&sys echo # field=%itemnumber% , type=alpha , name=[quote %itemn%] ,~
  f_length=%second% >> %outfile%
&sys echo # >> %outfile%
&call descript
&end
&when I
&do
/* INTEGER
&sys echo # field=%itemnumber% , type=numeric , name=[quote %itemn%]~
 , ffff=%missing% , bitlength=32 , f_type=integer ,~
  f_length=%second% >> %outfile%
&sys echo # >> %outfile%
&call descript
&end
&end
/*  avoids to many items in memory
&delvar %itemn%
```

```
/* end of main loop is this end here
&end
/* write last row in descriptuion file
/* have do to some tweaking to get it out
/* &messages &on
/* &echo &on
&if [length %t1%] > 0 &then
&do
&sys echo %t1%>> 1
&sys echo %t2%>> 2
&sys echo %t3%>> 3
&sys echo %t4%>> 4
&sys echo %t5%>> 5
&sys echo %t6% >> 6
&sys echo %t% >> 7
&sys paste 1 2 3 4 5 6 7 > 8
&sys tr -d '\t' < 8 > 9
/* &sys paste 9 6 > 10
&sys cat 9 >> %outfile%
/* &messages &on
&sys rm 1
&sys rm 2
&sys rm 3
&sys rm 4
&sys rm 5
&sys rm 6
&sys rm 7
&sys rm 8
&sys rm 9
&end
&else
&sys echo %t% >> %outfile%
&type done with itemdescriptions
&type create ascii table with data ...
&if [exists %file%.asc1 -file] &then &sys rm %file%.asc1
&if [exists %file%.asc.hd -file] &then &sys rm %file%.asc.hd
&if [exists %file%.asc -file] &then &sys rm %file%.asc
&type now
&messages &off
/* do renaming of tmp12345# in %file%#
tables
sel tmp12345.pat
alter
tmp12345#
%file%#
~
~
~
~
/* unload list
/* &messages &on
unload %file%.asc1 # columnar %file%.asc.hd
q
/* &messages &on
/* do replacing ',' to . ; placing header and data files beneath
&sys tr -d , . < %file%.asc1 > %file%.asc2
&sys cat %outfile%  %file%.asc2 > %file%.asc
/* &if [exists tmp12345 -cover] &then kill tmp12345
/* &if [exists %outfile% -file] &then &sys rm %outfile%
&if [exists %file%.asc1 -file] &then &sys rm %file%.asc1
&if [exists %file%.asc2 -file] &then &sys rm %file%.asc2
&if [exists %file%.asc.hd -file] &then &sys rm %file%.asc.hd
&if [exists %file%.hd -file] &then &sys rm %file%.hd
```

```
&if [exists itemfile -file] &then &sys rm itemfile
&sv sfile = [close %ifile%]
/* section needed for replacement of umlaut values
/* &sys tr -d , . < %file%.asc1 > %file%.asc2
&messages &on
&sys mv %file%.asc %file%.hd
&if [exists tmp12345 -cover] &then kill tmp12345
&type produced ascii file called %file%.hd
LOGFILE ON
LOG * add
arcisa %file% %file%.hd
&label error
&return
/* ##########################
/* writes data in isatis ascii format
&routine descript
&if %sign% = plus &then
&do
&do x := 1 &to %second%
&sv t = %t%+
&end
&sv sign = minus
&end
&else
&do
&do x := 1 &to %second%
&sv t = %t%-
&end
&sv sign = plus
&end
&return
/* ###########
&routine BailOut
&sev &error &fail
&type \--> AML ERROR: %aml$errorfile%
&type --> LINE : %aml$errorline%
&type --> ERROR: %aml$message%
LOGFILE ON
&return
```

ARCISA.TMPL (needed for ARCISA.AML)

```
#
# FILE SAVING: Directory:   File:
#
# structure=free , x_unit=m , y_unit=m
#
#
# field=1 , type=xg , name="X coord" , bitlength=32 ;
#         f_type=Decimal , f_length=12 , f_digits=3
#
# field=2 , type=yg , name="Y coord" , bitlength=32 ;
#         f_type=Decimal , f_length=12 , f_digits=3
#
```

SRADAGGREG.AML

```
/* INPUT: srad *.bin files, parameters to extract from srad data set..
/* DEVELOPER: hannes isaak reuter, zalf muencheberg, germany
/* CONTACT:  hreuter@zalf.de / gisxperts@web.de
/* CHANGES: HIR 1999-2002
/*    HIR 23/06/2003 cleaning for arcsripts upload....
```

```
/* COMMENTS:
/* aml to prepare srad bin files in arcinfo grids , just looks for a wildcard to process all file in ther
&args wildcard parameter1 parameter2
/* &echo &on
/* file setting and so on
&s covcount = [filelist %wildcard%.bin xx12345.tmp -file]
&if %covcount% = 0 &then
  &type \--> No files match specification %wildcard%
&else &do
  &s file = [open xx12345.tmp openstat -r]
  &s cov = [read %file% readstat]
  &do &while %readstat% = 0
&do
&if [exists n%cov% -grid ] &then &type Grid %cov% already exist- skipping
&else &do
&type work on %cov%
&call convert
&end
&end
   &s cov = [read %file% readstat]
  &end
  &s close = [close %file%]
&end
&if [exist xx12345.tmp -file] &then &s del [delete xx12345.tmp -file]
&return
/* real work is done here
&routine convert
/* &messages &off &all
/* &echo &on
&sv infile = %cov%
/* generate good name with before
&s outfile = [before %infile% .]
&if [exists tmp1234 -cover] &then kill tmp1234
&if [exists tmp1234 -grid] &then kill tmp1234
&if [exists tmp1235 -cover] &then kill tmp1235
&if [exists tmp1235 -grid] &then kill tmp1235
&r tapestoarc %infile% %parameter1% tmp1234 A
&type %parameter1%
&r tapestoarc %infile% %parameter2% tmp1235 A
&type %parameter2%
grid
tmp1234gk1 = tmp1234 + tmp1235
q
&echo &off
copy tmp1234gk1 %outfile%%parameter1%%parameter2%p
/* &if [exists *.flt -file] &then &sys rm *.flt
&if [exists tmp1234gk1 -grid] &then kill tmp1234gk1
&if [exists tmp1234 -grid] &then kill tmp1234
&if [exists tmp1235 -grid] &then kill tmp1235
&messages &on
&return
```

TAPESTOARC.AML

```
/* INPUT: tapes file, attribute number, output grid name, grid type
/* DEVELOPER: hannes isaak reuter, zalf muencheberg, germany
/* CONTACT: hreuter@zalf.de / gisxperts@web.de
/* CHANGES: HIR 1999-2002
/*    HIR 23/06/2003 cleaning for arcsripts upload....
/* COMMENTS:
&sev &error &routine BailOut
&args tapesfile att_num att_grid format
/* Call tapestoarc, load the attribute att_num into att_grid if specified.
```

```
/* tapestoarc writes the attribute file to the current directory, which
/* when called from Arc is the current workspace.
/* added routine for automatically flipping grid
/* routine uses ascii grid function !!!
/* Base structure of the aml from CSIRO.
display 0
/* &echo &off
/* calling environment
&if [quote [show program]] = 'ARCPLOT' &then &do
 &sv shell = arc
&end
&else &if [quote [show program]] = 'GRID' &then &do
 &sv shell = arc
&end
&else &do
 &sv shell =
&end
/* usage, arguments
&if [null %tapesfile%] &then &do
 &type Usage: tapestoarc <tapesfile> {<attribute> <outgrid>}
 &return
&end
&if ^ [exists %tapesfile% -file] &then &do
 &type %tapesfile% does not exist
 &return
&end
/* get the first part of the tapes file name, excluding the pathname.
&sv tap_head = [before [entryname %tapesfile%] .]
/* 1st usage, return information about the tapes file and exit
&if [null %att_grid%] &then &do
 &sys tapestoarc -I %tapesfile%
 &return
&end
/* 2nd usage, copying out an attribute to a target grid name
&if [exists %att_grid% -grid] &then &do
 &type %att_grid% allready exists
 &return
&end
/* Enforce the rule that the target grid must not be a pathname.
&if [entryname %att_grid%] <> %att_grid% &then &do
 &type target grid must be in current workspace, and specified without a pathname.
 &return
&end
&sv typ = -%format%
&if [exists testtt5.txt -file] &then &sys rm testtt5.txt
&sys (tapes5arc -I %tapesfile% -n %att_num% %typ%) >& testtt5.txt
&sv filunit = [open testtt5.txt openstatus -read]
&sv text = [read %filunit% readstatus]
&if [search %text% W] = 1 &then &sv flip = 1
&sv test = [close %filunit%]
/* &lv
&if ( %format% = A ) &then &do
&if [exists %tap_head%_%att_num%.asc] &then
&do
/* add 07.02.2002 remove one number from column for case luettewitz
/* &sys te %tap_head%_%att_num%.asc
/* added one line in tapes5arc to do the number reduction there
/* just need to lower one count, sed not working for me !
%shell% asciigrid %tap_head%_%att_num%.asc %att_grid% FLOAT
&end
&end
&else &do
&if [exists %tap_head%_%att_num%.flt] &then
```

```
%shell% floatgrid %tap_head%_%att_num%.flt %att_grid%
&end
/* &pause
&if %flip% = 1 &then &do
&if [exists tmp12345 -grid] &then kill tmp12345
rename %att_grid% tmp12345
grid
%att_grid% = flip ( tmp12345 )
q
&end
/* add to the log
%shell% log %att_grid% add
Attribute %att_num% from TAPES-G file %tapesfile%
/* clean up
&type cleaning
&sys rm %tap_head%_%att_num%.flt %tap_head%_%att_num%.hdr  %tap_head%_%att_num%.asc
&type done
&return
/* ##########################################################
/* routine bailout start
/* ##########################################################
&routine BailOut
&sev &error &fail
&type \--> AML ERROR: %aml$errorfile%
&type --> LINE : %aml$errorline%
&type --> ERROR: %aml$message%
&r logoff on
&messages &on
&return
/* ##########################################################
/* routine bailout end
/* ##########################################################
```

RSME.AML

```
/* NAME: rsme3.aml
/* WHAT: determine optimum resolution for a raster data set
/* INPUT: contour line cover with  height, start and end resolution to be tested
/* DEVELOPER: hannes isaak reuter, zalf muencheberg, germany
/* CONTACT:  hreuter@zalf.de / gisxperts@web.de
/* CHANGES: HIR 1999-2002
/* COMMENTS:
/* uses 2^x for resolution increase
/* developement for futher parameter
/* USAGE: rsme3 <dem> <item> <statsfile> <resolution start> <resolution end> <interval> <type>
&sev &error &routine BailOut
&args dem item stats start ende interval type
/* Überprüfen ob alle Parameter festgelegt wurden:
  &do I &list %dem% %item% %stats% %start% %ende% %interval% %type%
   &if [null %I%] &then
     &do
      &type Usage: rsme <DEM/COVER><item><statsfile><resolution start> <resolution end> <interval> <slope/plan/prof/
aspect/curvature>
      &return; &return
     &end
   &end

&if not [exists %dem% -cover] &then
/* check for dem
&if [exists %dem% -grid] &then
&do
/* &describe %dem%
```

```
grid
%dem%1234 = contour ( %dem%, interval, 1 )
q
&sv dem = %dem%1234
&sv item = contour
/* &sv type = grd
/* &else
/* &do
&end
&else &return --> NO Cover avaiable EXIT
/* check for item in cover
&if not [iteminfo %dem%.aat -INFO %item% -EXISTS] &then &return --> Item does not exist - EXIT
/* &type = cover
/* &end
/* &display 0
/* &messages &off &all
/* ########################
/* logfile and grid testing
/* ########################
&sv programold = %:program%
&if %:program% = ARC &then &do
&r logoff off
GRID
&end
&else
&r logoff OFF
/* delete any existing working files
&call cleanup
&sv stats = [upcase %stats%]
/* ##############
/* logfile off end
/* ##############
/* creating statistics file
&if [exists %stats% -info] &then
&do
&messages &on
&return STATS file already exist -> EXIT
&end
/* q
&data ARC INFO > /dev/null
ARC
DEFINE %stats%
RESOLUTION
8
8
f
4
RSME
8
8
f
4
~
Q STOP
&end
/* ##############################################################
/* processing starts
/* reso contains resolution
/* ##############################################################
&sv x = 1
out = scalar( pow ( 2, %interval% ) )
&sv reso1 = [show out]
&sv reso = [calc 1 * %reso1%]
```

```
&do &until %reso% > %ende%
&if %reso% > %start% &then
/* &type runn
&call rsme
&else
&Type increase resolution %reso% to meet start criteria %start%
/* &end
&sv x = %x% + 1
&sv y = %interval% * %x%
out = scalar( pow ( 2, %y% ) )
&sv resol = [show out]
&sv reso = [calc 1 * %resol%]
/* &call cleanup
/* &pause
&end
&if [exists %dem%1234 -cover] &then kill %dem%1234
/* ################################################################
/* processing ends
/* ################################################################
&if %programold% = GRID &then
&r LOGOFF ON
&else
&do
&r logoff ON
q
&end
LOG * add
RSME3 %dem%, %stats%, %start%, %ende%, %interval%, %type%
&messages &on
&return
/* ################################################################
/* program ends
/* ################################################################
/* ############################################################
/* routine rsme start
/* ############################################################
&routine rsme
&type process resolution: %reso%
/* &if %type% = cover &then &do
&data arc topogrid tmp10 %reso%
datatype contour
contour %dem% %item%
end
&end
&if not [exists tmp10 -grid] &then
&do
&r logoff ON
&return TOPOGRID FAILURE
&end
/* end of cover run
/* &end
tmp11 = curvature (tmp10, tmp12, tmp13, tmp14, tmp15)
/* tmp12 = slope ( tmp11 )
&select %type%
&when slope
rename tmp14 tmp20
&when aspect
rename tmp15 tmp20
&when plan
rename tmp13 tmp20
&when planform
rename tmp13 tmp20
&when prof
```

```
rename tmp12 tmp20
&when profile
rename tmp12 tmp20
&when asp
rename tmp15 tmp20
&when slp
rename tmp14 tmp20
&when curv
rename tmp11 tmp20
&when curvature
rename tmp11 tmp20
&otherwise
rename tmp14 tmp20
&end
tmp21 = tmp20 * tmp20
&describe tmp21
&sv rsme = %GRD$Mean%
&type RSME is: %rsme%
/* ##################
/* save result to file
/* ###################
&data ARC INFO > /dev/null
ARC
sel %stats%
add
%reso%
%rsme%
~
Q STOP
&end
&call cleanup
&return
/* ######################################################
/* routine rsme end
/* ######################################################
/* ######################################################
/* routine cleanup start
/* ######################################################
&routine cleanup
&if [exists tmp10 -grid] &then kill tmp10
&if [exists tmp11 -grid] &then kill tmp11
&if [exists tmp12 -grid] &then kill tmp12
&if [exists tmp13 -grid] &then kill tmp13
&if [exists tmp14 -grid] &then kill tmp14
&if [exists tmp15 -grid] &then kill tmp15
&if [exists tmp20 -grid] &then kill tmp20
&if [exists tmp21 -grid] &then kill tmp21
&return
/* ######################################################
/* routine cleanup start
/* ######################################################
/* ######################################################
/* routine bailout start
/* ######################################################
&routine BailOut
&sev &error &fail
&type \--> AML ERROR: %aml$errorfile%
&type --> LINE : %aml$errorline%
&type --> ERROR: %aml$message%
&r logoff on
&messages &on
&return
/* ######################################################
```

```
/* routine bailout end
/* ###########################################################
```

EXCEL MAKRO FOR SRAD PARAMETER CLACULATION

```
Sub SRAD()
' Active Cell conatins julian day
' neext to it the daily global irradiance in J/cm^2
Dim x As Double
Let Pi = 3.14
ABC = True:
' Let LAT = InputBox(Prompt:= _"Geben Sie die Latitude in Grad ein.")
' kassow 53.57
' luette 51.14
' beckum 48 ?
' Let LAT = 51.14
' Let LAT = 51.14
Let LAT = ActiveCell.Offset(-2, 1).Value
If LAT = „“ Then Let LAT = InputBox(Prompt:="Please provide the  Latitude in Degree.")
ActiveCell.Offset(-1, 2).Value = „RAD(meas)“
ActiveCell.Offset(-1, 3).Value = „DRC(max)“
ActiveCell.Offset(-1, 4).Value = „SUND“
ActiveCell.Offset(-1, 5).Value = „DL“
ActiveCell.Offset(-1, 6).Value = „AT“
ActiveCell.Offset(-1, 7).Value = „SF“
ActiveCell.Offset(-1, 8).Value = „CIRC“
ActiveCell.Offset(-1, 9).Value = „ALPHA“
' loop which calculates everything up to the end
Do
Let Tag = ActiveCell.Value
'  5000 conversion  j/cm^2 in j/m^2
Let rad = ActiveCell.Offset(0, 1).Value * 5000
'     -------- Calculation of Daylength and irradiance -----
'!    ---------------------- DECLINATION ----------------------
    Let DEC = -23.4 * Cos(2 * Pi * (Tag + 10) / 365)
    Let SINLD = Sin(DEC * Pi / 180) * Sin(LAT * Pi / 180)
    Let COSLD = Cos(DEC * Pi / 180) * Cos(LAT * Pi / 180)
  '  !   -------------------- ASTRONOMICAL DAY LENGTH  -----------
    Let x = SINLD / COSLD
    Let dl = 12 * (Pi + 2 * asin2(x)) / Pi
  '  !   -------------------- EFFECTIVE DAY LENGTH----------------
    Let x = (-Sin(8 * Pi / 180) + SINLD) / COSLD
    Let DLE = 12 * (Pi + 2 * asin2(x)) / Pi
    Let x = (-Sin(-6 * Pi / 180) + SINLD) / COSLD
    Let DLP = 12 * (Pi + 2 * asin2(x)) / Pi
  '  !   ----- AVERAGE PHOTOSYNTHETIC ACTIVE IRRADIANCE ------
    Let RDN = 3600 * (SINLD * dl + 24 / Pi * COSLD * Sqr(1 - (SINLD / COSLD) ^ 2))
  '  !   ------------IRRADIANCE CLEAR DAY (in Joule/m^2)-----------
    Let drc = 0.5 * 1300 * RDN * Exp(-0.14 / (RDN / (dl * 3600)))
  '  !   --------------IRRADIANCE CLOUDY DAY ---------------------
    Let DRO = 0.2 * drc
  '  !   --------------Calculation duration of sun shine ( h)
    Let sund = (rad - DRO) / (drc - DRO) * dl
    If sund < 0 Then Let sund = 0
    If sund > dl Then Let sund = dl
'   !    ------------  sunshine fraction (sf)
    Let quot1 = sund / dl
'   !    ----------- athmospheric transmittance (AT)
'   implemented idso (69) by using only clear sky conditions, large values and !! the number of values is very small.
    ' if quot1 = 1 then
    Let quot = rad / drc
'   !    ---------------- neue berechnung von at
'   Let rths = ((rad / 5000) * 0.36) / (0.35 + (0.61 * quot1))
```

```
'     Let rths2 = rths / 1354
'     !     ------------- Berechnung von CIRC daily
'      0.36 is conversion  j/cm^2 in w/m^2 (daily)
      Let circ = ((rad / 5000) * 0.36) / 1354
      'problems with circ - limit values ?
      'If circ < 0.15 Then cird = 0.15
      ' Else: If circ > 0.65 Then circ = 0.65
'     !     ------------- calculation of alpha / or beta e.g. cloudiness
'            if quuotl = 0 than clear day
      If quotl > 0 Then
      Let alpha = (rad * (1 - quotl)) / (drc * quotl)
      Else: Let alpha = „“
      End If
      ' if alpha larger 1 some error
      If alpha > 1 Then Let alpha = „“
'     Convert back
      Let rad1 = rad / 5000
      Let drc1 = drc / 5000
'     OUTPUT
' meas rad
ActiveCell.Offset(0, 2).Value = rad1
' calculated max rad
ActiveCell.Offset(0, 3).Value = drc1
' calculated sunshine duration
ActiveCell.Offset(0, 4).Value = sund
'  day length
ActiveCell.Offset(0, 5).Value = dl
' calculated AT
ActiveCell.Offset(0, 6).Value = quot
' calculated SF
ActiveCell.Offset(0, 7).Value = quotl
' calculated CIRC
ActiveCell.Offset(0, 8).Value = circ
'berechnet alpha
ActiveCell.Offset(0, 9).Value = alpha
'  next row
ActiveCell.Offset(1, 0).Select
If ActiveCell.Value = „“ Then ABC = False
Loop Until ABC = False
End
End Sub
```

Appendix Geostatistics

Table 45. Significant cross correlations between different variables at the field site „Bei Lotte“. Additionally some signifcant cross correlations for the field site „Sportkomplex“ are shown which are specially signed with SK. The abbrevitions used are described below the table. Length indicates the number of significant lag classes (lag distance 27 m, except for the SK protein content, here 54 m).

Heading and Tailing property	CC Lag 0	Len gth	Heading and Tailing property	CC Lag 0	Len gth
Clay 030 FAT 030	0.95	1	EC 030 1998 Meas10slp	0.36	1
EC 060 1998 EC 030 1998	0.88	1	EC 060 1998 Meas10slp	0.34	1
EM38m Meas10max	0.20	2	EM38m Meas10mean	0.19	3
EM38m Meas10min	0.28	4	EM38m Meas10pctg	-0.22	3
EM38m Meas10sd	-0.42	4	EM38m Meas10slp	-0.25	1
EUF-K1 1999 EUF-K1 2000	0.75	2	EUF-K1 1999 EUF-K1 2001	0.77	1
EUF-K1 1999 EUF-P1 2000	0.57	1	EUF-K1 1999 Meas10dev	-0.37	1
EUF-K1 1999 Meas10flacc	0.41	1	EUF-K1 1999 Meas10m	-0.45	3
EUF-K1 1999 Meas10max	-0.33	2	EUF-K1 1999 Meas10mean	-0.37	2
EUF-K1 1999 Meas10min	-0.28	1	EUF-K1 1999 Meas10range	-0.30	2
EUF-K1 1999 Meas10sd	-0.26	2	EUF-K1 1999 Meas10spi	0.37	1
EUF-K1 1999 NSensor 04 05 2000	0.29	1	EUF-K1 1999 Penm 010	0.25	1
EUF-K1 1999 Penm 120	0.29	2	EUF-K1 1999 Penm 560	0.26	1
EUF-K1 1999 Penm 670	0.19	1	EUF-K1 1999 SSM 12 05 2000	0.26	1
EUF-K1 2000 EC 030 1998	0.29	1	EUF-K1 2000 EUF-P1 2000	0.60	1
EUF-K1 2000 Meas10dev	-0.35	1	EUF-K1 2000 Meas10flacc	0.28	1
EUF-K1 2000 Meas10m	-0.31	2	EUF-K1 2000 Meas10max	-0.22	1
EUF-K1 2000 Meas10mean	-0.24	1	EUF-K1 2000 Meas10range	-0.20	2
EUF-K1 2000 Meas10spi	0.32	1	EUF-K1 2000 Meas10stc	0.34	1
EUF-K1 2000 Penm 010	0.26	1	EUF-K1 2000 Penm 120	0.25	1
EUF-K1 2001 EUF-K1 2000	0.69	1	EUF-K1 2001 EUF-P1 2000	0.54	1
EUF-K1 2001 Meas10flacc	0.36	1	EUF-K1 2001 Meas10m	-0.36	2
EUF-K1 2001 Meas10max	-0.28	2	EUF-K1 2001 Meas10mean	-0.30	2
EUF-K1 2001 Meas10min	-0.23	1	EUF-K1 2001 Meas10range	-0.26	4
EUF-K1 2001 Meas10sd	-0.25	2	EUF-K1 2001 NSensor 04 05 2000	0.29	1
EUF-K1 2001 Penm 010	0.23	1	EUF-K1 2001 Penm 120	0.26	2
EUF-K2 1999 EUF-K1 1999	0.92	2	EUF-K2 1999 EUF-K1 2000	0.71	2
EUF-K2 1999 EUF-K1 2001	0.69	1	EUF-K2 1999 EUF-K2 2000	0.72	1
EUF-K2 1999 EUF-K2 2001	0.65	1	EUF-K2 1999 EUF-P1 2000	0.56	1
EUF-K2 1999 Meas10dev	-0.35	1	EUF-K2 1999 Meas10flacc	0.43	1
EUF-K2 1999 Meas10m	-0.38	2	EUF-K2 1999 Meas10max	-0.27	2
EUF-K2 1999 Meas10mean	-0.31	2	EUF-K2 1999 Meas10min	-0.23	1
EUF-K2 1999 Meas10range	-0.24	1	EUF-K2 1999 Meas10sd	-0.20	1
EUF-K2 1999 Meas10spi	0.39	1	EUF-K2 1999 NSensor 04 05 2000	0.28	1
EUF-K2 1999 Penm 120	0.28	1	EUF-K2 1999 Penm 560	0.22	1
EUF-K2 1999 SSM 12 05 2000	0.24	1	EUF-K2 2000 EC 030 1998	0.35	1
EUF-K2 2000 EUF-K1 1999	0.73	1	EUF-K2 2000 EUF-K1 2000	0.89	1
EUF-K2 2000 EUF-K1 2001	0.66	1	EUF-K2 2000 EUF-P1 2000	0.60	1

Table 45. Significant cross correlations between different variables at the field site „Bei Lotte". Additionally some signifcant cross correlations for the field site „Sportkomplex" are shown which are specially signed with SK. The abbrevitions used are described below the table. Length indicates the number of significant lag classes (lag distance 27 m, except for the SK protein content, here 54 m).

Heading and Tailing property	CC Lag 0	Length	Heading and Tailing property	CC Lag 0	Length
EUF-K2 2000 Meas10dev	-0.27	1	EUF-K2 2000 Meas10m	-0.25	1
EUF-K2 2000 Penm 120	0.22	1	EUF-K2 2001 EUF-K1 1999	0.71	1
EUF-K2 2001 EUF-K1 2000	0.69	1	EUF-K2 2001 EUF-K1 2001	0.88	1
EUF-K2 2001 EUF-K2 2000	0.69	1	EUF-K2 2001 EUF-P1 2000	0.52	1
EUF-K2 2001 Meas10sd	-0.26	2	EUF-K2 2001 Penm 120	0.21	1
EUF-P1 2000 Meas10m	-0.31	1	EUF-P1 2000 Meas10mean	-0.27	1
EUF-P1 2000 Meas10min	-0.22	1	EUF-P1 2000 SSM 25 03 1999	0.26	1
NSensor 11 05 1999 Meas10range	-0.21	2	K 030 Meas10asp	-0.22	2
K 360 1999 Meas10m	-0.47	1	K 360 Meas10dev	-0.58	1
K 360 Meas10m	-0.47	1	KPS 2000 Meas10M	0.31	6
Meas10dev Meas10cur	0.55	1	Meas10flacc Meas10dev	-0.48	2
Meas10m Meas10dev	0.34	1	Meas10m Meas10flacc	-0.32	2
Meas10max Meas10flacc	-0.19	2	Meas10max Meas10m	0.90	5
Meas10mean Meas10flacc	-0.21	2	Meas10mean Meas10m	0.97	6
Meas10mean Meas10max	0.95	5	Meas10min Meas10flacc	-0.16	1
Meas10min Meas10m	0.91	6	Meas10min Meas10max	0.90	5
Meas10min Meas10mean	0.95	6	Meas10pctg Meas10asp	0.41	3
Meas10pctg Meas10cur	0.34	1	Meas10pctg Meas10dev	0.57	2
Meas10pctg Meas10m	-0.33	3	Meas10pctg Meas10max	-0.67	4
Meas10pctg Meas10mean	-0.51	4	Meas10pctg Meas10min	-0.60	4
Meas10plan Meas10dev	0.50	1	Meas10prof Meas10dev	-0.43	1
Meas10prof Meas10pctg	-0.35	1	Meas10range Meas10m	0.60	5
Meas10range Meas10max	0.82	5	Meas10range Meas10mean	0.65	5
Meas10range Meas10min	0.48	4	Meas10range Meas10pctg	-0.56	4
Meas10sd Meas10m	0.21	2	Meas10sd Meas10max	0.28	2
Meas10sd Meas10mean	0.25	2	Meas10sd Meas10prof	0.20	1
Meas10sd Meas10range	0.56	4	Meas10slp Meas10dev	-0.31	1
Meas10slp Meas10sd	0.44	2	Meas10spi Meas10dev	-0.52	2
Meas10spi Meas10flacc	0.91	1	Meas10spi Meas10m	-0.31	3
Meas10spi Meas10max	-0.21	2	Meas10spi Meas10mean	-0.18	2
Meas10spi Meas10min	-0.17	2	Meas10spi Meas10range	-0.19	1
Meas10spi Meas10slp	0.18	1	Meas10stc Meas10cur	-0.63	1
Meas10stc Meas10dev	-0.56	2	Meas10stc Meas10flacc	0.67	1
Meas10stc Meas10m	-0.21	2	Meas10stc Meas10plan	-0.69	1
Meas10stc Meas10slp	0.51	2	Meas10stc Meas10spi	0.88	2
Meas10wi Meas10asp	-0.26	1	Meas10wi Meas10pctg	-0.42	2
NDVI 04 1999 EC 030 1998	-0.52	1	NDVI 04 1999 EC 060 1998	-0.41	1
NDVI 04 1999 FAT 030	-0.64	1	NDVI 04 1999 FAT 360	-0.54	1
NDVI 04 1999 Meas10asp	-0.29	1	NDVI 04 1999 Meas10m	0.28	2
NDVI 04 1999 Meas10max	0.47	3	NDVI 04 1999 Meas10mean	0.35	2
NDVI 04 1999 Meas10min	0.36	2	NDVI 04 1999 Meas10pctg	-0.50	3
NDVI 04 1999 Meas10range	0.46	3	NDVI 04 1999 Meas10slp	-0.23	1
NDVI 04 1999 Meas10wi	0.27	1	NDVI 04 1999 NDVI 04 2000	0.21	1
NDVI 04 1999 NIR	-0.52	3	NDVI 04 1999 NPROT 1999	0.27	2
NDVI 04 1999 NSensor 04 05 2000	-0.37	2	NDVI 04 1999 Nsensor 11 05 1999	-0.32	1
NDVI 04 1999 Penm 010	-0.34	2	NDVI 04 1999 Penm 340	0.32	1
NDVI 04 1999 SM 030 1998	0.25	2	NDVI 04 1999 SM 030 2001	-0.26	1

Table 45. Significant cross correlations between different variables at the field site „Bei Lotte“. Additionally some signifcant cross correlations for the field site „Sportkomplex“ are shown which are specially signed with SK. The abbrevitions used are described below the table. Length indicates the number of significant lag classes (lag distance 27 m, except for the SK protein content, here 54 m).

Heading and Tailing property	CC Lag 0	Len gth	Heading and Tailing property	CC Lag 0	Len gth
NDVI 04 1999 YPS 1999	-0.29	2	NDVI 04 2000 Meas10m	0.17	3
NDVI 04 2000 Meas10max	0.19	3	NDVI 04 2000 Meas10mean	0.19	3
NDVI 04 2000 Meas10range	0.33	3	NDVI 04 2000 Meas10sd	0.39	2
NDVI 04 2000 NSensor 04 05 2000	-0.61	2	NDVI 04 2000 Nsensor 11 05 1999	-0.42	1
NDVI 04 2000 SM 030 1998	0.32	1	NDVI 05 1998 Meas10m	-0.44	3
NDVI 05 1998 Meas10max	-0.37	4	NDVI 05 1998 Meas10mean	-0.45	3
NDVI 05 1998 Meas10min	-0.52	3	NDVI 05 1998 Meas10pctg	0.26	3
NDVI 05 1998 NDVI 2001	0.59	2	NDVI 05 1998 NIR	0.25	1
NDVI 05 1998 Nsensor 11 05 1999	-0.56	2	NDVI 05 1998 Nsensor 18 04 2000	-0.25	1
NDVI 05 1998 Penm 120	0.22	1	NDVI 05 1998 Penm 230	0.16	1
NDVI 05 1998 SM 030 1999	0.41	1	NDVI 05 1998 SM 030 2000	-0.35	2
NDVI 05 1999 EUF-P1 2000	0.48	1	NDVI 05 1999 Meas10m	-0.39	1
NDVI 05 1999 Meas10mean	-0.35	1	NDVI 05 1999 Meas10min	-0.39	1
NDVI 05 1999 Meas10slp	-0.25	1	NDVI 05 1999 NDVI 04 1999	0.47	2
NDVI 05 1999 NDVI 05 1998	0.65	2	NDVI 05 1999 NDVI 2001	0.47	1
NDVI 05 1999 Nsensor 11 05 1999	-0.67	1	NDVI 05 1999 Nsensor 18 04 2000	-0.44	1
NDVI 05 1999 SM 030 1999	0.56	1	NDVI 05 1999 SM 030 2000	-0.24	1
NDVI 05 1999 SpiKeN 2000	0.42	2	NDVI 05 1999 SSM 10 05 1999	0.32	1
NDVI 05 1999 SSM 29 04 1999	0.41	1	NDVI 05 2000 KPS 1999	-0.20	6
NDVI 05 2000 Meas10m	0.46	3	NDVI 05 2000 Meas10max	0.38	3
NDVI 05 2000 Meas10mean	0.46	3	NDVI 05 2000 Meas10min	0.31	2
NDVI 05 2000 Meas10range	0.35	3	NDVI 05 2000 Meas10sd	0.39	2
NDVI 05 2000 NDVI 04 2000	0.84	6	NDVI 05 2000 NIR	-0.37	3
NDVI 05 2000 NSensor 04 05 2000	-0.69	3	NDVI 05 2000 Penm 010	-0.17	2
NDVI 05 2000 Penm 120	-0.19	2	NDVI 05 2000 Penm 560	-0.23	1
NDVI 05 2000 Penm 670	-0.16	2	NDVI 05 2000 SM 030 1998	0.42	3
NDVI 05 2000 SM 030 2000	0.28	2	NDVI 05 2000 YPS 1999	-0.24	6
NDVI 05 2002 NSensor 11 05 1999	-0.26	1	NDVI 05 2002 KPS 1999	-0.31	6
NDVI 05 2002 Meas10m	0.30	4	NDVI 05 2002 Meas10max	0.41	4
NDVI 05 2002 Meas10mean	0.26	4	NDVI 05 2002 Meas10range	0.63	4
NDVI 05 2002 Meas10sd	0.48	3	NDVI 05 2002 Meas10wi	0.23	1
NDVI 05 2002 NDVI 04 1999	0.29	6	NDVI 05 2002 NDVI 2001	-0.39	6
NDVI 05 2002 NDVI 2001	-0.49	1	NDVI 05 2002 NIR	-0.32	4
NDVI 05 2002 NPROT 2000	-0.26	6	NDVI 05 2002 Penm 010	-0.44	4
NDVI 05 2002 YPS 1999	-0.22	6	NDVI 0502 NSensor11 05 1999	-0.27	3
NDVI 0502 Penm 670	-0.21	1	NDVI 0502 PH 030	0.42	4
NDVI 0502 PH 360	0.47	4	NDVI 1997 EM38m	-0.43	3
NDVI 1997 EUF-K1 1999	-0.36	3	NDVI 1997 EUF-K1 2000	-0.33	1
NDVI 1997 EUF-K1 2001	-0.32	3	NDVI 1997 EUF-K2 1999	-0.32	1
NDVI 1997 EUF-K2 2000	-0.34	1	NDVI 1997 EUF-K2 2001	-0.34	2
NDVI 1997 EUF-P1 2000	-0.23	1	NDVI 1997 KPS 1999	-0.33	6
NDVI 1997 Meas10m	0.36	4	NDVI 1997 Meas10max	0.34	3
NDVI 1997 Meas10mean	0.38	4	NDVI 1997 Meas10prof	0.21	1
NDVI 1997 Meas10range	0.44	3	NDVI 1997 Meas10sd	0.73	5
NDVI 1997 NDVI 04 2000	0.22	2	NDVI 1997 NDVI 05 2000	0.32	6
NDVI 1997 NDVI 05 2002	0.65	6	NDVI 1997 NDVI 05 2002	0.48	5
NDVI 1997 NIR	-0.47	5	NDVI 1997 NSensor 04 05 2000	-0.44	4

Table 45. Significant cross correlations between different variables at the field site „Bei Lotte". Additionally some signifcant cross correlations for the field site „Sportkomplex" are shown which are specially signed with SK. The abbrevitions used are described below the table. Length indicates the number of significant lag classes (lag distance 27 m, except for the SK protein content, here 54 m).

Heading and Tailing property	CC Lag 0	Length	Heading and Tailing property	CC Lag 0	Length
NDVI 1997 Penm 010	-0.43	4	NDVI 1997 Penm 120	-0.39	4
NDVI 1997 Penm 230	-0.22	1	NDVI 1997 Penm 670	-0.19	1
NDVI 1997 SM 030 1998	0.57	5	NDVI 1997 SM 030 2000	0.45	5
NDVI 1997 SM 360 1998	0.30	4	NDVI 1997 SM 360 1999	0.34	3
NDVI 1997 SM 360 2000	0.19	2	NDVI 1997 SM 690 1998	0.34	4
NDVI 1997 SM 690 2001	0.30	1	NDVI 1997 SSM 2904 1999	0.31	3
NDVI 1997 TKG 2000	0.19	6	NDVI 1997 YPS 1999	-0.29	6
NDVI 2001 EC 060 1998	-0.47	1	NDVI 2001 NIR	0.23	1
NDVI 2001 NPROT 1999	0.43	6	NDVI 2001 Nsensor 11 05 1999	-0.39	1
NDVI 2001 Nsensor 18 04 2000	-0.34	1	NDVI 2001 Penm 340	0.24	1
NDVI 2001 SM 030 1999	0.41	1	NDVI 2001 SM 030 2000	-0.25	1
NDVI 2001 SM 360 2000	-0.19	1	NDVI04 2000 Kernel 1999	0.28	2
NDVI04 2000 NPROT 1999	0.34	2	NDVI04 2000 YPS 1999	-0.31	2
NIR Meas10asp	0.17	1	NIR Meas10m	-0.69	4
NIR Meas10max	-0.76	5	NIR Meas10mean	-0.75	5
NIR Meas10min	-0.66	4	NIR Meas10pctg	0.48	4
NIR Meas10range	-0.66	5	NIR Meas10sd	-0.44	3
NIR NSensor 04 05 2000	0.45	4	NIR NSensor1804 2000	-0.36	3
NIR PH 360	-0.45	2	NIR SSM 10 05 1999	-0.33	1
NIR SSM 2904 1999	-0.43	3	Nmin 030 1998 EUF-K1 1999	0.41	1
Nmin 030 1998 Meas10m	-0.36	2	Nmin 030 1998 Meas10max	-0.25	2
Nmin 030 1998 Meas10mean	-0.32	2	Nmin 030 1998 Meas10min	-0.27	2
Nmin 030 1998 NDVI 2001	0.47	1	Nmin 030 1998 NIR	0.19	1
Nmin 030 1998 SM 030 1999	0.36	1	Nmin 030 1999 EC 060 1998	-0.36	1
Nmin 030 1999 Meas10mean	-0.25	1	Nmin 030 1999 Meas10min	-0.25	1
Nmin 030 1999 NDVI 05 1998	0.38	1	Nmin 030 1999 NDVI 05 1999	0.50	1
Nmin 030 1999 NDVI 2001	0.48	1	Nmin 030 1999 Nmin 030 1998	0.39	1
Nmin 030 1999 Nsensor 11 05 1999	-0.43	1	Nmin 030 1999 Nsensor 18 04 2000	-0.42	1
Nmin 030 1999 SM 030 1999	0.46	1	Nmin 030 2000 Meas10m	-0.24	1
Nmin 030 2000 Penm 340	0.26	1	Nmin 030 2000 Penm 450	0.27	1
Nmin 030 2000 SM 360 2000	-0.21	1	Nmin 030 2001 Meas10m	-0.45	4
Nmin 030 2001 Meas10max	-0.38	3	Nmin 030 2001 Meas10mean	-0.41	3
Nmin 030 2001 Meas10min	-0.34	3	Nmin 030 2001 Meas10range	-0.31	3
Nmin 030 2001 NDVI 05 2000	-0.31	2	Nmin 030 2001 NIR	0.31	3
Nmin 030 2001 NSensor 04 05 2000	0.16	3	Nmin 030 2001 Penm 010	0.23	2
Nmin 030 2001 Penm 120	0.24	2	Nmin 360 1998 NDVI 2001	0.39	1
Nmin 360 1999 Meas10slp	-0.27	1	Nmin 360 1999 NDVI 2001	0.50	1
Nmin 360 1999 SM 030 2000	-0.18	1	Nmin 360 2000 Meas10max	0.19	3
Nmin 360 2000 Meas10mean	0.18	3	Nmin 360 2000 Meas10sd	0.19	1
Nmin 360 2000 NDVI 1997	0.19	2	Nmin 360 2000 SM 030 1998	0.30	2
Nmin 360 2001 Meas10dev	-0.23	1	Nmin 360 2001 Meas10m	-0.41	3
Nmin 360 2001 Meas10max	-0.33	3	Nmin 360 2001 Meas10mean	-0.37	3
Nmin 360 2001 Meas10min	-0.32	3	Nmin 360 2001 Meas10range	-0.25	3
Nmin 360 2001 NDVI 05 2000	-0.17	1	Nmin 360 2001 NIR	0.33	1
Nmin 360 2001 Penm 010	0.22	2	Nmin 360 2001 Penm 120	0.19	1
Nmin 360 2001 Penm 560	0.18	1	Nmin 360 2001 SM 030 1998	-0.14	1
Nmin 360 2001 SM 030 2000	-0.18	1	Nmin 690 1998 NDVI 2001	0.35	1

Table 45. Significant cross correlations between different variables at the field site „Bei Lotte". Additionally some signifcant cross correlations for the field site „Sportkomplex" are shown which are specially signed with SK. The abbrevitions used are described below the table. Length indicates the number of significant lag classes (lag distance 27 m, except for the SK protein content, here 54 m).

Heading and Tailing property	CC Lag 0	Len gth	Heading and Tailing property	CC Lag 0	Len gth
Nmin 690 1999 EC 060 1998	-0.31	1	Nmin 690 1999 NDVI 2001	0.45	1
Nmin 690 1999 NIR	0.18	1	Nmin 690 1999 Nmin 030 1999	0.69	1
Nmin 690 1999 Nsensor 18 04 2000	-0.44	1	Nmin 690 2000 Meas10max	0.20	3
Nmin 690 2000 Meas10mean	0.19	2	Nmin 690 2000 Meas10min	0.17	2
Nmin 690 2000 Meas10pctg	-0.20	1	Nmin 690 2000 NIR	-0.16	2
Nmin 690 2000 NSensor 04 05 2000	-0.20	1	Nmin 690 2001 Meas10dev	-0.26	1
Nmin 690 2001 Meas10m	-0.30	2	Nmin 690 2001 Meas10max	-0.21	1
Nmin 690 2001 Meas10mean	-0.25	2	Nmin 690 2001 Meas10min	-0.21	1
Nmin 690 2001 NDVI 05 2000	-0.17	1	Nmin 690 2001 SM 030 1998	-0.15	1
NPROT 2000 Meas10M	-0.41	6	NSensor 04 05 2000 EC 030 1998	0.30	1
NSensor 04 05 2000 NSensor 11 05 1999	0.15	1	NSensor 04 05 2000 Meas10m	-0.47	5
NSensor 04 05 2000 Meas10max	-0.53	5	NSensor 04 05 2000 Meas10mean	-0.54	5
NSensor 04 05 2000 Meas10min	-0.45	5	NSensor 04 05 2000 Meas10pctg	0.35	2
NSensor 04 05 2000 Meas10range	-0.47	4	NSensor 04 05 2000 Meas10sd	-0.40	4
NSensor 04 05 2000 NIR	0.45	4	Nsensor 11 05 1999 Meas10range	-0.30	2
Nsensor 11 05 1999 Meas10slp	0.32	2	Nsensor 18 04 2000 EC 060 1998	0.32	1
Nsensor 18 04 2000 Meas10m	0.32	1	Nsensor 18 04 2000 Meas10max	0.30	4
Nsensor 18 04 2000 Meas10mean	0.33	3	Nsensor 18 04 2000 Meas10min	0.30	2
Nsensor 18 04 2000 Meas10pctg	-0.14	2	Nsensor 18 04 2000 Meas10range	0.20	3
Nsensor 18 04 2000 NIR	-0.36	3	Penm 010 EUF-K1 1999	0.25	1
Penm 010 EUF-K1 2000	0.26	1	Penm 010 EUF-K1 2001	0.23	1
Penm 010 KPS 1999	0.28	6	Penm 010 Meas10m	-0.48	5
Penm 010 Meas10max	-0.51	5	Penm 010 Meas10mean	-0.49	5
Penm 010 Meas10min	-0.41	4	Penm 010 Meas10pctg	0.26	2
Penm 010 Meas10range	-0.49	5	Penm 010 Meas10sd	-0.27	3
Penm 010 Meas10wi	-0.16	1	Penm 010 NIR	0.48	5
Penm 010 NSensor 04 05 2000	0.30	4	Penm 010 SpiK eN 1999	-0.22	6
Penm 010 SSM 29 04 1999	-0.39	2	Penm 010 YPS 1999	0.19	6
Penm 120 EUF-K1 1999	0.29	2	Penm 120 EUF-K1 2000	0.25	1
Penm 120 EUF-K1 2001	0.26	2	Penm 120 EUF-K2 1999	0.28	1
Penm 120 EUF-K2 2000	0.22	1	Penm 120 EUF-K2 2001	0.21	1
Penm 120 EUF-P1 2000	0.25	1	Penm 120 KPS 1999	0.20	6
Penm 120 Meas10m	-0.48	4	Penm 120 Meas10max	-0.44	5
Penm 120 Meas10mean	-0.48	5	Penm 120 Meas10min	-0.42	4
Penm 120 Meas10range	-0.32	4	Penm 120 Meas10sd	-0.22	3
Penm 120 NIR	0.33	4	Penm 120 NSensor 04 05 2000	0.30	3
Penm 120 Penm 010	0.56	4	Penm 230 Meas10m	-0.25	1
Penm 230 Meas10max	-0.20	1	Penm 230 Meas10mean	-0.25	1
Penm 230 Meas10min	-0.22	1	Penm 230 Penm 120	0.61	1
Penm 340 NSensor11 05 1999	-0.24	1	Penm 340 SSM 29 04 1999	0.25	1
Penm 450 FAT 360	-0.51	1	Penm 450 Meas10dev	-0.33	1
Penm 450 Meas10flacc	0.19	1	Penm 450 Meas10m	-0.24	1
Penm 450 NPROT 2000	0.38	2	Penm 450 Penm 340	0.73	1
Penm 560 EUF-K1 1999	0.26	1	Penm 560 EUF-K2 1999	0.22	1
Penm 560 KPS 1999	0.26	6	Penm 560 Meas10dev	-0.34	2
Penm 560 Meas10flacc	0.22	1	Penm 560 Meas10m	-0.40	3

Table 45. Significant cross correlations between different variables at the field site „Bei Lotte". Additionally some signifcant cross correlations for the field site „Sportkomplex" are shown which are specially signed with SK. The abbrevitions used are described below the table. Length indicates the number of significant lag classes (lag distance 27 m, except for the SK protein content, here 54 m).

Heading and Tailing property	CC Lag 0	Length	Heading and Tailing property	CC Lag 0	Length
Penm 560 Meas10max	-0.31	3	Penm 560 Meas10mean	-0.34	3
Penm 560 Meas10min	-0.28	3	Penm 560 Meas10range	-0.26	3
Penm 560 Meas10spi	0.21	1	Penm 560 Meas10stc	0.22	1
Penm 560 NIR	0.32	1	Penm 560 NPROT 2000	0.32	6
Penm 560 Penm 120	0.18	1	Penm 560 Penm 450	0.80	1
Penm 560 Sand 360	0.31	2	Penm 670 EUF-K1 1999	0.19	1
Penm 670 KPS 1999	0.23	6	Penm 670 Meas10dev	-0.25	1
Penm 670 Meas10m	-0.37	4	Penm 670 Meas10max	-0.31	3
Penm 670 Meas10mean	-0.32	3	Penm 670 Meas10min	-0.26	3
Penm 670 Meas10range	-0.29	3	Penm 670 NIR	0.33	1
Penm 670 NPROT 2000	0.26	6	Penm 670 Penm 010	0.16	2
Penm 670 Penm 560	0.88	1	Penm 670 PH 030	-0.20	1
Penm 780 Meas10dev	-0.23	1	Penm 780 Meas10m	-0.26	3
Penm 780 Meas10max	-0.21	2	Penm 780 Meas10mean	-0.22	2
Penm 780 Meas10min	-0.18	1	Penm 780 Meas10range	-0.19	2
Penm 780 NIR	0.26	1	Penm 780 NPROT 2000	0.26	2
Penm 780 Penm 560	0.70	1	Penm 780 Penm 670	0.86	1
PH 030 Meas10max	0.39	4	PH 030 Meas10range	0.47	5
PH 030 Penm 010	-0.57	4	PH 030 SSM 29 04 1999	0.53	2
PH 360 FAT 030	-0.44	1	PH 360 Meas10range	0.44	2
PH 360 NIR	-0.45	2	PH 360 SSM 29 04 1999	0.57	2
PH30 Meas10sd	0.33	3	PH30 NIR	-0.48	4
PH30 NSensor11 05 1999	-0.49	1	PH60 NSensor11 05 1999	-0.39	2
Sand 030 Meas10dev	-0.62	2	Sand 030 Meas10m	-0.42	2
Sand 360 Meas10dev	-0.60	2	Sand 360 Meas10m	-0.44	2
SG 1999 Meas10SLP	-0.13	2	SK NDVI 04 1999 NDVI 04 2000	0.47	2
SK NDVI 05 1998 NDVI 04 1999	0.34	6	SK NDVI 05 1998 NDVI 04 2000	0.47	2
SK NDVI 05 1999 NDVI 04 1999	0.60	2	SK NDVI 05 1999 NDVI 04 2000	0.44	2
SK NDVI 05 2000 NDVI 04 1999	0.43	2	SK NDVI 05 2000 NDVI 04 2000	0.72	2
SK NDVI 07 1997 NDVI 04 1999	-0.33	2	SK NDVI 07 1997 NDVI 04 2000	-0.38	2
SK NDVI 07 1997 NDVI 05 1999	-0.46	2	SK NDVI 07 1997 NDVI 05 2000	-0.51	6
SK NPROT 2000 Penm 670	-0.01	2	SK NPROT 2000 Penm 780	0.00	1
SK NPROT 2002 Penm 560	0.26	2	SK NPROT 2002 Penm 670	0.26	2
SK YIE 1997 NDVI 05 1998	0.31	6	SK YIE 1998 Clay 030	-0.16	6
SK YIE 1998 Clay 360	-0.27	6	SK YIE 1998 FAT 030	-0.19	6
SK YIE 1998 FAT 360	-0.28	6	SK YIE 1998 Meas10CUR	0.19	6
SK YIE 1998 Meas10FLAC	-0.25	6	SK YIE 1998 Meas10M	0.60	9
SK YIE 1998 Meas10PCT1	0.59	6	SK YIE 1998 Meas10PCTG	0.35	6
SK YIE 1998 Meas10PROF	-0.24	6	SK YIE 1998 Meas10SLP	-0.08	2
SK YIE 1998 Meas10SPI	-0.31	6	SK YIE 1998 Meas10WI	-0.16	6
SK YIE 1998 NDVI 04 1999	0.24	2	SK YIE 1998 NDVI 04 2000	0.42	2
SK YIE 1998 NDVI 05 1998	0.69	6	SK YIE 1998 NDVI 05 1999	0.30	2
SK YIE 1998 Sand 690	-0.28	6	SK YIE 1998 Silt 030	0.17	2
SK YIE 1998 Silt 360	0.32	6	SK YIE 1998 Silt 690	0.25	6
SK YIE 1998 YIE 1997	0.19	2	SK YIE 1999 Meas10SLP	-0.27	6
SK YIE 1999 NDVI 04 1999	0.45	2	SK YIE 1999 NDVI 04 2000	0.34	2
SK YIE 1999 NDVI 05 1998	0.42	2	SK YIE 1999 NDVI 05 1999	0.51	2

Table 45. Significant cross correlations between different variables at the field site „Bei Lotte". Additionally some signifcant cross correlations for the field site „Sportkomplex" are shown which are specially signed with SK. The abbrevitions used are described below the table. Length indicates the number of significant lag classes (lag distance 27 m, except for the SK protein content, here 54 m).

Heading and Tailing property	CC Lag 0	Len gth	Heading and Tailing property	CC Lag 0	Len gth
SK YIE 1999 NDVI 07 1997	-0.26	2	SK YIE 1999 YIE 1997	0.36	2
SK YIE 1999 YIE 1998	0.29	2	SK YIE 1999 YIE 2002	0.37	6
SK YIE 2000 Meas10M	-0.35	9	SK YIE 2000 Meas10M	-0.21	6
SK YIE 2000 NDVI 04 1999	0.42	2	SK YIE 2000 NDVI 04 2000	0.52	2
SK YIE 2000 NDVI 05 1999	0.42	2	SK YIE 2000 NDVI 05 2000	0.68	2
SK YIE 2000 NDVI 07 1997	-0.41	6	SK YIE 2000 Sand 030	-0.22	6
SK YIE 2000 Sand 030	-0.40	2	SK YIE 2000 Sand 360	-0.34	2
SK YIE 2000 Sand 690	-0.30	2	SK YIE 2000 Silt 030	0.23	2
SK YIE 2000 Silt 360	0.17	2	SK YIE 2000 Silt 690	0.25	2
SK YIE 2000 YIE 2000	0.57	6	SK YIE 2001 NDVI 04 1999	0.27	2
SK YIE 2001 NDVI 04 2000	0.40	2	SK YIE 2001 NDVI 05 1999	0.34	6
SK YIE 2001 NDVI 05 2000	0.60	6	SK YIE 2001 NDVI07 1997	-0.49	6
SK YIE 2001 YIE 2000	0.57	6	SK YIE 2002 NDVI 04 1999	0.32	2
SK YIE 2002 NDVI 04 2000	0.44	2	SK YIE 2002 NDVI 05 1999	0.45	2
SK YIE 2002 NDVI 05 2000	0.50	1	SK YIE 2002 NDVI07 1997	-0.19	2
SK YIE 2002 YIE 2000	0.53	2	SK YIE 2002 YIE 2001	0.49	2
SM 030 1998 EM38m	-0.17	2	SM 030 1998 Meas10m	0.46	5
SM 030 1998 Meas10max	0.44	5	SM 030 1998 Meas10mean	0.50	5
SM 030 1998 Meas10min	0.36	3	SM 030 1998 Meas10range	0.40	5
SM 030 1998 Meas10sd	0.58	4	SM 030 1998 Meas10slp	0.31	1
SM 030 1998 NIR	-0.46	4	SM 030 1998 NSensor 04 05 2000	-0.45	4
SM 030 1998 Penm 010	-0.28	4	SM 030 1998 Penm 120	-0.34	4
SM 030 1998 Penm 230	-0.22	2	SM 030 1998 Penm 560	-0.20	1
SM 030 1998 Penm 670	-0.24	1	SM 030 1998 SM 030 2000	0.54	4
SM 030 1999 Meas10m	-0.38	2	SM 030 1999 Meas10max	-0.29	1
SM 030 1999 Meas10mean	-0.34	2	SM 030 1999 Meas10min	-0.34	2
SM 030 1999 Nsensor 18 04 2000	-0.33	1	SM 030 2000 Meas10asp	0.21	1
SM 030 2000 Meas10m	0.45	3	SM 030 2000 Meas10max	0.37	4
SM 030 2000 Meas10mean	0.48	4	SM 030 2000 Meas10min	0.37	3
SM 030 2000 Meas10range	0.26	5	SM 030 2000 Meas10sd	0.47	4
SM 030 2000 Meas10slp	0.43	2	SM 030 2000 NIR	-0.45	4
SM 030 2000 NSensor 04 05 2000	-0.33	4	SM 030 2000 Nsensor 11 05 1999	0.30	1
SM 030 2000 Penm 010	-0.18	3	SM 030 2000 Penm 120	-0.37	3
SM 030 2000 Penm 230	-0.36	2	SM 030 2001 Meas10asp	0.19	1
SM 030 2001 Meas10max	-0.20	2	SM 030 2001 Meas10mean	-0.16	1
SM 030 2001 Meas10range	-0.21	2	SM 030 2001 Meas10slp	0.18	1
SM 030 2001 Meas10wi	-0.17	1	SM 030 2001 Nsensor 11 05 1999	0.23	1
SM 030 2001 Penm 010	0.25	1	SM 360 1998 Meas10sd	0.29	3
SM 360 1998 Meas10slp	0.18	1	SM 360 1998 SM 030 1998	0.44	2
SM 360 1998 SM 030 2000	0.31	2	SM 360 1999 Meas10range	0.21	1
SM 360 1999 Meas10sd	0.20	2	SM 360 1999 NIR	-0.27	1
SM 360 1999 Penm 010	-0.16	2	SM 360 1999 SM 030 1998	0.22	1
SM 360 2000 Meas10m	0.29	2	SM 360 2000 Meas10max	0.22	1
SM 360 2000 Meas10mean	0.27	2	SM 360 2000 Meas10min	0.21	1
SM 360 2000 Meas10sd	0.20	1	SM 360 2000 Meas10slp	0.28	1
SM 360 2000 NIR	-0.27	2	SM 360 2000 Nsensor 11 05 1999	0.21	1
SM 360 2000 Penm 340	-0.41	1	SM 360 2000 Penm 450	-0.35	1

Table 45. Significant cross correlations between different variables at the field site „Bei Lotte". Additionally some signifcant cross correlations for the field site „Sportkomplex" are shown which are specially signed with SK. The abbrevitions used are described below the table. Length indicates the number of significant lag classes (lag distance 27 m, except for the SK protein content, here 54 m).

Heading and Tailing property	CC Lag 0	Length	Heading and Tailing property	CC Lag 0	Length
SM 360 2000 Penm 560	-0.30	1	SM 360 2000 SM 030 1998	0.35	3
SM 360 2000 SM 030 2000	0.40	2	SM 360 2001 Meas10slp	0.15	1
SM 360 2001 Nsensor 11 05 1999	0.26	1	SM 360 2001 Penm 340	-0.37	1
SM 360 2001 SM 360 2000	0.36	1	SM 690 1998 Meas10m	0.25	3
SM 690 1998 Meas10max	0.26	3	SM 690 1998 Meas10mean	0.30	3
SM 690 1998 Meas10min	0.25	1	SM 690 1998 Meas10range	0.19	3
SM 690 1998 Meas10sd	0.25	3	SM 690 1998 NIR	-0.31	3
SM 690 1998 NSensor 04 05 2000	-0.22	3	SM 690 1998 SM 030 1998	0.40	3
SM 690 1998 SM 030 2000	0.39	3	SM 690 1998 SM 360 2000	0.38	1
SM 690 1998 SM 690 2001	0.58	1	SM 690 2000 Meas10m	0.24	1
SM 690 2000 Meas10mean	0.21	1	SM 690 2000 NIR	-0.29	1
SM 690 2000 SM 030 2000	0.42	2	SM 690 2000 SM 360 2000	0.56	1
SM 690 2001 Meas10mean	0.19	1	SM 690 2001 SM 030 2000	0.27	2
SM 690 2001 SM 360 2000	0.33	1	SM 690 2001 SM 690 2000	0.43	1
SpikeN 1999 Meas10M	0.22	6	SpikeN 2000 Meas10M	-0.26	6
SSM 10 05 1999 Meas10asp	-0.20	1	SSM 10 05 1999 Meas10max	0.24	1
SSM 10 05 1999 Meas10pctg	-0.42	3	SSM 10 05 1999 Meas10range	0.27	2
SSM 10 05 1999 Meas10wi	0.37	1	SSM 10 05 1999 NIR	-0.33	1
SSM 10 05 1999 NSensor 04 05 2000	-0.24	1	SSM 12 05 2000 EUF-K1 1999	0.26	1
SSM 12 05 2000 EUF-K2 1999	0.24	1	SSM 12 05 2000 K 360	0.46	1
SSM 12 05 2000 Meas10dev	-0.31	1	SSM 12 05 2000 Meas10pctg	-0.29	2
SSM 25 03 1999 EC 060 1998	-0.47	1	SSM 25 03 1999 EUF-P1 2000	0.26	1
SSM 25 03 1999 Meas10asp	-0.17	1	SSM 25 03 1999 Meas10pctg	-0.33	2
SSM 25 03 1999 SSM 10 05 1999	0.56	1	SSM 25 03 1999 SSM 12 05 2000	0.36	1
SSM 2503 1999 Meas10wi	0.43	1	SSM 29 04 1999 Meas10max	0.33	3
SSM 29 04 1999 Meas10mean	0.21	2	SSM 29 04 1999 Meas10range	0.46	3
SSM 29 04 1999 Meas10sd	0.29	2	SSM 29 04 1999 NSensor 04 05 2000	-0.28	3
SSM 29 04 1999 SSM 25 03 1999	0.51	1	SSM 2904 1999 Meas10pctg	-0.44	3
SSM 2904 1999 Meas10wi	0.26	1	SSM 2904 1999 NIR	-0.43	3
SSM 2904 1999 Nsensor 11 05 1999	-0.33	1	SSM 2904 1999 PH 360	0.57	2
SSM 2904 1999 SSM 10 05 1999	0.54	1	TKG 2000 Meas10M	0.33	6
WR 2000 Sand 030	-0.04	2	WR 2000 Sand 360	-0.03	2
WR 2000 Sand 390	-0.04	2	WR 2000 SG 1999	0.03	2
WR 2000 Silt 030	0.15	2	WR 2000 Silt 360	0.17	2
WR 2001 Sand 030	0.05	2	WR 2001 SG 1999	-0.15	2
WR 2001 WR 2000	-0.09	2	YPS 1999 Meas10M	-0.20	6
YPS 2000 Meas10M	0.24	6			

Table 46. Abbreviations used in Table 45

Abbrev.	Description	Abbrev.	Description
010	sampling depth 0-10 cm	030	sampling depth 0-30 cm
060	sampling depth 0-60 cm	120	sampling depth 10-20 cm
230	sampling depth 20-30 cm	340	sampling depth 30-40 cm
360	sampling depth 30-60 cm	450	sampling depth 40-50 cm
560	sampling depth 50-60 cm	670	sampling depth 60-70 cm
690	sampling depth 60-90 cm	780	sampling depth 70-80 cm
ASP	aspect	Clay	Clay Content
DEV	Deviation in filter	EC	Electrical Conductivity Measurements
EM38m	EM-Measurments with VERIS technology	EUF	Electro Ulta Filtration Data
FAT	Fine Particle Content	FLACC	Flowaccumulation
K	Potassium obtained using DL method	K1	First Fraction of EUF-Potassium
K2	Second Fraction of EUF-Potassium	Kernel	Number of Kernels
KpS	Kernel Per Spike	M	Mean height in context of Meas10
MAX	Maximum height in filter	Meas10	indicates relief parameter
MIN	Minimum height in filter	NDVI	Near Difference Vegetation Index
NIR	Near Infrared Image Data	Nmin	Mineralized Nitrogen
Nprot	Grain Protein Content	Nsensor	Crop N status
P1	First Fraction of EUF-Phosphorus	PCTG(1)	Landscape Position (see Table 4)
Penm	Penetration Resistance	PH	ph
PLAN	Profile Curvature	PROF	Profile Curvature
RANGE	Range of height in filter	Sand	Sand content
SD	SD of height in filter	SG	Spring Barley Yield
Silt	Silt content	SK	indicates data for „Sportkomplex“
SLP	slope	SM	Soil Moisture
SM	Soil Moisture	SPI	Stream power Index
SpikeN	Number of Spikes	SSM	Soil surface moisture
STC	Sediment transport Capacity	TKM	Thousand Kernel Mass
TWI	Topographic Wetness Index	WR	Winter Rye Yield
YIE	Grain Yield	YpS	Yield per Spike

Appendix Landform

Table 47. N_{min} Content in $kgha^{-1}$ for Profile curvature in1/100m for 1998-2001 (Mean = Average, STD = Standard deviation, N = number of samples, CV = Coefficient of Variation, 030 0-30cm depth layer, 360 30-60cm depth layer, 690 60-90 cm depth layer). All samples taken at the field site „Bei Lotte".

PROFILE CURVATURE	-0.2--0.1	-0.1-0	0-0.1	0.1-0.2	0.3-0.4	FIELD
Mean - Nmin - 030 - 100898	38.10	43.05	42.45	43.23	59.40	42.65
Mean - Nmin - 360 - 100898	14.40	18.66	20.98	17.01	42.53	19.59
Mean - Nmin - 690 - 100898	5.72	8.61	8.24	8.94	20.89	8.41
Mean - Nmin - 030 - 160899	39.54	52.31	57.10	45.57	60.90	53.30
Mean - Nmin - 360 - 160899	10.01	14.53	16.40	13.41	31.84	15.21
Mean - Nmin - 690 - 160899	4.80	6.35	6.58	5.68	14.05	6.39
Mean - Nmin - 030 - 170800	27.47	28.71	29.67	26.68	52.89	29.17
Mean - Nmin - 690 - 170800	2.49	2.49	3.66	2.44	3.10	3.03
Mean - Nmin - 360 - 170800	4.03	5.36	5.18	5.28	13.71	5.28
Mean - Nmin - 030 - 130801	30.59	35.60	32.32	32.69	104.62	34.28
Mean - Nmin - 360 - 130801	6.59	9.24	7.51	8.43	30.30	8.44
Mean - Nmin - 690 - 130801	3.26	3.01	2.57	3.42	8.07	2.90
STD - Nmin - 030 - 100898	14.59	15.05	19.34	24.58	12.09	17.90
STD - Nmin - 360 - 100898	9.96	16.00	13.65	13.92	2.86	14.58
STD - Nmin - 690 - 100898	4.05	8.05	5.42	9.53	11.94	6.99
STD - Nmin - 030 - 160899	14.17	20.47	21.63	19.06	6.34	20.94
STD - Nmin - 360 - 160899	3.40	5.57	7.38	8.35	1.48	6.96
STD - Nmin - 690 - 160899	2.39	3.14	3.06	2.37	2.21	3.11
STD - Nmin - 030 - 170800	12.03	10.19	11.14	10.23	36.63	11.25
STD - Nmin - 690 - 170800	1.47	1.49	3.75	1.21	1.47	2.81
STD - Nmin - 360 - 170800	2.47	3.72	2.58	2.76	7.98	3.23
STD - Nmin - 030 - 130801	9.74	18.87	14.19	12.83	18.18	17.45
STD - Nmin - 360 - 130801	3.36	7.87	4.65	6.51	12.88	6.66
STD - Nmin - 690 - 130801	3.59	2.34	1.88	2.59	5.49	2.32
N	12	74	89	15	2	192
CV - Nmin - 030 - 100898	38.29	34.96	45.56	56.86	20.36	41.98
CV - Nmin - 360 - 100898	69.17	85.71	65.08	81.83	6.73	74.43
CV - Nmin - 690 - 100898	70.80	93.59	65.76	106.62	57.17	83.06
CV - Nmin - 030 - 160899	35.84	39.13	37.88	41.82	10.40	39.28
CV - Nmin - 360 - 160899	34.01	38.34	44.99	62.22	4.64	45.75
CV - Nmin - 690 - 160899	49.92	49.36	46.54	41.74	15.77	48.69
CV - Nmin - 030 - 170800	43.79	35.50	37.54	38.33	69.26	38.58
CV - Nmin - 690 - 170800	59.11	59.98	102.42	49.84	47.38	92.70
CV - Nmin - 360 - 170800	61.35	69.38	49.77	52.35	58.21	61.27
CV - Nmin - 030 - 130801	31.82	53.00	43.92	39.25	17.38	50.91
CV - Nmin - 360 - 130801	50.98	85.12	61.99	77.18	42.51	78.87
CV - Nmin - 690 - 130801	110.11	77.66	72.96	75.70	68.03	79.95

Table 48. N_{min} Content in $kgha^{-1}$ for Landforms for 1998-2001 (Mean = Average, STD = Standard deviation, N = number of samples, CV = Coefficient of Variation, 030 0-30cm depth layer, 360 30-60cm depth layer, 690 60-90 cm depth layer). All samples from the field site „Bei Lotte".

Landforms	SHOULDER	BACKSLOPE	FOOTSLOPE	LEVEL	FIELD
Mean - Nmin - 030 - 100898	40.50	40.25	40.56	44.35	42.65
Mean - Nmin - 360 - 100898	14.59	20.08	18.42	20.32	19.59
Mean - Nmin - 690 - 100898	5.44	7.90	10.46	8.80	8.41
Mean - Nmin - 030 - 160899	41.28	52.82	49.89	55.86	53.30
Mean - Nmin - 360 - 160899	11.71	13.68	15.93	16.32	15.21
Mean - Nmin - 690 - 160899	4.93	5.61	6.51	6.94	6.39
Mean - Nmin - 030 - 170800	27.75	27.97	28.75	29.99	29.17
Mean - Nmin - 690 - 170800	2.86	3.88	2.39	2.78	3.03
Mean - Nmin - 360 - 170800	4.17	5.14	6.31	5.36	5.28
Mean - Nmin - 030 - 130801	29.60	32.08	44.13	34.50	34.28
Mean - Nmin - 360 - 130801	7.95	7.46	11.43	8.51	8.44
Mean - Nmin - 690 - 130801	3.04	2.56	4.22	2.84	2.90
STD - Nmin - 030 - 100898	10.10	15.18	23.72	19.00	17.90
STD - Nmin - 360 - 100898	8.29	14.12	14.63	15.51	14.58
STD - Nmin - 690 - 100898	3.38	5.67	10.58	7.21	6.99
STD - Nmin - 030 - 160899	11.43	22.92	18.43	20.97	20.94
STD - Nmin - 360 - 160899	4.01	6.89	9.30	6.74	6.96
STD - Nmin - 690 - 160899	2.06	2.71	3.72	3.20	3.11
STD - Nmin - 030 - 170800	12.34	11.89	16.24	9.97	11.25
STD - Nmin - 690 - 17080	1.32	4.84	1.52	1.59	2.81
STD - Nmin - 360 - 17080	2.27	2.59	4.39	3.40	3.23
STD - Nmin - 030 - 130801	9.38	14.34	27.24	17.50	17.45
STD - Nmin - 360 - 130801	4.50	5.06	9.61	6.97	6.66
STD - Nmin - 690 - 130801	2.86	1.46	3.18	2.38	2.32
N	17	49	16	110	192
CV - Nmin - 030 - 100898	24.93	37.72	58.48	42.84	41.98
CV - Nmin - 360 - 100898	56.81	70.32	79.41	76.34	74.43
CV - Nmin - 690 - 100898	62.21	71.80	101.14	81.94	83.06
CV - Nmin - 030 - 160899	27.68	43.40	36.93	37.55	39.28
CV - Nmin - 360 - 160899	34.25	50.33	58.34	41.28	45.75
CV - Nmin - 690 - 160899	41.75	48.36	57.23	46.15	48.69
CV - Nmin - 030 - 170800	44.46	42.52	56.50	33.23	38.58
CV - Nmin - 690 - 17080	46.22	124.85	63.58	57.23	92.70
CV - Nmin - 360 - 17080	54.47	50.45	69.66	63.51	61.27
CV - Nmin - 030 - 130801	31.70	44.68	61.72	50.73	50.91
CV - Nmin - 360 - 130801	56.68	67.86	84.03	81.96	78.87
CV - Nmin - 690 - 130801	94.27	57.00	75.31	83.54	79.95

Table 49. Nutrient Content for potassium (K_{DL}) and phosphorus (P_{DL}) using the DL method and the EUF-method (Fractions are denoted with subscripts 1 and 2) seperated for Profile curvature in 1/100m for 1998-2001 (Mean = Average, STD = Standard deviation, N = number of samples, CV = Coefficient of Variation, 030 0-30cm depth layer, 360 30-60cm depth layer). All samples from the field site „Bei Lotte".

PROFILE CURVATURE	-0.2--0.1	-0.1-0	0-0.1	0.1-0.2	0.3-0.4	FIELD
Mean K_{DL} - 030 - 1998	4.43	6.55	5.36	4.61	10.00	5.83
Mean K_{DL} - 360 - 1998	2.46	4.25	3.97	3.29	10.59	4.05
Mean P_{DL} - 030 - 1998	8.32	12.20	8.16	8.26	22.02	10.19
Mean P_{DL} - 360 - 1998	6.29	7.29	4.91	5.66	11.81	6.24
Mean P_1 1999	1.12	1.39	1.20	1.45	2.75	1.31

Table 49. Nutrient Content for potassium (K_{DL}) and phosphorus (P_{DL}) using the DL method and the EUF-method (Fractions are denoted with subscripts 1 and 2) seperated for Profile curvature in 1/100m for 1998-2001 (Mean = Average, STD = Standard deviation, N = number of samples, CV = Coefficient of Variation, 030 0-30cm depth layer, 360 30-60cm depth layer). All samples from the field site „Bei Lotte".

Mean P_2 1999	0.45	0.58	0.52	0.59	1.00	0.55
Mean K_1 1999	4.69	5.50	4.54	6.56	11.27	5.15
Mean K_2 1999	2.20	2.69	2.25	3.05	4.43	2.50
Mean P_1 2000	1.02	1.20	1.07	1.20	2.16	1.14
Mean P_2 2000	0.53	0.72	0.61	0.65	1.44	0.66
Mean K_1 2000	4.48	4.56	4.14	5.72	9.57	4.50
Mean K_2 2000	2.58	2.66	2.41	2.89	4.18	2.57
Mean P_1 2001	1.10	1.19	1.07	1.19	1.96	1.14
Mean P_1 2001	0.54	0.60	0.57	0.60	1.17	0.59
Mean K_1 2001	5.18	5.39	4.49	5.02	11.67	5.00
Mean K_2 2001	2.53	2.55	2.30	2.36	4.40	2.44
STD K_{DL} - 030 - 1998	1.25	4.26	1.23	1.18		3.05
STD K_{DL} - 360 - 1998	0.42	2.60	2.11	0.99		2.40
STD P_{DL} - 030 - 1998	1.07	7.06	1.84	1.06		5.39
STD P_{DL} - 360 - 1998	0.24	3.99	1.17	0.88		3.01
STD P_1 - 1999	0.41	0.57	0.40	0.79	1.46	0.55
STD P_2 - 1999	0.19	0.24	0.21	0.37	0.64	0.25
STD K_1 - 1999	0.70	3.12	2.03	3.35	3.08	2.70
STD K_2 - 1999	0.73	1.04	0.84	1.38	1.67	1.02
STD P_1 - 2000	0.51	0.48	0.36	0.66	0.04	0.46
STD P_2 - 2000	0.31	0.31	0.25	0.40	0.02	0.30
STD K_1 - 2000	1.35	2.15	1.50	2.59	1.09	1.96
STD K_2 - 2000	0.61	1.05	0.72	1.20	0.71	0.92
STD P_1 - 2001	0.40	0.46	0.38	0.52	0.40	0.43
STD P_2 - 2001	0.23	0.25	0.24	0.31	0.04	0.25
STD K_1 - 2001	0.92	2.50	1.45	1.72	0.15	2.07
STD K_2 - 2001	0.40	0.87	0.66	0.67	0.37	0.77
N	12	74	89	15	2	192
CV K_{DL} - 030 - 1998	28.16	65.07	23.01	25.66	0.00	52.31
CV K_{DL} - 360 - 1998	17.07	61.17	53.29	30.03	0.00	59.21
CV P_{DL} - 030 - 1998	12.85	57.89	22.54	12.87	0.00	52.92
CV P_{DL} - 360 - 1998	3.76	54.67	23.79	15.63	0.00	48.16
CV P_1 - 1999	36.34	41.36	33.06	54.67	53.32	41.89
CV P_2 - 1999	41.86	41.86	40.59	61.89	63.64	44.77
CV K_1 - 1999	14.86	56.73	44.69	51.00	27.31	52.51
CV K_2 - 1999	33.33	38.67	37.34	45.03	37.67	40.69
CV P_1 - 2000	50.38	39.97	33.42	55.06	1.96	40.29
CV P_2 - 2000	57.37	42.93	41.08	61.61	1.48	46.04
CV K_1 - 2000	30.05	47.12	36.13	45.39	11.38	43.56
CV K_2 - 2000	23.51	39.50	29.85	41.42	16.92	35.62
CV P_1 - 2001	36.50	38.45	35.26	43.91	20.62	37.91
CV P_2 - 2001	43.24	41.98	41.95	52.12	3.03	43.30
CV K_1 - 2001	17.82	46.33	32.35	34.17	1.27	41.37
CV K_2 - 2001	15.71	34.13	28.82	28.20	8.36	31.48

Table 50. Nutrient Content for potassium (K_{DL}) and phosphorus (P_{DL}) using the DL method and the EUF-method (Fractions are denoted with subscripts 1 and 2) seperated for Landforms for 1998-2001 (Mean = Average, STD = Standard deviation, N = number of samples, CV = Coefficient of Variation, 030 0-30cm depth layer, 360 30-60cm depth layer, 690 60-90 cm depth layer). All samples from the field site „Bei Lotte".

Landforms	SHOULDER	BACKSLOPE	FOOTSLOPE	LEVEL	FIELD
Mean - K_{DL} -030 - 1998	6.21	5.32	6.19	5.95	5.83
Mean - K_{DL} -360 - 1998	3.86	4.54	5.10	3.69	4.05
Mean - P_{DL} - 030 - 1998	13.11	8.28	11.67	10.35	10.19
Mean - P_{DL} - 360 - 1998	8.38	5.40	7.06	6.13	6.24
Mean - P_1 - 1999	1.37	1.24	1.67	1.27	1.31
Mean - P_2 - 1999	0.54	0.51	0.67	0.55	0.55
Mean - K_1 - 1999	6.40	4.89	6.65	4.85	5.15
Mean - K_2 - 1999	3.07	2.47	3.02	2.35	2.50
Mean - P_1 - 2000	1.24	1.09	1.28	1.12	1.14
Mean - P_2 - 2000	0.72	0.59	0.75	0.66	0.66
Mean - K_1 - 2000	5.41	4.68	5.82	4.09	4.50
Mean - K_2 - 2000	2.98	2.63	2.91	2.44	2.57
Mean - P_1 - 2001	1.27	1.06	1.33	1.13	1.14
Mean - P_2 - 2001	0.65	0.52	0.68	0.59	0.59
Mean - K_1 - 2001	5.87	4.59	5.83	4.92	5.00
Mean - K_2 - 2001	2.80	2.34	2.58	2.40	2.44
STD - K_{DL} -030 - 1998	4.27	2.18	2.74	3.33	3.05
STD - K_{DL} -360 - 1998	2.65	2.72	3.79	2.00	2.40
STD - P_{DL} - 030 - 1998	5.08	2.20	6.98	6.17	5.39
STD - P_{DL} - 360 - 1998	3.58	1.18	3.27	3.39	3.01
STD - P_1 - 1999	0.56	0.53	0.94	0.46	0.55
STD - P_2 - 1999	0.25	0.24	0.39	0.22	0.25
STD - K_1 - 1999	2.35	1.94	3.82	2.76	2.70
STD - K_2 - 1999	1.17	0.85	1.49	0.94	1.02
STD - P_1 - 2000	0.55	0.44	0.70	0.41	0.46
STD - P_2 - 2000	0.38	0.31	0.46	0.25	0.30
STD - K_1 - 2000	2.00	1.61	2.93	1.80	1.96
STD - K_2 - 2000	0.90	0.79	1.28	0.89	0.92
STD - P_1 - 2001	0.43	0.41	0.51	0.42	0.43
STD - P_2 - 2001	0.30	0.23	0.33	0.24	0.25
STD - K_1 - 2001	1.98	1.43	2.82	2.15	2.07
STD - K_2 - 2001	0.73	0.59	0.97	0.80	0.77
N	17	49	16	110	192
CV - K_{DL} - 030 - 1999	68.76	41.03	44.28	55.99	52.31
CV - K_{DL} -360 - 1999	68.48	59.84	74.46	54.37	59.21
CV - P_{DL} - 030 - 1999	38.75	26.52	59.85	59.59	52.92
CV - P_{DL} - 360 - 1999	42.74	21.78	46.32	55.29	48.16
CV - P_1 - 1999	40.60	43.12	56.41	35.82	41.89
CV - P_2 - 1999	46.28	46.14	58.16	40.04	44.77
CV - K_1 - 1999	36.74	39.77	57.50	56.90	52.51
CV - K_2 - 1999	38.05	34.34	49.43	39.83	40.69
CV - P_1 - 2000	44.42	39.95	54.66	36.45	40.29

Table 50. Nutrient Content for potassium (K_{DL}) and phosphorus (P_{DL}) using the DL method and the EUF-method (Fractions are denoted with subscripts 1 and 2) seperated for Landforms for 1998-2001 (Mean = Average, STD = Standard deviation, N = number of samples, CV = Coefficient of Variation, 030 0-30cm depth layer, 360 30-60cm depth layer, 690 60-90 cm depth layer). All samples from the field site „Bei Lotte".

CV - P_2 - 2000	53.00	51.91	61.93	38.23	46.04
CV - K_1 - 2000	36.92	34.49	50.42	44.11	43.56
CV - K_2 - 2000	30.07	29.87	43.84	36.63	35.62
CV - P_1 - 2001	33.98	38.95	38.36	37.41	37.91
CV - P_2 - 2001	46.64	43.89	49.00	40.30	43.30
CV - K_1 - 2001	33.73	31.13	48.30	43.67	41.37
CV - K_2 - 2001	25.98	25.26	37.42	33.38	31.48

Table 51. Soil surface moisture content (SSM) in g(100g soil)$^{-1}$ for Landforms for 1999 and 2000 (Mean = Averag, STD = Standard deviation, N = number of samples, CV = Coefficient of Variation, Sampling depth 0-10cm). All samples from the field site „Bei Lotte".

Landforms	SHOULDER	BACKSLOPE	FOOTSLOPE	LEVEL	FIELD
Mean - SSM - 250399	25.99	26.65	27.46	27.08	26.90
Mean - SSM - 290499	24.86	25.85	25.70	25.76	25.70
Mean - SSM - 100599	22.33	22.64	23.14	22.72	22.70
Mean - SSM - 120500	11.82	11.31	11.90	11.40	11.45
STD - SSM - 250399	1.74	1.51	2.31	1.89	1.85
STD - SSM - 290499	0.81	0.73	1.15	1.01	0.97
STD - SSM - 100599	0.89	1.09	1.35	1.01	1.06
STD - SSM - 120500	1.29	1.38	1.49	1.34	1.36
N	17	49	16	110	192
CV - SSM - 250399	6.69	5.66	8.43	6.97	6.87
CV - SSM - 290499	3.26	2.82	4.49	3.93	3.79
CV - SSM - 100599	4.00	4.80	5.82	4.46	4.66
CV - SSM - 120500	10.92	12.23	12.49	11.79	11.91

Table 52. Soil moisture (SM) content in g(100g soil)$^{-1}$ for Landforms for 1999-2001 (Mean = Average, STD = Standard deviation, N = number of samples, CV = Coefficient of Variation, 030 0-30cm depth layer, 360 30-60cm depth layer, 690 60-90 cm depth layer). All samples from the field site „Bei Lotte".

Landforms	SHOULDER	BACKSLOPE	FOOTSLOPE	LEVEL	FIELD
Mean - SM - 030 - 160899	21.02	21.73	21.74	21.65	21.62
Mean - SM - 360 - 160899	20.10	20.51	19.78	20.08	20.17
Mean - SM - 690 - 160899	19.78	20.83	20.02	20.25	20.34
Mean - SM - 030 - 170800	14.02	18.07	15.61	13.64	14.97
Mean - SM - 360 - 170800	10.53	11.98	11.65	10.82	11.16
Mean - SM - 690 - 170800	11.78	13.36	12.09	12.19	12.44
Mean - SM - 030 - 130801	15.64	15.41	15.99	15.24	15.38
Mean - SM - 360 - 130801	16.15	16.44	16.59	15.90	16.12
STD - SM - 030 - 160899	1.45	1.85	2.41	1.37	1.61
STD - SM - 360 - 160899	1.11	0.81	1.27	0.90	0.95
STD - SM - 690 - 160899	1.82	1.10	1.83	0.99	1.23
STD - SM - 030 - 170800	4.16	5.05	4.79	3.82	4.64
STD - SM - 360 - 170800	2.42	2.92	2.73	2.33	2.57
STD - SM - 690 - 170800	2.01	2.29	1.88	1.73	1.99
STD - SM - 030 - 130801	2.42	1.68	1.47	1.67	1.73
STD - SM - 360 - 130801	1.68	1.96	1.80	1.55	1.70
STD - SM - 690 - 130801	2.11	1.71	1.38	1.40	1.57
N	17	49	16	110	192

Table 52. Soil moisture (SM) content in g(100g soil)$^{-1}$ for Landforms for 1999-2001 (Mean = Average, STD = Standard deviation, N = number of samples, CV = Coefficient of Variation, 030 0-30cm depth layer, 360 30-60cm depth layer, 690 60-90 cm depth layer). All samples from the field site „Bei Lotte".

CV - SM - 030 - 160899	6.89	8.51	11.09	6.34	7.46
CV - SM - 360 - 160899	5.53	3.93	6.41	4.49	4.71
CV - SM - 690 - 160899	9.20	5.27	9.14	4.89	6.04
CV - SM - 030 - 170800	29.69	27.93	30.65	28.00	31.03
CV - SM - 360 - 170800	23.00	24.35	23.46	21.56	23.05
CV - SM - 690 - 170800	17.08	17.11	15.56	14.23	15.97
CV - SM - 030 - 130801	15.49	10.93	9.17	10.93	11.24
CV - SM - 360 - 130801	10.42	11.90	10.87	9.73	10.56
CV - SM - 690 - 130801	146.02	92.47	57.33	101.84	97.38

Table 53. Soil surface moisture content (SSM) in g(100g soil)$^{-1}$ for Profile curvature in1/100m for 1999 and 2000 (Mean = Average, STD = Standard deviation, N = number of samples, CV = Coefficient of Variation, sampling depth 0-10cm). All samples from the field site „Bei Lotte".

PROFILE CURVATURE	-0.2--0.1	-0.1-0	0-0.1	0.1-0.2	0.3-0.4	FIELD
Mean - SSM - 250399	25.30	26.65	27.22	27.01	31.19	26.90
Mean - SSM - 290499	24.69	25.52	25.95	25.78	26.39	25.70
Mean - SSM - 100599	21.81	22.53	22.84	23.36	23.35	22.70
Mean - SSM - 120500	11.54	11.35	11.40	12.18	11.76	11.45
STD - SSM - 250399	1.50	1.59	1.90	1.64	2.09	1.85
STD - SSM - 290499	0.48	0.93	0.96	0.98	0.13	0.97
STD - SSM - 100599	0.78	0.86	1.13	1.18	0.47	1.06
STD - SSM - 120500	1.29	1.33	1.31	1.82	0.61	1.36
N	12	74	89	15	2	192
CV - SSM - 250399	5.94	5.97	6.99	6.09	6.71	6.87
CV - SSM - 290499	1.96	3.63	3.71	3.79	0.48	3.79
CV - SSM - 100599	3.57	3.82	4.96	5.05	2.03	4.66
CV - SSM - 120500	11.15	11.74	11.51	14.96	5.21	11.91

Table 54. Soil moisture content (SM) in g(100g soil)$^{-1}$ seperated for Profile curvature in1/100m for 1999 and 2000 (Mean = Average, STD = Standard deviation, N = number of samples, CV = Coefficient of Variation, 030 0-30cm depth layer, 360 30-60cm depth layer, 690 60-90 cm depth layer). All samples from the field site „Bei Lotte".

PROFILE CURVATURE	-0.2--0.1	-0.1-0	0-0.1	0.1-0.2	0.3-0.4	FIELD
Mean - SM - 030 - 160899	20.39	21.57	21.79	21.77	22.80	21.62
Mean - SM - 360 - 160899	19.76	20.07	20.31	20.18	19.61	20.17
Mean - SM - 690 - 160899	19.77	20.22	20.50	20.66	19.01	20.34
Mean - SM - 030 - 170800	13.89	14.23	15.35	17.32	13.93	14.97
Mean - SM - 360 - 170800	11.51	10.92	11.15	12.08	11.48	11.16
Mean - SM - 690 - 170800	12.25	12.54	12.39	12.66	10.90	12.44
Mean - SM - 030 - 130801	15.75	15.54	15.05	15.96	17.85	15.38
Mean - SM - 360 - 130801	16.67	16.02	15.95	16.98	17.45	16.12
Mean - SM - 690 - 130801	17.86	17.69	17.92	18.68	18.96	17.90
STD - SM - 030 - 160899	1.91	1.34	1.57	2.44	0.90	1.61
STD - SM - 360 - 160899	1.52	0.79	0.94	1.18	0.56	0.95
STD - SM - 690 - 160899	1.57	1.24	1.01	1.70	2.52	1.23
STD - SM - 030 - 170800	3.98	4.09	4.99	5.20	0.89	4.64
STD - SM - 360 - 170800	2.11	2.35	2.78	2.88	0.00	2.57
STD - SM - 690 - 170800	1.52	1.92	2.14	1.84	1.50	1.99
STD - SM - 030 - 130801	1.50	1.80	1.69	1.31	1.20	1.73
STD - SM - 360 - 130801	1.22	1.62	1.77	1.78	1.44	1.70
STD - SM - 690 - 130801	1.29	1.67	1.54	1.33	0.74	1.57

Table 54. Soil moisture content (SM) in g(100g soil)$^{-1}$ seperated for Profile curvature in1/100m for 1999 and 2000 (Mean = Average, STD = Standard deviation, N = number of samples, CV = Coefficient of Variation, 030 0-30cm depth layer, 360 30-60cm depth layer, 690 60-90 cm depth layer). All samples from the field site „Bei Lotte".

N	12	74	89	15	2	192
CV - SM - 030 - 160899	9.39	6.21	7.21	11.22	3.94	7.46
CV - SM - 360 - 160899	7.69	3.91	4.61	5.86	2.83	4.71
CV - SM - 690 - 160899	7.96	6.13	4.91	8.24	13.25	6.04
CV - SM - 030 - 170800	28.66	28.71	32.53	30.03	6.37	31.03
CV - SM - 360 - 170800	18.35	21.49	24.90	23.84	0.01	23.05
CV - SM - 690 - 170800	12.45	15.30	17.29	14.51	13.80	15.97
CV - SM - 030 - 130801	9.53	11.59	11.25	8.21	6.75	11.24
CV - SM - 360 - 130801	7.34	10.13	11.09	10.47	8.28	10.56
CV - SM - 690 - 130801	7.23	9.44	8.57	7.15	3.89	8.78

Table 55. Texture, C_{org} and pH for profile curvature in1/100m (Mean = Average, STD = Standard deviation, N = number of samples, CV = Coefficient of Variation, 030 0-30cm depth layer, 360 30-60cm depth layer, 690 60-90 cm depth layer). All samples from the field site „Bei Lotte".

PROFILE CURVATURE	-0.2--0.1	-0.1-0	0-0.1	0.1-0.2	0.3-0.4	FIELD
Mean - Sand - 030	2.82	3.22	3.59	3.05	7.15	3.45
Mean - Sand - 360	2.28	2.87	3.27	2.45	8.45	3.13
Mean - Sand - 690	2.81	2.75	2.97	2.20	9.08	3.00
Mean - Silt - 030	78.41	79.76	80.92	79.92	75.99	80.02
Mean - Silt- 360	76.60	78.46	80.07	78.33	75.42	78.86
Mean - Silt - 690	76.61	77.31	77.98	78.20	73.87	77.49
Mean - Clay - 030	18.78	17.02	15.49	17.04	16.87	16.53
Mean - Clay - 360	21.12	18.67	16.66	19.23	16.13	18.01
Mean - Clay - 690	20.58	19.93	19.05	19.60	17.05	19.51
STD - Sand - 030	0.55	1.20	2.25	1.00	2.67	1.81
STD - Sand - 360	0.75	1.76	2.44	0.79	3.72	2.22
STD - Sand - 690	2.28	3.30	3.66	0.80	6.45	3.45
STD - Silt - 030	3.67	2.25	3.52	1.84	1.30	3.02
STD - Silt - 360	1.63	2.34	4.45	2.39	1.93	3.48
STD - Silt - 690	3.12	2.41	6.06	1.26	7.33	4.37
STD - Clay - 030	3.67	2.25	2.32	2.37	1.37	2.52
STD - Clay - 360	1.45	2.49	3.31	2.19	1.79	3.04
STD - Clay - 690	2.49	2.67	3.78	1.77	0.88	3.09
N	5	26	26	5	2	64
Mean - SOC - 030	1.03	1.14	1.08	1.03	1.22	1.10
Mean - SOC - 360	0.37	0.55	0.51	0.43	0.96	0.53
Mean - SOC - 690	0.18	0.26	0.26	0.20	0.70	0.26
Mean - PH - 030 - 1998	6.22	6.54	6.55	6.63	6.86	6.53
Mean - PH - 060 - 1998	6.18	6.45	6.44	6.39	6.25	6.42
STD - SOC - 030	0.12	0.13	0.11	0.06	0.08	0.12
STD - SOC - 360	0.12	0.14	0.13	0.02	0.05	0.16
STD - SOC - 690	0.05	0.07	0.06	0.02	0.14	0.10
STD - PH - 030	0.37	0.34	0.21	0.12		0.29
STD - PH - 360	0.27	0.38	0.25	0.13		0.31
Mean - PH - 030 - 0800	6.50	6.53	6.34	6.43	6.74	6.47
STD - PH - 030 - 0800		0.04	0.25	0.00		0.18
N - PH - 030 - 0800	1	6	5	2	1	15
CV - Sand - 030	19.64	37.34	62.61	32.74	37.34	52.55
CV - Sand -360	33.04	61.40	74.56	32.11	44.03	70.86
CV - Sand -690	81.09	119.79	123.20	36.54	71.04	115.04

Table 55. Texture, C_{org} and pH for profile curvature in1/100m (Mean = Average, STD = Standard deviation, N = number of samples, CV = Coefficient of Variation, 030 0-30cm depth layer, 360 30-60cm depth layer, 690 60-90 cm depth layer). All samples from the field site „Bei Lotte".

CV - Silt -030	4.68	2.82	4.35	2.30	1.71	3.77
CV - Silt - 360	2.13	2.98	5.56	3.06	2.56	4.41
CV - Silt - 690	4.07	3.11	7.77	1.61	9.92	5.64
CV - Clay - 030	19.56	13.25	14.99	13.88	8.10	15.27
CV - Clay - 360	6.87	13.33	19.87	11.39	11.09	16.86
CV - Clay - 690	12.08	13.41	19.86	9.04	5.14	15.85
CV - SOC - 030	11.35	11.50	10.36	5.81	6.26	11.13
CV - SOC - 360	31.93	25.87	24.83	4.98	5.52	29.96
CV - SOC - 690	27.38	26.97	22.17	7.81	20.49	39.26
CV - PH - 030	5.96	5.14	3.24	1.86	0.00	4.46
CV - PH - 360	4.37	5.97	3.81	1.98	0.00	4.83

Table 56. Texture, C_{org} and pH for landform elements for 1998-2001 (Mean = Average, STD = Standard deviation, N = number of samples, CV = Coefficient of Variation, 030 0-30cm depth layer, 360 30-60cm depth layer, 690 60-90 cm depth layer). All samples from the field site „Bei Lotte".

Landforms	SHOULDER	BACKSLOPE	FOOTSLOPE	LEVEL	FIELD
Mean - Sand - 030	3.87	3.56	4.88	3.09	3.45
Mean - Sand - 360	3.20	3.35	5.04	2.69	3.13
Mean - Sand - 690	3.40	3.18	5.11	2.50	3.00
Mean - Silt - 030	77.96	80.39	78.55	80.39	80.02
Mean - Silt - 360	77.03	79.62	77.05	79.01	78.86
Mean - Silt - 690	76.72	77.37	76.57	77.84	77.49
Mean - Clay - 030	18.17	16.05	16.57	16.52	16.53
Mean - Clay - 360	19.77	17.03	17.91	18.30	18.01
Mean - Clay - 690	19.88	19.45	18.32	19.66	19.51
STD - Sand - 030	0.92	2.48	2.62	1.15	1.81
STD - Sand - 360	0.85	2.78	3.69	1.57	2.22
STD - Sand - 690	2.10	4.19	4.89	2.89	3.45
STD - Silt - 030	1.71	3.58	3.00	2.72	3.02
STD - Silt - 360	1.71	4.64	2.96	2.82	3.48
STD - Silt - 690	2.52	6.37	4.57	3.07	4.37
STD - Clay - 030	2.13	2.19	2.29	2.77	2.52
STD - Clay - 360	1.84	3.27	2.85	3.00	3.04
STD - Clay - 690	1.24	3.36	1.95	3.34	3.09
Mean - SOC - 030	1.13	1.07	1.12	1.11	1.10
Mean - SOC - 360	0.51	0.51	0.63	0.52	0.53
Mean - SOC - 690	0.21	0.25	0.40	0.26	0.26
Mean - PH - 030	6.19	6.58	6.72	6.55	6.53
Mean - PH - 360	6.25	6.46	6.37	6.43	6.42
STD - CORG- 030	0.17	0.10	0.12	0.13	0.12
STD - CORG -360	0.15	0.14	0.30	0.14	0.16
STD - CORG - 690	0.06	0.06	0.28	0.07	0.10
STD - PH - 030	0.46	0.23	0.12	0.26	0.29
STD - PH - 360	0.54	0.27	0.15	0.30	0.31
N - SOC - 030	6	20	5	33	64
Mean - PH - 030 - 0800		6.44	6.59	6.45	6.47
STD PH - 030 - 0800		0.12	0.22	0.20	0.18
N - PH - 030 - 0800		4	2	9	15
CV - Sand - 030	23.78	69.73	53.76	37.26	52.55
CV - Sand - 360	26.58	83.09	73.15	58.26	70.86

Table 56. Texture, C_{org} and pH for landform elements for 1998-2001 (Mean = Average, STD = Standard deviation, N = number of samples, CV = Coefficient of Variation, 030 0-30cm depth layer, 360 30-60cm depth layer, 690 60-90 cm depth layer). All samples from the field site „Bei Lotte".

CV - Sand - 690	61.76	131.80	95.81	115.59	115.04
CV - Silt - 030	2.19	4.46	3.81	3.38	3.77
CV - Silt - 360	2.22	5.83	3.84	3.58	4.41
CV - Silt - 690	3.29	8.24	5.97	3.94	5.64
CV - Clay - 030	11.74	13.64	13.80	16.78	15.27
CV - Clay - 360	9.33	19.19	15.91	16.40	16.86
CV - Clay - 690	6.26	17.30	10.65	16.97	15.85
CV - SOC - 0 30	14.99	9.27	10.35	11.61	11.13
CV - SOC - 360	29.97	27.74	47.39	27.02	29.96
CV - SOC - 690	28.96	24.67	69.87	26.20	39.26
CV - PH - 030 - 1998	7.46	3.54	1.85	3.97	4.46
CV - PH - 360 - 1998	8.61	4.21	2.29	4.65	4.83

Table 57. Correlation between SOC in 0-30cm and 30-60cm and EUF-P content and $P_{(DL)}$ content. All samples from the field site „Bei Lotte" (n= 64).

Type		SOC 0-30cm	SOC 30-60cm	SOC 60-90cm
SOC 0-30cm	Pearson	1.000	.580	.410
	signif	.	.000	.001
SOC 30-60cm	Pearson	.580	1.000	.758
	signif	.000	.	.000
SOC 60-90cm	Pearson	.410	.758	1.000
	signif	.001	.000	.
$P_{(DL)}$ 0-30cm 1998	Pearson	.663	.442	.330
	signif	.000	.001	.021
$P_{(DL)}$ 30-60cm 1998	Pearson	.486	.334	.205
	signif	.000	.019	.158
EUF-P_1 1999	Pearson	.534	.561	.523
	signif	.000	.000	.000
EUF-P_2 1999	Pearson	.490	.577	.528
	signif	.000	.000	.000
EUF-P_1 2000	Pearson	.447	.567	.491
	signif	.000	.000	.000
EUF-P_2 2000	Pearson	.437	.643	.556
	signif	.000	.000	.000
EUF-P_1 2001	Pearson	.470	.528	.392
	signif	.000	.000	.001
EUF-P_2 2001	Pearson	.451	.613	.471
	signif	.000	.000	.000

** 0.01 (2-side) significant.

* 0.05 (2-side) significant.

Curriculum Vitae

Name	Hannes Isaak Reuter
Profession	Diplom-Geoecologist
Birthday	21.05.1973 in Dessau / Saxonia-Anhalt / Germany
Marital Status	single
Parents	Dr. Hans Reuter (Chemist, retired), Anette Reuter (parish priest, retired)
Education	09.1980-08.1990 Polytechnical schools in Dessau and Dranske - elementary school
	09.1990-08.1992 Ernst-Moritz-Arndt Gymnasium / Bergen / Rügen - high school
	01.1993-03.1994 Evangelisches Rüstzeitheim Lohme / Rügen - civil service
	04.1994-03.1999 Universität Potsdam/ Potsdam / Study of Geoecology
	10.1996-09.1997 University of Guelph / Guelph / Canada / Exchange Student at LRS
Technical expierience	06.1998-03.1999 PIK / Potsdam - Technical assistant in the AGREC working group
	05.1999-08.2002 ZALF / Müncheberg / Germany
	Scientific assistant in the german research foundation (DFG) project „Analyse der räumlichen und zeitlichen Variabilität ausgewählter Boden und Pflanzenbestandesmerkmale und ihrer Wechselwirkung auf Ackerflächen zur Vorbereitung des teilflächenspezifischen Managements“
	since 06.2003 ZALF / Müncheberg / Germany
	Scientific assistant in the DEKLIM network / part: wind erosion from agricultural fields